## About the Author

LEE KENNETT is professor emeritus of history at the University of Georgia and the author of *Marching Through Georgia: The Story of Soldiers and Civilians During Sherman's Campaign* and *G.I.: The American Soldier in World War II.* He lives in Pleasant Garden, North Carolina.

ALSO BY LEE KENNETT

*Marching Through Georgia*

*Gettysburg, 1863*

*G.I.: The American Soldier in World War II*

# SHERMAN
## A SOLDIER'S LIFE

### LEE KENNETT

Perennial

*An Imprint of* HarperCollins*Publishers*

First Perennial edition published 2002.

*Designed by Jackie McKee*

*Maps by Michael Caron*

The Library of Congress has catalogued the hardcover edition as follows:

Kennett, Lee B.
Sherman: a soldier's life/Lee Kennett.–1st ed.
   p. cm.
Includes bibliographical references (p. ) and index.
ISBN 0-06-017495-1 (alk. paper)
 1. Sherman, William T. (William Tecumseh), 1820–1891. 2. Generals–United
States–Biography. 3. United States–Army–Biography. 4. United States–History–
Civil War–1861–1865–Campaigns. I. Title.

E467.1.S55 K36 2001

355'.0092–dc21

[B]

2001016687

ISBN 0-06-093074-8 (pbk.)

02 03 04 05 06 ❖/RRD 10 9 8 7 6 5 4 3 2 1

*Undertaking the biography of a man with the character and stature of William Tecumseh Sherman is somewhat like taking in a forceful celebrity as a long-term houseguest. The general was with us for five years, during which time his sayings and doings frequently figured in our conversations at table; books, notes, maps, photographs, photocopies, and microfilms relating to him gradually filled our library. He dictated our travel plans: we gave up customary tourist sites for a succession of libraries and archives. My wife, Anne-Marie Durand Kennett, submitted to this regime with unfailing patience and good humor. To her, therefore, this work is dedicated.*

# Contents

*Illustrations follow page 302.*

# MAPS

# PREFACE

William Tecumseh Sherman first came to the attention of the American public 140 years ago. Since then his has been an enduring presence—not a surprising fate for a man whom a whole generation of Americans was taught to look upon as either heroic or infamous. In today's population at large and among students of our most celebrated war there is a still a considerable range of opinion, some of it quite vehement—in history, as in Sherman's life, rarely could anyone be indifferent about the man.

My own interest in the general began as simple curiosity, stirred in the preparation of a previous work and encouraged by my editors at HarperCollins. As I progressed in research I found that conclusions about the man I had formed in the course of my earlier work had to be modified—a sign of fallibility, no doubt, but also, I hope, one of an essentially open mind.

The general once said, "I must be judged as a soldier," and I have taken him at his word. The focus of this study, then, is Sherman's military career; it accords well with my interest and work in military history over the past forty years. Necessarily, other aspects of the general's life can receive only limited coverage, but this choice too Sherman would have endorsed. His military career was central to his being; his marriage, his domestic and social life—all else, in fact, had to be fitted in where the army left room.

With few exceptions I have forsaken traditional campaign history, the dissection of battles and the chronicling of campaigns; these matters merit books of their own, and in many cases have them. Then, more importantly, the man himself has claimed almost all the pages available, and I think rightly so.

From the day Sherman entered West Point until his retirement from the army forty-seven years later, virtually every aspect of his service life was documented in army records, and most of those records have survived; therefore this study rests to a considerable degree on quarrying done in the National Archives. For the period of the Civil War I have tried to go beyond archival materials printed in that vast compilation known to Civil War historians as the *Official Records*. To some extent I have succeeded, though my admiration for the compilers of the "*ORs*" has grown in the process. The general's twenty-three wartime letter books, lent to the government and never reclaimed by him, were a particularly rich source in the National Archives; so were the hundreds of orders he issued on the most varied subjects.

The treatment is in the main chronological, though I have paused here and there for closer looks at the nature of the man and his notions of waging war. The portrait presented here incorporates more somber hues than the reader may have encountered before. The man had his eccentricities, his not always laudable behavior patterns. These and the question of his supposed insanity seem to call for some sort of informed judgment on the general's psyche and its possible anomalies.

Finally I have concluded the book with a treatment of what might be called Sherman's afterlife; what he contributed to the evolution of warfare and how his hero-villain image has fared over the last century, both in our nation's history and in its folklore.

# Acknowledgments

In the half-decade it took for this work to progress from an idea to a bound volume I benefited at every turn from the advice and assistance of others. The initial proposal was framed with the help of Michael Condon, my literary agent, and Cynthia Barrett, who had edited an earlier book of mine at HarperCollins; subsequently I profited from the keen and incisive critique of my manuscript made by Paul McCarthy of HarperCollins's editorial department; subsequently Leslie Stern, Tim Duggan, and Kristen Schmidt ably guided the manuscript on through the editorial process. Thanks are also due to Eleanor Mikucki for her fine, thorough copyediting.

Much of the research for this work was done in Washington. At the National Archives, Michael Musick and Michael Meier directed me to the wealth of sources upon which this book largely rests. At the Library of Congress John Sellers shared his consummate knowledge of the Library's manuscript collections.

I found able and helpful people at several repositories with large collections of Sherman materials. At West Point Suzanne Christoff and Allen Aimone of the United States Military Academy Library; at the Ohio Historical Society in Columbus, Gary J. Arnold and Tom House; at the U.S. Army Historical Research Center, Carlisle Barracks, Pennsylvania, John

A. Slonaker, David Keough and Richard Sommers; at the Archives of the University of Notre Dame, Charles Lamb, Peter J. Lysy, and Marlene Wasikowsi.

I am also grateful for the help of persons in a number of other libraries and research institutions: Marjorie McNinch, at the Hagley Museum and Library, Wilmington, Delaware; Olga Tsapina, the Huntington Library; Connie Leitnaker, the Sherman House, Lancaster, Ohio; Henry G. Fulmer, South Caroliniana Library, Columbia; Lynn Hollingsworth, Kentucky Historical Society, Frankfort; Jill Costill and Stephen E. Towne, Indiana State Archives, Indianapolis; Terry Cook, California State Archives, Sacramento; Nicole Wells, New-York Historical Society; James J. Holmberg, the Filson Club Historical Society, Louisville; Diane B. Jacob, Archives of the Virginia Military Institute, Lexington; Brandon Sloane, Department of Military Affairs, Frankfort, Kentucky; Patricia Keats, California Historical Society, San Francisco; Ellen B. Thomasson, Missouri Historical Society, St. Louis; Patricia M. La Pointe, Memphis–Shelby County Library and Information Center; Sibille Zemitis, California State Library, Sacramento; Jennifer J. Bryan, Maryland Historical Society, Baltimore; Judy Bolton and Jo Jackson, LSU Libraries, Baton Rouge; William M. Fowler Jr., and Nicholas Graham, Massachusetts Historical Society, Boston; E. Lee Shepard, Virginia Historical Society, Richmond; Barbara E. Benson, Historical Society of Delaware, Wilmington; Joseph A. Horn, Boston Public Library; Pat Medert, Ross County Historical Association, Chillicothe, Ohio; and Edward Gaynor, Alderman Library, University of Virginia, Charlottesville.

Several fellow historians helped by directing me to Sherman materials, sharing with me items they had come upon or by reading portions of the manuscript and offering critiques: Fred Edmiston, Bill Kaan, Charles Wynes, Nash Boney, Archer Jones, and David Evans. Robert David Dawson was kind enough to share with me some of the countless Sherman letters he has tracked down and transcribed in archives all over the country.

I benefited from observations on Sherman by two men who shared with him the profession of arms: Lieutenant General Ormond Simpson, USMA (Ret.) and Colonel Charles Payne, United States Army. Dr. Walter Pharr looked over various texts and from them provided answers to a number of questions I had about Sherman's health and final illness. Dr.

Martha Simpson examined innumerable writings from the general's pen with the knowledge and insights of a psychologist. She guided me through the world of mental disorders and helped me find a meaningful pattern in the general's quirks and foibles, providing an answer to the charge of insanity that would dog him for the last thirty years of his life. Finally, Kendall Bennett and Michael Smith were of great assistance in preparing the final manuscript.

# SHERMAN

# 1

## LANCASTER

AMONG THE COUNTLESS paths and trails that crisscrossed pre-Columbian America there was a track called the Mingo Trail that wound through the woodlands of southern Ohio. This area was then the home of the Wyandots. One of their principal settlements lay some thirty miles south of modern-day Columbus, where the trail met a river the Indians called the Hockhocking; a ford or "ripple" there made for an easy crossing and spawned a village they called Cranetown. Well into the eighteenth century the area remained isolated from the British colonies to the east; the momentous struggle that began at Lexington and Concord produced no reverberations along the Hockhocking. Frontiersmen visiting Cranetown in the era of the American Revolution found a settlement with a hundred wigwams and perhaps five times that number of Wyandots.

The land of the Wyandots would shortly lose both its isolation and its identity. In the Northwest Ordinance of 1787 Congress laid down the rules for the settlement of the area; nine years later the Federal government struck a bargain with one Ebenezer Zane, who undertook to lay out an access route running from Wheeling, in what was then Virginia, to Maysville, Kentucky. Zane and his men were soon at work, cutting underbrush and marking or "blazing" trees along their way.

Zane followed preexisting trails when he could, so it is not surprising that his route took the Mingo Trail to the river crossing at Cranetown. Zane's Trace, as it came to be known, could not at first accommodate wheeled traffic, but later it was transformed to Zane's Road and survives today as Highway 22.

The first log cabin in the vicinity of Cranetown went up in April 1798; before the summer was over it was joined by a dozen others. These first homesteaders were farmers whose overriding concern was to get a crop of grain in the ground so they would be able to feed their livestock through the winter. Theirs was hard living, isolated and closely bound to the soil, without benefit of clergy or physicians; they doctored themselves with what was at hand, herbs, doses of gunpowder, and alcohol, taking the last "in the winter to sustain animal heat and in the summer to counteract the same."

What this burgeoning community most needed were the goods and services of a town, so Ebenezer Zane founded one. With a real estate agent's eye for development he took up a quarter-section, 160 acres, at the Cranetown site and in 1800 he had it surveyed out into lots. Several lots he set aside for a courthouse and other public buildings, and others were offered to settlers; they found ready buyers at prices ranging from $5 to $50. Named New Lancaster, shortened in 1805 to Lancaster, the town became the seat of Fairfield County.

The new town, like the countryside around it, continued to attract inhabitants, and as it grew it changed. In 1800 it was a cluster of log cabins, by 1805 it had ninety buildings, and by 1810 there were three structures in brick. It had a post office from its founding, and soon after it became a regular stop on the stage line; in 1836 it became linked to distant markets by a canal—whose construction gave the future Union general his first employment at fifty cents a day.

A class structure emerged. With government came officialdom; with law courts had come judges and lawyers to serve the town and surrounding county. Several physicians set up practice; clerics of various denominations had come to minister to their flocks and a succession of schoolmasters set up shop. In the town's early days it was the commercial and financial sector that seemed to lag behind; most of the business ventures launched in Lancaster's first decade foundered. Money was always in short supply and barter was common; credit was tight. Not until 1816 did the town have a bank.

In new communities such as Lancaster, where everybody was from somewhere else, a number of inhabitants gave themselves a fresh start by hiding their past or by taking on a new and more glamorous identity; in this period frontier communities played host to more than one "long-lost Dauphin of France." Lancaster had a celebrity in Mrs. Bilderbeck, who said that for many months she had been held prisoner by the Miami Indians; the stories she related were probably true. There was Augustus Witter, who claimed to have fought in the Battle of Waterloo, which was quite possible; and there was an old black woman named Aunt Disa who said she had served as wet nurse to the infant George Washington, which was problematical.

The population of Lancaster and Fairfield County was a rich mixture of ethnic strains and cultures; many of the inhabitants were a logical "spillover" from more settled areas to the east, notably from western Virginia and Pennsylvania; but New England was also represented, particularly Connecticut. Recent immigrants from the German states were so numerous that for a time Lancaster had a German-language newspaper. There was a sprinkling of Frenchmen and Spaniards. While the bulk of the population was white and Protestant, other races and creeds were represented. Free blacks seem to have been there from the first, though relatively few in number. They were often craftsmen; the town's tinsmith, for example, was a black man; finally, there was a seasoning of Wyandots, some of whom stayed on until the 1840s.

As time passed these people coexisted, mingled, and somehow became one. According to an early Ohio historian the process was a simple one: "The families married and intermarried and a race of native Buckeyes was the result, combining all that was good in the different races." Certain it is that over time the new settlers' welter of dialects and accents was transformed into a common way of speaking, instantly recognizable to the Southerner or New Englander who hears it. And by the mid-nineteenth century the "Westerner" had a discernible culture to match his speech, announcing himself fully as much by what he said as by how he spoke. He was more frank, more direct in his discourse; Easterners sometimes called him "blunt." The product of a simpler, less stratified society, he seemed relatively insensitive to nuances of rank and status. Where the New Yorker or Philadelphian would be properly deferential, the Buckeye or the Hoosier would appear brash.

Lancaster's upper class, a cluster of families in spacious homes, could be found on East Main Street where the air was less noxious than along the river. They had at first little to do with commerce and even less with industry; they had distinguished and often enriched themselves with public service, wise investments, and the practice of law. In the early nineteenth century they dominated the town's political life and they produced its most famous son, William Tecumseh Sherman.

In the first decades of the nineteenth century four allied families formed part of Lancaster's elite: the Beechers, the Boyles, the Ewings, and the Shermans. The first on the scene was Philemon Beecher from Litchfield, Connecticut, who arrived in 1801. Trained in the law, he came in the very year the local courts began functioning. Like most prominent men of his day he became an officer in the Ohio militia; though his knowledge of military matters was not great, he rose to the rank of major general. His imposing figure and splendid uniform loomed large in General Sherman's childhood memories: "I thought Napoleon a very small soldier when compared to him."

Hugh Boyle arrived in Lancaster one year after Philemon Beecher. Irish and Catholic, he had been obliged to flee his homeland because of some "trouble with the British authorities." In post-Revolutionary America such a past was virtually a letter of recommendation. Boyle brought his wife, sister-in-law, and two small daughters to Lancaster and settled in as clerk of Court of Common Pleas, a position he would occupy for three decades. Boyle forged a strong friendship with Philemon Beecher, its bonds being strengthened when Beecher married Boyle's sister-in-law. Then Eleanor Gillespie Boyle died while her daughters were still young, so her sister stepped in to take care of them. The Beechers and the Boyles essentially merged into one family.

Thomas Ewing's entrance into the circle of Lancaster's elite was anything but easy. It was the consequence of tremendous effort, both physical and mental, by an extraordinary young man. Ewing was born in 1789 in what is today West Virginia, the son of a Continental Army officer who had been impoverished by the Revolution. In 1798 George Ewing moved his family to the Hocking Valley. There were no schools in the vicinity, but an elder sister taught Thomas to read when he was six, and in so doing she transformed his life. He was enthralled by the written word and would remain so to his death; it was said that as a youth he once walked forty miles to borrow a book. In reading Thomas Ewing was an omnivore—poetry, navigation, geometry, husbandry, he devoured them all. Though he digested a library of English literature, Euclid was always his favorite author.

Ewing's physical development was as remarkable as his intellectual growth. Contemporaries regarded him as a colossus. With massive chest and powerful limbs, Ewing in his prime stood well over six feet and weighed 260 pounds. The young man's stamina found ample use on the Ewing farm, but Ewing also hired himself out to the owner of some salt wells. Salt was scarce on the frontiers and work at the wells, or "salines," well-paid, but exhausting and unrelenting. When a workman was not drawing brine to the surface he was hauling it to the boiling cauldrons, or he was stoking the fires under them, and when he was not tending the fires he was in the forest nearby felling trees and turning them into more fire-wood.

Were this not occupation enough, Ewing found time to attend Ohio University, which opened in 1809, and got his degree in 1815. By then he knew what would be his life's work. One day when he'd had some time to kill he wandered into a courtroom; when he left he knew that he wanted to practice law. In those days legal training was done by an on-the-job apprenticeship, one "read" law with some established attorney, and in due course the legal community admitted him to its ranks. One day Ewing appeared in Lancaster and asked Philemon Beecher to accept him as a clerk and reader. Beecher was impressed with the earnestness of the giant in country clothes and he agreed. Ewing plunged into law as he had plunged into a dozen other subjects; a year later he was admitted to the bar. Ewing came to know the Beecher-Boyle clan; his friendship with Hugh Boyle's daughter Maria blossomed into love and led to their marriage in 1820.

Among Ewing's close friends was another attorney, Charles R. Sherman. Considering the prominence that Judge Charles Sherman enjoyed and the esteem in which he was held by his contemporaries, the records chronicling his passage through life are surprisingly few: an unflattering portrait, legal opinions hidden in law books nearly two centuries old, and a few papers at the University of Notre Dame. The major written tribute to him is a sixteen-page biography by his son-in-law, William J. Reese. Even if one adds what was said about him by contemporaries, it is by no means easy to draw the man out of the shadows.

About the Sherman family a good deal is known, thanks to genealogical researches. The Shermans can be fixed in the English county of Essex, particularly around the town of Dedham, in the fifteenth century. Their name is probably a derivative of "shearman," for they were often associated with the wool trade. There were Shermans in New England by the early seventeenth century. They were religious dissenters, and General Sherman himself would be saluted as a "superb specimen of the pure Puritan stock," though the general's own religious leanings were and still are a matter of conjecture. The direct ancestors of Charles Sherman settled in Connecticut, where they were often attorneys, judges, and legislators. Charles, who was born in 1788, followed the family tradition. He joined his father's law firm and was admitted to the bar in 1810.

That same year he married Mary Hoyt, who may have brought an unwelcome dowry: a tendency to mental illness that would supposedly become hereditary in the line of Charles R. Sherman. In his later years Mary Hoyt's brother Charles became sufficiently deranged to be confined, and a couple of similar cases were noted, with depression apparently the chief manifestation. These stories would surface understandably enough in 1861–62, when General Sherman's own sanity was being questioned, but they were so scattered and imprecise that it is impossible to make much out of them clinically. A historian who recently sifted the evidence concluded that one could "only guess at the nature of the affliction." As for the matter of incidence, it would appear that numerous as the clan was, it probably had no more than its share of mental and behavioral problems. And they were a prolific and hardy strain: all of General Sherman's ten brothers and sisters would reach adulthood and marry, a remarkable record for the era.

Charles Sherman soon left his bride and took off alone to investigate job and land possibilities beyond the Ohio, where the advice of his father, Taylor Sherman, would follow him: he should take his time in settling and pick a place after careful consideration. Charles seems to have followed this counsel; his travels eventually took him along Zane's Trace to Lancaster, where he tested the waters by opening a law office, then returned to Connecticut for his wife, who had in the meantime borne him a son. The couple showed up in Lancaster in 1811; they had come on horseback, taking turns holding their infant son on a pillow.

Charles and Mary Sherman soon moved into a frame house on Main Street and entered the society of the town. Charles's legal background helped him make his way, but his friendship with the Beechers, Boyles, and Ewings helped even more. Once admitted to the circle the young attorney made friends everywhere. By all accounts he was an engaging young man who moved in Lancaster society with ease and grace. Like his son the general, Judge Sherman was good company.

Charles Sherman gave every appearance of a man on the way up. He soon had a thriving practice; in 1817 he began to ride circuit, first as a lawyer, and ultimately as justice of the Ohio Supreme Court. His domestic life was also filled with activity. Between 1811 and 1828 he sired eleven children and apparently took a hand in caring for them. In his leisure hours he indulged a strong literary bent. According to his biographer his mind was "too generous and refined, too expansive, exploring and versatile, to exist upon the exclusive pabulum of the law." He wrote at least one play and composed sermons. He had some notions of medicine; when the town's doctors were all out on call the sick knew they could turn to the obliging Judge Sherman. He was an avid gardener who introduced asparagus to Lancaster.

One suspects that Judge Sherman was something of a free spirit; the catholicity of his interests suggests this and so does the naming of his third son. Born February 8, 1820, the baby was a redhead. Charles named him Tecumseh for the Shawnee chief who had allied himself with the British in the War of 1812 and posed a grave threat to the settlers in the old Northwest until his death in battle in 1813. According to William Reese the choice of names caused some comment in Lancaster and came up in a social gathering at the Sherman home: "Judge Sherman was remonstrated with, half in play and half in earnest, against perpetuating in his family this savage Indian name. He only replied, but it was with seriousness, 'Tecumseh was a great warrior' and the affair of the name was settled."

With all his qualities Judge Sherman had some flaws. From time to time, according to William Reese, he was a prey to self-doubt. When the Ohio legislature elected him to the state's Supreme Court he accepted the position with "great personal diffidence and apprehension." But he accepted, and in doing so he added to an already heavy burden of cares and duties. For the judge was spread thin; with functions and interests in so many directions it was perhaps inevitable that he should have neglected some of them. That is the most charitable explanation for the catastrophe that struck him in 1817.

Early in 1816 the U.S. Treasury Department announced that after February 1817 money paid into it by its collectors of revenue would have to be either in specie or in notes of the Bank of the United States or one of its branches. In the Northwest, where money was always in short supply, payments had customarily been made in local bank notes. Sherman was collector for the Third District of Ohio, and somehow he let his deputies in the various counties be caught by the deadline with large quantities of local bank notes; he, as their chief, assumed the debt and undertook to reimburse the Treasury. In so doing he assumed a burden that he would carry the rest of his life.

Sometime after 1815 Charles's sister Betsy came out to Lancaster with their mother, Elisabeth Stoddard Sherman. Grandmother Stoddard, as she was called, would become a fixture in the Sherman household. By all accounts a vigorous and impressive woman, she shouldered part of the burden of maintaining the large Sherman household, including the disciplining of the children. Her grandson Tecumseh said of her, "she never spared the rod or broom, but she had more square, hard sense to the yard than any woman I ever saw. From her Charles, John and I inherit what little sense we possess." She was to be a tower of strength when tragedy struck the family.

In the spring of 1829 Judge Sherman was forty-one years old. He was holding court in the town of Lebanon on June 18 when he was suddenly taken ill. Friends sent to Cincinnati for the best medical help and dispatched an express messenger to Lancaster, a hundred miles away. As soon as she got word Mary Sherman started to Lebanon by coach. She had gone about halfway when a second messenger met her with word that her husband was dead.

For Charles Sherman's widow and children 1829 was a watershed year. Family and friends stepped in to help; money was taken up, and the Masons were particularly helpful. Eldest son Charles Hoyt Sherman, who had come to Ohio on a pillow, was finishing his studies at Ohio University and would soon be on his own; the same was true with Elisabeth, who would shortly marry William Reese. The three youngest children Mary Sherman would keep with her. The six in the middle she had to confide to relatives and friends. This diaspora would be permanent. By 1870 there was not a single Sherman in Lancaster. As for Tecumseh, he would keep ties with his elder sister Elisabeth and even closer ones with John, three years younger.

Nine-year-old Tecumseh left home but did not leave Lancaster: Thomas Ewing came for him and took him to his brick mansion just up the street. As Charles Sherman had sunk into financial ruin, his friend and neighbor had prospered. Thomas Ewing never lost the habits of hard work and discipline; with the proceeds of a large and lucrative practice he made judicious investments, real estate mostly. He kept his hand in the salt business, perhaps because it had been part of his youth, certainly because it was profitable. Ewing had started his family later than Judge Sherman. He and his wife had seven children, one of whom died in infancy. The six others would figure prominently in General Sherman's life: there were Philemon Beecher, born in 1820, Eleanor Boyle (1824), Hugh Boyle (1826), Thomas Jr. (1829), Charles (1835), and Maria Theresa (1837).

The family was a close, harmonious one, far more so than the Sherman brood. For all his formidable exterior Thomas Ewing was a kind and attentive father. Maria seems to have been the children's chief disciplinarian, but rigid and uncompromising chiefly in matters of religion; all the children were raised as Catholics. It was Maria who insisted that young Sherman add a Christian name; she had a priest perform the baptism, though "William" proved to be an unused appendage. At West Point and in the military generally, colleagues called him "Sherman"; in his own family he had been nicknamed "Cumpy" and the Ewings called him "Cump."

As a substitute family the Ewings would have been hard to surpass. To the degree that they could, they made him one of them. Thanks to Thomas Ewing young Cump would make a privileged entry into the profession of arms; Philemon would be his closest friend during his last years in Lancaster; Hugh would be his favorite "brother" after he was grown; Eleanor, universally called Ellen, would be his wife.

Two conflicting portraits of the youthful Sherman emerge. The general saw himself as a mischievous, adventurous boy: trying to ride his father's horse and falling off, stealing a neighbor's kindling instead of chopping wood himself, drawing the ire of his teacher with his pranks. Yet the stronger evidence suggests that he was a shy and sensitive youngster. Thomas Ewing worried that the boy he brought home with him was "disposed to be bashful, not quite at home." He urged his wife to do what she could to make him one of the family. Later Ewing recalled his youthful charge as "transparently honest, faithful and reliable, studious and correct in habits." John Sherman described the brother he knew in the Lancaster days as "a steady student, quiet in his manner and easily moved by sympathy and affection. I was regarded as a wild reckless lad, eager in controversy and ready to fight. No one could then anticipate that he would be a great warrior and I a plodding lawyer and politician."

In the first edition of his *Memoirs* Sherman said nothing of his infancy and youth, beginning the narrative with his departure for California in 1846. Even when he filled in this gap in a later edition he covered those sixteen Lancaster years in six pages; he claimed that his memory did not go back beyond about 1827. But the Lancaster years left their mark. It is possible that experiencing the shipwreck of his family at a vulnerable age had psychological consequences that could be seen in the serious emotional crisis of Brigadier General Sherman thirty-two years later. He had suffered two distinct traumas in 1829, with the loss of his father probably a lesser blow than his separation from his mother. Sherman's references to the judge are fond and respectful, though along with them there were intimations that he might have managed his affairs better. Charles Sherman's third son would have a horror of indebtedness: "Don't borrow," he would tell his wife. "It is more honest to steal."

Through all the years that they were separated Sherman would worry about his mother's welfare, send her what money a threadbare lieutenant could spare, and lecture his brothers about her needs. According to his brother John his devotion to his mother was his most salient trait. It was she whom he would have called back to life to see the wonders of the West, to witness his triumphs. Shortly after his death in 1891 his aide-de-camp made the surprising revelation that "Gen. Sherman was proud of tracing his powers of endurance to his mother, to whom he also frequently ascribed the heritage of other soldierly characteristics."

By birth Sherman was a man of the West, though he would not become aware of that distinction until he went East. Far from rejecting this Western heritage, as others would do, he would glory in it. For him the "Great West," as he called it, held the future and the destiny of America.

Then it is pretty clear that the young man who left Lancaster for West Point in 1836 was a burgeoning conservative, certainly in the political and social sense. The message of conservatism pervaded the comfortable houses along Main Street; it was proclaimed in the newspapers that were read in those houses; it seasoned the table talk of the Beechers, the Boyles, and the Ewings. The message was clear: the best government was one run for the benefit of all—by the better sort of people.

Finally, the milieu in which Sherman was brought up was permeated by law. Most of the adult males in the world of his childhood were attorneys or judges. Between his father and his foster father he must have heard endlessly about venues and estoppels and nolo contendere. This exposure did not "take" in the usual sense; he came to detest lawyers, along with journalists and politicians. But the law itself fascinated him and appealed to him. In his life he would do many things that were unorthodox, controversial, and to some people criminal, but always he would find the law on his side.

# 2

## WEST POINT

CHARLES AND MARY Sherman chose the career of their third son; Thomas Ewing said as much in a letter to Secretary of War Lewis Cass, and Thomas Ewing was not one to trifle with the truth: "It was his father's wish, often expressed before his death," Ewing wrote in the summer of 1835, "that he should receive an education that would fit him for the public service of his country in the army or navy." Mary Sherman said that she did not want her son to go to sea, so Ewing was writing about the boy's admission to West Point. Cass wrote back that the deadline for 1835 admissions was past, but young Sherman would be considered the following year.

A successful applicant for admission to the United States Military Academy had to be at least sixteen years old but not more than twenty-one, with a minimum height of four feet, nine inches. He could not be deformed "or afflicted with any disease or infirmity which would render him unfit for the military service or with any disorder of an infectious or moral character." He had to demonstrate sufficient education to pursue the academy's course of study with some hope of success. Each June authorities at West Point would test the applicants' physical condition and mental capacities. Much of the nation's youth might have shown up, then, save for this proviso: academy officials examined only applicants who came armed with letters of appointment from the secretary of war.

The letters were awarded by a complicated system: most were distributed among the various congressional districts; two were awarded to each state "at large," and a dozen or so "at large" slots were accorded to the president. In practice the appointments were often political plums, and much sought ones; after all, four years of education at government expense was no small thing. Sherman discovered that as a result of political horse-trading some of the six other "Ohio boys" were in fact from other states. (He was to remember this precedent. Twenty-nine years later, when his forces occupied much of Mississippi, he chose her quota of appointees from the ranks of his Midwestern regiments.)

The academy's Board of Visitors asserted that it was "open to the sons of all classes of our citizens," which was technically true enough, but a recent study of the academy in the pre–Civil War period shows that most cadets came from "families with governmental connections or political influence." One has only to look through cadet applications for 1836: one applicant buttressed his dossier with a letter of recommendation from President Andrew Jackson. A young man from Indiana supported his candidacy with a petition on his behalf signed by twenty-five state senators. It is a measure of the weight of Thomas Ewing in these circles of power and influence that he had already secured the appointment of a relative of his wife, was able to place Sherman in 1836, and later sent his own sons Philemon and Hugh.

Such things lent credence to the arguments of political figures who denounced the academy as "an aristocratic excrescence," inimical to the democratic principles of the nation. Andrew Jackson was one of the academy's enemies; though the institution survived his administration, it suffered from a chronic lack of funds—when Sherman came the institution was short twenty instructors, their place being taken by officers on temporary duty or even some of the more advanced cadets.

When the examiners assembled at West Point in June 1836 they had 121 dossiers before them. Each candidate submitted to a somewhat perfunctory physical exam; to test his vision a dime was held up at a distance of fourteen feet and he had to tell whether it was heads or tails. The academy's examiners might ask him to work a math problem at the blackboard, and would ask questions on a number of subjects, including geography (one candidate described the Pacific Ocean as beginning at Gibraltar). Of ninety-nine candidates who showed up, the examiners accepted ninety-four. Of that number about twenty percent would drop out within six months, this attrition being attributed to "want of inclination for the service"; others would fall by the wayside, so that in June 1840 a total of forty-two cadets would graduate.

For young William T. Sherman the examination presented no problems, nor did the life he embraced that June. The first night he spent as a cadet he shared a tent with a second-year man who introduced himself simply as "Ord." Thus began his lifelong friendship with Edward Otho Cresap Ord. Their paths would cross innumerable times in the coming years. Each would get an invitation to the other's wedding; a half-century after their first meeting Sherman would be present at the interment of his old comrade-in-arms. And what was true with Ord was true with most of the cadets Sherman would meet; though their relations would not always be so cordial, their lives and fates would be entwined.

In September summer camp ended, the cadets regained their barracks, classes began, and routine was fixed by an academic timetable that accorded ten hours per weekday to class work and study. Reveille was at 5 A.M. (six in the winter), followed by breakfast and infantry drill; classes began at eight and ran until two in the afternoon, with an hour for dinner. In the afternoon there was more drill, both infantry and artillery this time; if there was still daylight when drill ended at 4 P.M. the cadets could use the time for games or other outdoor activities. The time after supper was dedicated to study. Tattoo was at nine and taps at ten. This was the routine Monday through Friday, but the weekend was by no means free. On Sunday there was chapel and, to the chaplain's distress, dress parade and inspection.

A distinctive characteristic of the academy was the seclusion that enveloped its students. This isolation was intentional. In 1828 the academy's Board of Visitors reported that thanks in large part to its remoteness, "the moral discipline of the institution is perfect." They announced with some exaggeration that "the avenues to vice are closed." With the exception of occasional furloughs granted on an individual basis and a "free" summer after the second year, cadets would stay with their comrades in this martial sequestration.

In some ways the system of instruction at West Point was remarkably modern. Teaching was done with courses broken into small sections of fifteen or so, with these sections containing students of similar abilities. Student participation was constant, with cadets called on to speak or "recite" almost daily. The cadets' progress was closely monitored, with daily and weekly grades and monthly "consolidations"; extensive examinations took place in January and June. Grades were made public to spur the students' efforts to excel; the names of the five cadets who stood highest in each class were printed each year in the *Army Register*. Deportment was also closely monitored, with demerits given for each failing (Sherman once received a demerit because he was not holding the butt of his musket in correct alignment with his body). The demerits reduced the cadet's class standing, and 200 of them in a single year could mean expulsion.

Wherever feasible the learning was "hands on." Cadets studying electricity got a small shock from a galvanic device. Those learning about "pyrotechny" fabricated rockets and rolled cartridges. Then they learned the duties and functions of common soldiers and noncommissioned officers—executing the manual of arms, standing guard, making the rounds, and so on—by performing these themselves. Artillery drill required them not only to learn loading and firing procedures, but also placing and moving the guns. Until 1839, when the academy acquired horses for this purpose, the cadets did the work themselves, hitched to the cannon with rope harnesses.

In the classroom the program of study began with basic skills, chiefly language and mathematics. In the 1830s French was a major ingredient of first- and second-year work. The professors were satisfied if their charges could translate freely into English, acknowledging that "but few can be expected to speak fluently." In the second year math studies turned to geometry and calculus; drawing and surveying were added, along with geography, rhetoric, and English grammar.

The third year, which cadets judged to be the hardest, was dedicated to the sciences: chemistry and physics, especially mechanics, optics, magnetism, electricity, and astronomy. The fourth year, which Sherman called "the most important and the most interesting," seems particularly charged. Military subjects now dominated the schedule: infantry and artillery tactics, fortification and siegecraft, military and civil engineering, including the construction of roads, bridges, railways, and canals, along with the improvement of rivers and harbors. Then there was more science (geology and mineralogy), rhetoric and moral philosophy, including constitutional and international law.

The cadets were presented with an orderly exposition of tactics, both artillery and infantry (in the 1830s and 1840s the army had no cavalry strictly speaking). Strategy was little studied. Dennis Hart Mahan, perhaps the institution's best-known professor and the author of *Outposts*, the cadet's Bible, was an extravagant admirer of Napoleon; in *Outposts* the men who would lead the Civil War armies were offered as model a commander whose mastery was innate, whose actions were intuitive: "No futilities of preparation; no uncertain feeling about in search of the key point; the whole field of view taken in by one eagle glance; what could not be seen divined by an unerring military instinct."

The curriculum was overly ambitious, trying to convey knowledge in too many fields for the time available, so some subjects were slighted. Then too, the quality of the faculty was uneven. A roster for the fall of 1839 shows that the professors heading some departments were well-grounded academics with significant publications to their credit, but in almost every instance the assistant professors were first or second lieutenants on temporary duty; they had varying degrees of experience, and three of them, Henry Halleck, Henry L. Smith, and J. F. Gilmer, were fresh academy graduates with no teaching experience. The chaplain doubled as professor of moral philosophy and complained about the difficulty of filling two positions. The texts were the critical element in instruction. According to Dr. Jasper Adams, chaplain in the 1839–40 period and a critic of the system, the texts were "carefully explained and illustrated by examples and the comments of professors. An exact knowledge of these text books is held to be of the greatest importance, long and patient examinations are held upon them, and the relative standing of the cadets in the Academy is made to depend upon their acquaintance with them."

One is struck by the fact that almost all of the cadets' work was oral. If certain graduates later revealed a masterful prose style—here one thinks immediately of Grant and Sherman—they did not acquire it at the academy. Then too, the cadets' performance in class essentially involved regurgitating the material absorbed from the text or supplying explanation thereof. There is no evidence of their being encouraged to challenge or critique the texts placed before them; that would have been foreign to the system.

Perhaps the most significant failing of the educational program in its military dimension was the failure to go very far beyond theories and principles, a deficiency particularly regrettable in military engineering. By the same token, the cadets had no practical experience of any unit larger than the infantry company formed from the corps of cadets. And then they graduated and entered an army that was distributed about the country in penny-packets, and never assembled for extensive field exercises or maneuvers. At best, some of those who would command Civil War armies got a practical grasp of small unit operations during the Mexican War; a smaller number who saw service in Mexico had positions on the staff of some general, and thus had a glimpse of how war was directed on a larger scale.

While the academy was creating soldiers in the classrooms and on the parade ground, in the barracks it was forming young men, this by a process that was largely unplanned and uncontrolled. In this hothouse environment a distinct society flourished. A hierarchy had crystallized quite early, with each class regarding itself as superior in status to the ones following it. New arrivals were referred to as "conditional things" until formally admitted; thereafter and through their first year they were "plebes." Hazing does not seem to have been a fixed custom, and when practiced was mild in form, with plebes doing chores for upperclassmen. Sherman played that role, then later had a plebe of his own who fetched water, cleaned his gun, and the like; Sherman repaid him "in the usual cheap coin—advice."

In this era there was no "beast barracks" for the newcomers; although cadets were usually with classmates during hours of instruction, members of two or three classes might room together. With enrollment under 300, everybody knew everybody.

The corps of cadets was a homogeneous group: they were all white, all male, and almost all arrived with the self-assurance of those who move into the mainstream of life from the bosom of the great American middle class; the sectional antagonisms that would eventually tear the country apart provoked no violent scenes while Sherman was a cadet. There is ample evidence of a general bonding. Sherman later claimed that the gathering of youths from so many different states in a common effort had a unifying effect. He told Dennis Hart Mahan that West Point graduates grew so attached to one another that they were "the very Siamese twins of society."

In this little world Cadet Sherman thrived. His surviving letters reveal no homesickness, impatience with the regimented life he led, or any indication he wanted to be anywhere else than where he was. He had some difficulty at the very outset since the cadets thought he was a close kinsman of Thomas W. Sherman, who had graduated in 1836. That Sherman had been extremely unpopular with his classmates since he had insisted on strictly observing every rule and regulation of the institution—a significant commentary on the mindset of the corps of cadets.

It appears from Sherman's letters that he was well-liked and considered good company, and this is confirmed by the reminiscences of those who knew him in his cadet days. William S. Rosecrans, class of 1842, recalled that he was easily enlisted for any lark or illicit enterprise. With all his popularity, Cadet Sherman did not appear to be a young man with an innate "gift" of command, the sort who imposed himself naturally as a leader of his willing fellows; he was not one of the cadet officers, for example. But as Sherman himself pointed out, those who stood tallest at the academy rarely achieved distinction later. He assured Hugh Ewing, who was also to enter the academy, that "Napoleon did not stand at the head of his class, nor did Wellington, though both were great men."

The life of the cadet as seen in the academy's rules and regulations was sober and spartan: he was prohibited from consuming alcohol, using tobacco, playing cards, leaving his room during study hours, or leaving the academy without permission. This did not prevent Cadet John M. Schofield from making a wager that he could travel to New York City and back between two roll calls without being detected—and he collected on his bet. Though cadets' quarters were inspected three times a day, all sorts of things were concealed in them, including food for illegal "hashes." The patterns of behavior, the "code" prevailing in the barracks, came from adherence to the rules, and also from violation of them.

Within the cadet community serious problems or altercations were uncommon. These were young gentlemen who did not steal; from 1833 to

1860 only three cases of theft were reported. On the other hand the cadets all seem to have been in collusion to violate consistently certain academy regulations, especially those regarding tobacco and alcohol. After a popular cadet named Heath died suddenly just before Christmas of 1839 the cadets went on a massive binge. The administration tried to hush it up, but it made the newspapers, one of which proclaimed, "so far as morality is concerned, the institution is going to the devil with quick and long strides."

There were cadets who were easily provoked and quick to feel a slight. One morning on the parade ground Cadet Lieutenant Tilghman told Cadet Private Steere to take his hands out of his pockets; Cadet Private Steere replied, "Go to hell," and the two came to blows. In 1836 one cadet stabbed another. If a cadet acted in a spontaneous outburst of anger, as in protecting his "honor," his transgression might not be taken very seriously by the authorities; if the violence was premeditated, the cadet involved would be dismissed—such was the case with Cadet Rogers, who attacked Cadet Hughes with a dirk a full week after they had had an altercation during infantry drill.

In September 1838 Major Richard Delafield succeeded Colonel René de Russy as superintendent. "Old Dick," as the cadets called him, showed considerable interest in what went on in the barracks; he made a number of changes that the cadets liked; installing iron bedsteads, for example, so cadets would no longer have to sleep on the floor.

Delafield believed that by intervening in cases of cadet misconduct he could build esprit de corps. If a cadet was found to have committed a fairly serious offense—playing cards, for example—Delafield would offer to reduce the culprit's punishment if he could obtain pledges from all his classmates that they would renounce card games. By banding together to help their comrade, by acting in solidarity, they would thus build esprit de corps—and card playing would cease to be a disciplinary problem. Major Delafield was proud of this strategy.

In the spring of 1840 a cadet named Story was caught in the act of writing obscenities on the barracks wall; this was frequently done, but no one had ever been caught at it. Delafield applied his esprit de corps strategy. Cadet Story was threatened with the direst punishment, then told he could get off lightly if he got all his classmates to take a pledge to stop writing obscene graffiti. Story made the rounds of his classmates, asked them to take the pledge, and came back to report that in a show of group solidarity they had all refused.

The name of William Tecumseh Sherman does not figure in the various causes célèbres that Major Delafield had to deal with. But his existence is documented now; from 1836 on the biographer is no longer limited to what he can infer from oft-told anecdotes and reminiscences. Thus it is recorded that Cadet Sherman reported to the academy's hospital eight times in his last two years for a variety of minor ailments, "sore foot," "inflamed eye," and the like—but he had no major illnesses.

Far more significant is Sherman's name as it appears on the academy's "Merit Rolls"; in a compilation of June 1837 his standing in his studies was ninth in a class of seventy-six, a level of performance that would remain fairly constant, so that he graduated sixth in a class of forty-two. French, math, and drawing were his strong suits. As for comportment, at the end of his first year he was ranked number 124 in a corps of cadets of 211. This position, if less praiseworthy, also did not change much. Each year he had what he called a "respectable" number of demerits—in the neighborhood of a hundred. These, factored in with his scholastic rating, inevitably lowered his general standing. The demerits seem to have been mainly for peccadilloes; when he was caught at something more serious, "absent from quarters" in June 1840, on the eve of graduation, Major Delafield seems to have quietly intervened on his behalf.

The young Sherman can be seen another way: he conducts a sizable correspondence, some of which has survived. His extant letters from the period before he went to West Point are few, but those still in existence have a certain bland, colorless quality, as though written by someone unsure of himself, one whose character had not "jelled."

But the letters written by Cadet Sherman are emphatically his own. His characteristic script, smoothly flowing and slanted, is appearing; while initially there are some awkward constructions in his West Point letters, no one could describe them—or anything else Sherman wrote over the next half-century—as bland or colorless. A contemporary who read some of Sherman's youthful letters at the end of the century pronounced them "sprightly, vivacious and a trifle eccentric, written by a young man little likely to be ever much retarded by any doubts as to the wisdom of his own opinions." While he was an early and omnivorous reader, his remarkable facility with words seems to have simply bloomed within him, through what genetic inheritance is unclear.

The most voluminous correspondence surviving is the one with Ellen Ewing, and it tells a good bit about the youth who was gradually falling in love with his foster sister. When she sends him a pencil he tells her that he will keep it in its wrapper, and only use it "on special occasions." A pair of slippers she made for him he also vows to keep in the wrapper, for they are too beautiful to wear; "it would be a sacrilege." When she sends him some candy she made, he gives his "Eastern friends" some, and reports that they had never tasted anything like it.

In everything he strives to show her his best side. After a brief furlough in New York City he pens a lyrical description of "the richly cultivated fields of Long Island." When Ellen writes about her life in the Academy of the Visitation, where she is a student, it is a worldly Sherman who replies: her "nunnery" sounds like a rather dull place, he writes, then adds, "but different persons are of different tastes, you know."

And as their correspondence continues Ellen becomes his confidante, a function she will fill the rest of her life. He confesses to her the ambivalent feelings he sometimes has over his relations with the Ewing family. He has written to Maria Ewing but has had no answer: "very often," he confesses, "I feel my insignificance and inability to repay the many kindnesses and favors received at her hands and those of her family."

Sherman also corresponded with the increasingly scattered members of his own family, though here few letters have survived. In letters to members of the Ewing family he mentions hearing regularly from his mother and from his sister Elisabeth; he seems to have had little contact with his brothers, with the important exception of John. The brothers' correspondence, like Sherman's exchange of letters with Ellen, would continue steadily in years to come; John, like Ellen, would become a confidant; when things were not going well for Cadet Sherman, and later for Lieutenant Sherman, he would sometimes unburden himself to both. But in his early letters he tended to play the older brother, counselor, and dispenser of advice, in which he would often take an emphatic tone. He did not think much of John's entering the legal profession: "For my part, it would be the last choice." As for himself he told John confidently that he would never study another profession: "If I ever resign it would be to turn farmer, if ever I can raise enough money to buy a good farm in Iowa."

Sherman was already a voracious reader, as he would remain the rest of his life (curiously, academy library records indicate he rarely checked out books there). It's clear from references in his letters from West Point that he was reading things that had no relation to his studies. The James Fenimore Cooper novels probably excited the interest he developed in the Far West. He became a regular reader of newspapers. The Ewings sent them to him and so did his sister Elisabeth; he no doubt read others that were making the rounds of the barracks. He boasted to Philemon Ewing, "I see papers from every portion of the Union." His criticism of the press antedates his own personal tribulations with it. To Maria Ewing he wrote that the excesses of the Lancaster newspapers ought to make their editors "blush."

The year 1840 would mark the end of Sherman's military apprenticeship. The question of what to do when it was over had occupied him for some time, and here he seems to have been troubled with an irresolution that was uncommon with him. Into his third year his choice was the infantry, chiefly because it manned outposts on the Western frontier. But he found himself in something of a quandary because of Thomas Ewing, or what he perceived Ewing wanted him to do. He wrote one of Ewing's sons at the end of his third year: "I am almost confident that your father's wishes and intentions will clash with my inclinations. In the first place, I think he wishes me to strive and graduate in the engineer corps. This I can't do. Next to resign and become a civil engineer . . ." His own preference was infantry and the Far West, "out of the reach of what is termed *civilization*."

Then a boundary crisis flared up with the British and Sherman caught the fever. If he were to fight "a civilized people" he would prefer to do it in the artillery, and this became his choice. For a time the barracks were awash with visions of epic battles with the Redcoats, and the cadets were fired with martial ardor. Sherman has left a description of the scene, and it should serve as a reminder that these young men had not yet left adolescence behind. They were not poring over maps, charting the best invasion routes to Canada. They were dashing about the halls and stairways, parrying and thrusting: "Books are thrown in the corner, and broadswords and foils supply their place. Such lunging, cutting and slashing—enough to dispose of a thousand British a day . . . "

In January 1840 Sherman wrote John that he still had not decided which arm he wanted to serve with and was becoming fatalistic. He would graduate sixth in his class, not high enough to have one of the coveted engineer slots. In the end he listed no preference—he would take what the army assigned him. He ordered his class ring, sword, epaulets, and chapeau complete with feathers. When the tailor came he was obliged to indicate the arm for which the uniform was intended, so he ordered one for an infantry officer. It proved to be a bad choice, for when he received his orders he found that he had been assigned to the Third Artillery.

# 3

## The Southern Years

IN NORMAL TIMES AN artillery officer could expect assignment to one of some seventy fortifications guarding the nation's ports and harbors from Penobscot, Maine, to the mouth of the Mississippi. But in 1840 these installations were largely abandoned while their garrisons did service as infantry in the Second Seminole War. In 1835 President Andrew Jackson had begun the removal of the eastern Indian tribes beyond the Mississippi, provoking disorders in many states and war in Florida.

Initially spokesmen for the Seminoles indicated their willingness to be moved to lands west of Arkansas, but the time for departure came and passed, as did later dates the government agents fixed. The authorities finally lost patience and began to move in troops. By the end of 1835 they had some 500 in Florida and more were on the way. The flashpoint came on December 28 of that year, when the Seminoles killed a government official and fell upon a hundred-man column of troops, killing all but two.

The government responded by calling in more troops; Company A, Third Artillery–Sherman's unit–had left Fort Moultrie, South Carolina, in 1836. Ultimately some 5,000 men–half the strength of the army–were committed to the conflict. Six successive commanders struggled to find a way out of the morass.

It was only in October 1840 that Lieutenant Sherman reached

Florida. He had taken full advantage of the ninety-day furlough that followed graduation; much of the time he spent in Lancaster, visiting the Ewings and what was left there of his own family: his mother and two sisters and a brother who barely recognized the tall young man in blue uniform. In the time that he spent with Ellen each found confirmation of the feeling that had blossomed between them; thereafter their correspondence would touch on the two subjects that might stand in the way of an eventual marriage: her religious convictions and his profession.

The war Sherman found in Florida would today be classified a low-intensity conflict. The army had the wherewithal to defeat the Seminoles in open battle, but the Seminoles wouldn't fight that way. For the most part the struggle was a war in the shadows, with the Seminoles relying on raids and ambushes, which the army found difficult to prevent; to identify and punish the perpetrators was well nigh impossible. No sustained campaign could be mounted, since in the hot summer season the army ceased operations in the interest of preserving the soldiers' health (only one out of four army fatalities would be the result of combat). There were intermittent negotiations that seemed to make progress—until the day the Seminoles were to show up for removal. As Sherman put it, "Indians plenty, always coming but none come." Then when negotiations failed and the army resumed operations, it found itself attacking an enemy whose chief weapons were ferocity and guile, one who had no supply line to cut, no vital center to be taken.

Sherman went out on a few patrols, or "scouts," but he saw no real combat. He did see enough to make some trenchant comments on his first war, particularly in letters addressed to his brother John. The white settlers of Florida he described as a "cowardly" lot, who helped provoke the Indians, then fled at the first sign of danger. He also had harsh words for the militia from neighboring states that had been summoned. As a child he had been impressed by militia reviews, but now, as a soldier in the field, he wrote John that in times of danger they were prone to panic and would abandon their positions to "gather together like sheep." What he saw of the militia in Florida was the beginning of a disdain for the "citizen-soldier" that he would carry into the Civil War. Only the "regulars" could be counted on: "The army, if properly directed, can bring these rascals to a sense of their destiny."

Sherman's solution, which he confided to John, was to carry the fight to the enemy, using columns or detachments of a hundred or more troops that would penetrate to the Seminole settlements and burn them, destroying boats and canoes. Writing in March 1841 he told John that in his district such a policy was already beginning to make a difference. In fact, Sherman's arrival in Florida coincided with the launching of a new strategy put into effect by Colonel William J. Worth. When he took command he resolved to keep the army in the field all year, reasoning that in the summer the destructive raids would be most effective, since they would keep the Seminoles from growing and harvesting crops. This campaigning across Florida in the summer heat was hard on the troops, but ultimately it was even harder on the Seminoles. By 1842, with starvation in prospect, most of the surviving Seminoles agreed to leave; the minority who refused took refuge deeper in the Everglades, where Colonel Worth was content to leave them.

Yet in Colonel Worth's command all was not well. The artillerymen were particularly aggrieved. They missed their comfortable quarters and were tired of playing infantrymen. So they deserted en masse. In Company G, the unit to which Sherman transferred when he was promoted to first lieutenant late in 1841, the desertion rate rose to sixteen percent per month in the summer of 1842. The officers departed too, simply resigning: fifty-five junior officers out of the artillery's four regiments. This exodus probably played a major role in Sherman's promotion to first lieutenant, several years before it might have been expected.

Company G was soon transferred to duty in St. Augustine, then a town of about 3,000, and a welcome change from the spartan existence in the field. St. Augustine also offered diversion; many years later, in 1879, Sherman visited the town again and reported to a good friend that all of his "old sweethearts" had died save one. For a time he was posted with a score of men at Picolata, a checkpoint some miles outside St. Augustine. This duty, humdrum as it was, constituted a milestone: for the first time in his military career he had a command of his own.

For this period a single official letter from Lieutenant Sherman to the adjutant general survives, but it is worth quoting. He had received a stern message from Washington about his last monthly return, which had not been sent in on the regulation form. He sent a reply that was probably passed around in the adjutant general's office: "The reason why a printed form was not used on that occasion is simply because there were none at hand at the post. Should the information it contains be deemed of sufficient importance at HDQRS, by sending me a blank printed return I will cause it to be made out and forwarded at once."

With the Second Seminole War winding down, the War Department began moving units out of Florida. Company G was soon packing and on its way to Fort Morgan, which guarded the entrance to Mobile Bay; then, after a few weeks' stay, back to its former billet, Fort Moultrie. This bastion stood at the end of an arm of land and covered the entrance to Charleston. Army engineers were erecting another fortification on an island in the middle of Charleston harbor, though it would not be finished two decades later when it had its brief moment of fame; named for a Revolutionary War hero, it was to be called Fort Sumter.

After the stir and action of the Seminole War, then the interlude in Fort Morgan, at Fort Moultrie Company G would finally come to rest; Sherman would eventually look upon it as home. At Moultrie he would get to know in all its details the close little world that was the peacetime army, a world he would inhabit for the next ten years. The army world was indeed a small one. In 1842 it contained 10,780 officers and men. The officers' corps numbered 781, about the number of students one would find today in a small college; more than half of those officers had a common background at West Point. And as in a small college, everyone knew everyone else, or at very least knew of them.

Much of the talk and gossip was about other officers, but it was not all idle. In the matter of promotion what happened to others was intimately linked to one's own professional future. With the army's authorized strength varying very little, and with promotion based essentially on seniority, any rumor of a resignation, death, or even serious illness caused great excitement. The departure of a single officer of considerable rank could produce a long chain of advancements.

Since there was no retirement system in the army of those days, officers sometimes served until they literally dropped. George Gibson was the army's commissary general of subsistence and Thomas S. Jesup was its quartermaster general when Sherman graduated from West Point, both having joined the army before he was born; both were still in their respective positions when he turned forty.

The disabled continued to serve alongside the superannuated.

William L. McClintock, the major of Sherman's regiment, continued to hold that position long after his condition made it impossible for him to exercise its functions. A report by the adjutant general in 1846 noted that McClintock "cannot walk; and could not when he was promoted in June, 1843." Not surprisingly, promotion in such a static organization was painfully slow. According to an estimate prepared by the adjutant general in 1836, it took eight years to reach the rank of first lieutenant, ten more to reach captain, and then twenty more to major; a colonelcy could be expected after fifty-eight years of service.

The base pay Sherman received was $25 a month—a salary that had not changed since 1798, though it is true that allowances had been increased, so that total annual income could be figured at $700. Still, a junior officer without private fortune had to give considerable thought to the expenses entailed in having a wife and family.

Given these manifold disadvantages, one may wonder why any man would opt for a military career. First of all, with it came status and respectability. To his delight Sherman discovered that in Charleston, as in the South generally, his calling was particularly ennobling. He found all doors open to him, including those of the most exclusive residences. He knew the town's most celebrated belles and danced with them at the most fashionable cotillions. Trying to sell Ellen on the advantages of army life, he assured her that "in the military one meets only ladies and gentlemen."

Then an army career offered job security in an age when the term was unknown. In its way the army took care of its own—Major McClintock offers a good example. Being an officer in peacetime was something like being a fireman. It mattered not that much of his time he was doing nothing in particular, as long as he was available when duty called. Leave policy was very generous; there were cases in which an officer was accorded a year and a half, and in 1853 the army would give Sherman six months off so that he could try another profession.

Finally, if a man could not make his fortune in the army, there was a chance he could win fame, and fame of the most lasting and satisfying kind; any number of young officers dreamed of that, and Lieutenant Sherman was one of them.

But at bottom the army simply suited Sherman, just as the academy had; from the very first he seems to have felt at home there. His letters and diary entries over the years are remarkably free of that sort of institutional griping we tend to think of as almost a prerogative of those in military service. All the army's rules and regulations made eminent sense to Lieutenant Sherman. Twenty years later General Sherman was of the same view. In 1863 he issued a general order informing the officers and men of his command that the United States Army "is, in fact, the most perfect machine which the wisdom of man has ever devised."

In the 1840s one problem the army could do little about was the ennui of peacetime service. This was most serious on the tiny, isolated frontier posts scattered throughout the West, but it proved to be something of a problem at Moultrie, whose garrison was composed of some 250 soldiers, fifteen officers, and a physician. During the week the routine did not vary: reveille at daylight, followed by drill, either artillery or infantry. In artillery drill the men went through the cadenced steps of loading, aiming, and firing the cannon as an officer bawled out the commands. In fact they did none of these actions, but only simulated them, for artillerymen were authorized to fire only a very small number of live rounds per year in gunnery exercises. An officer described the drill as "loading in twelve times, and firing imaginary balls at invisible men. We do it every day to the destruction of our lungs." Then breakfast was at seven, dress parade at eight, and changing of the guard at eight-thirty. Unless they had the duty, by 9 A.M. most of the officers were free for the rest of the day and could leave the fort. Thereafter, Sherman wrote John, "some read, some write, some loaf, and some go into the city."

Those officers who stayed at the fort often congregated in Sherman's quarters. One day they arrived when he was writing a letter to Ellen, so he put away pen and paper; when his guests had departed he resumed his letter: "Some half dozen of them interrupted me. Some tumbled on my bed, others filled the rocking chair and other chairs and for the past two hours we have taken a cursory view of the world in general."

Politics was a perennial subject. It seemed increasingly turbulent and demagogic as the decade of the forties progressed; the conservatism and professionalism of the officers' corps inclined its members to regard the world of Washington with "elitist disdain." Sherman said the carping and bickering of political factions in the nation's capital reminded him of "two pelicans quarreling over a dead fish."

Of Sherman's duties at Fort Moultrie we have a single vignette, but it is a revealing one. He had taken on the duties of post treasurer, which meant he kept the books and made disbursements for the installation. Along with the job he inherited a soldier named George Smith, who was recommended as a reliable assistant. One day in November 1844 Sherman sent Smith to Charleston to pay the garrison's bakery bill and also to pay Sherman's tailor for work done; in all Smith left with a little over a hundred dollars. He did not come back that day or the next. On November 15 he was declared a deserter.

Sherman had already started a search for the man, quizzing people in Charleston and offering a reward for information. Finally Smith was picked up, still in uniform. He said he had never intended to desert but to return as soon as he finished his "spree." During the court-martial the judge advocate asked how much money was owing to the post fund. Sherman answered, "not a cent." He had already made up the deficit with his own money. Private Smith was put to hard labor wearing a ball and chain, with his pay stopped until Sherman had been reimbursed.

If the enticements of Charleston had proved the downfall of Private Smith, they were a safety valve for much of the garrison, and particularly for the officers. It was only four miles by water from the fort to the docks along Charleston's "Battery," with boats crossing the harbor frequently. Charleston would give Sherman his first solid impressions of the South and Southern ways. Their officer's epaulets gave Sherman and his companions easy entrée to Charleston society. In the winter "season" they were showered with invitations to balls and receptions, including the cotillion of the Saint Cecilia Society. In the summers the Charlestonians became next-door neighbors, moving into summer houses in nearby "Moultrieville," so the contacts continued. After four years at Fort Moultrie Sherman could say, "I'm pretty well acquainted with all the rich people."

He seems to have accepted many of these invitations, particularly early in his stay. As time passed he became more selective about the functions he attended. He danced, but apparently derived no great pleasure from doing so. "I never had nor now have much inclination for ladies' society," he confessed. "To sit down and spend a comfortable evening with little diversions I enjoy much, but a fixed ball or soiree, where one contemplates in advance the pleasure he has to enjoy and measure by rules, to me is a bore." There is no doubt that he was attracted to some of the young women; in later years he sometimes spoke of his "old attachment to Mary Lamb"–when he was on his celebrated campaign through the South two decades later he sent her word by an intermediary that he would be happy to serve her in any way. She had married by then, and there is no evidence he ever saw or heard from her thereafter.

From the first Sherman seems to have been repelled by the snobbery, pretension, and ancestor worship he encountered in Charleston society. Among themselves the officers ridiculed the hollow self-importance they found. Sherman's good friend Captain Braxton Bragg was apparently the most talented in poking fun at the more haughty Charlestonians they knew; years later Sherman wrote an acquaintance from those days: "You and I remember how Bragg used to ridicule the pretensions of the 'MacTabs' of Charleston to dominance over us." Those whom Sherman disliked the most were the broken-down aristocrats whose indolence had brought them into genteel impoverishment, with nothing to boast of but crumbling mansions and long-dead forebears. On the other hand he admired the hardworking and enterprising low-country planters who kept their estates flourishing. He made fast friends among them; he visited in their homes, hunted with them, and listened to them as they told him how they ran their plantations. They explained to him the business of "managing negroes." He listened attentively; some of their ideas he made his own. Slavery did not shock him. Black people he accepted for what they were–or seemed to be.

Sherman's stay at Fort Moultrie was increasingly interrupted. He was sent on various missions to other parts of the South: to Key West to look into the conduct of a captain stationed there, to Augusta to check the books of the arsenal and to resolve a quarrel among the officers there, to North Carolina and Louisiana for other hearings. One trip he made for the army was particularly significant because it took him into areas of Georgia and Alabama where he would campaign twenty years later. In February 1844 he was ordered to join Colonel Sylvester Churchill and a Lieutenant R. P. Hammond in a mission to examine claims submitted by Georgia and Alabama militiamen for horses they had lost while on government service in Florida.

Sherman's trip took him across Georgia, passing over an area he would see again in his March to the Sea. He passed close to a little settlement that would take the name of Atlanta three years later and made rendezvous with his colleagues at Marietta. From there the three officers moved through North Georgia, questioning militiamen and their officers. They found evidence of wholesale fraud. In some cases the horse for which compensation was claimed was alive and well, in others the militiamen had put their animals to death in prospect of gain; in almost every case the horses were grossly overvalued. The commission pared down the claims mercilessly. Sherman noted in his diary with a certain satisfaction: "These deductions will about consume the good claims and bring the bad claimants into debt to the government."

Sherman had already seen the militia on campaign in Florida; now this experience reinforced the already bad impression he had of the country's "citizen-soldiers." In his diary he made an ironic entry on the subject of "patriotic citizens . . . defending the firesides of their fellows, not from any sordid motive but from the *promptings* of a generous heart." The three officers carried on with their work into northern Alabama, staying for a time in a settlement named Bellefonte. Sherman was taken by the natural beauty of the area and by the cordiality with which the locals welcomed them, holding a fish fry in honor of their guests. At the same time he was struck by many of the young women: "I have never seen so great proportion of pretty, well-looking, modest and intelligent country girls." And through his stay he continued to meet "strapping girls." The officers soon wrapped up their business and Sherman took the road back to Charleston. Their work was highly praised in Washington since they had saved the government a considerable amount of money.

During his stay at Fort Moultrie Sherman was able to obtain two leaves sufficiently long for him to visit Lancaster. The army had an unwritten rule that a new officer should not apply for his first leave until he had completed three years of service. In July 1843 his three years were up, so he made application for a leave of three months, then asked for two additional months to take care of what he described as pressing family business; eventually the extension was granted. There were in fact important matters to settle in the Sherman family. Sherman's mother was then residing in the family home in Lancaster and in doing so was living beyond her means. Sherman and some of the other sons were giving her financial support. Some wanted their mother to stay in Lancaster, others wanted her to move to Mansfield, where it would be easier to look after her; Sherman was in this latter group, and anxious to settle the matter.

Then there was Ellen. Their correspondence had continued, and now, when he arrived in Lancaster in the summer of 1843, their relationship advanced further, for they talked at length about a life together. The obstacles were the same: her commitment to her church and her family, his pursuit of a military career. Though they did not resolve anything, they seem to have parted with a belief that things would work out for them. This is clear in Sherman's first letter to Ellen after his return to Fort Moultrie. His customary salutation "Dear Ellen" is now replaced by "Dearest Ellen." For the first time he speaks openly of love, telling her how happy he is in "the assurance of your love," and the habitual "Very Affectionately Yours" he replaced with "One who loves you dearly." On Monday, March 5, 1844, he crossed the Rubicon, writing a letter to Thomas Ewing, asking for Ellen's hand in marriage. Apparently he was not sure of the response, for he wrote in his diary that the letter would settle his fate as a bachelor "one way or another." Thomas Ewing gave his assent.

Few of Ellen's letters for this period survive, but from those Sherman wrote her one can see the outlines of a compromise of sorts. Sherman has promised his intended that he will examine "with an honest heart and a wish to believe, if possible," the doctrines of the Catholic Church–this from a man who described himself as "not scrupulous in matters of religion." He will also look into the possibility of a civil career. Strictly speaking, neither of these promises is kept. Not long after his return he was ordered to join Colonel Churchill to investigate the militia claims, so he wrote Ellen that he would have to postpone his research into religious matters for want of books, the backcountry of Georgia and Alabama having "no religion except the crudest species."

But Ellen was apparently relentless in asking to know his plans. In a letter of June 14, 1844, he gave a response of sorts. On the matter of changing careers, he seems to be fighting a delaying action. He has been reading law, and will continue to do so, but he confides to her: "Somehow or other I do not feel as though I would make a good lawyer." As for his staying in the service, "I want you to see something of the Army before it is rejected." He assures her that he has known officers' wives "to form a singular attachment to the life." He tells her that he has been tantalized by a fanciful scenario, a "phantasy": he is transferred to Fort McHenry, while Henry Clay is elected president and appoints Thomas Ewing to his cabinet; Ewing and his family move to Washington. "Now Fort McHenry is not very far from Washington," he told her, "and a Railroad connects the two places." Finally, the Whig-dominated Congress votes a pay increase for army officers.

Sherman then takes up the religious issue, outlining his own credo: he believed that God "made the universe and afterward permitted his son to be massacred to display his interest in the human family." In his view the relationship of God and man was essentially contractual: "He will enable us, if we exercise properly our judgments, with due charity and sincerity, to attain a fair share of worldly happiness." Beyond this point he is obviously loath to go. He interjects that it has been "more than an hour since the sentinel called out 'twelve o'clock and all's well,'" and "midnight will have its influence on all." Thereafter he talks about letters he was expecting and a Mexican ship in the harbor. He makes inquiries about her health and gives assurance that his own is "most excellent." Then he signs himself, "yours in love and obedience, W. T. Sherman."

Ellen must have returned to the charge, for he took up religion again in a letter of September 17. He said he had so far drawn little profit from his readings in doctrinal matters and suggested that Ellen should "catechise" him by mail and he would answer her fully and honestly; it would be an even better idea to postpone the catechizing until they were together. At this point he wrote that the call to dinner had sounded, so he would continue later; when he did, it was to tell her that he had a new boat and must decide what to name it.

In January 1845 Sherman suffered a fall with his horse while hunting deer on the plantation of his friend James Poyas, dislocating his shoulder; this qualified him for another leave, which he took in Lancaster. Though their union had Thomas Ewing's blessing, this time the couple made little progress in resolving their differences; whether the catechizing took place is not clear. Ellen's desire to stay in Lancaster, or at least in close proximity to her parents, seems to have become more manifest. Thereafter their correspondence resumed, but things did not go smoothly. Sherman said openly what had long been in his mind: he never wanted to live in Ohio again. Ellen apparently suggested that Iowa might be acceptable, but her future husband objected that the winters were too severe. Since she had made frequent complaints about her health, he suggested the harsh winters of the Midwest were to blame. The South, he decided, would be better for her.

In November 1845 he wrote to lay before Ellen a new proposal: they would settle in northern Alabama, which he assured her was "one of the most beautiful parts of the United States." The following spring he would get a leave to explore job possibilities in that region; if they looked good he would resign and find a place for them to live. They could marry in September and move to their new home. But the couple was fated never to live in northern Alabama. Less than a month after Sherman's letter Congress voted to annex Texas, setting the stage for war with the country's neighbor to the South. In September 1846 he was not settled in Alabama, as he had projected; he was in the South Atlantic, headed around the Horn and bound for the battlefields of the Mexican War.

# 4

## CALIFORNIA

LIKE MOST SOLDIERS, Sherman had watched the war clouds gather in 1845. In June he wrote Ellen that he was tempted to arrange a transfer with an infantry officer in Louisiana, putting him closer to the future theater of war—a pretty clear indication that for the moment he was more interested in combat than in matrimony; that December he put in an application to accompany a detachment of recruits to Corpus Christi, Texas, which would put him closer still to the Mexican border. His request got the endorsement of his superiors, but in Washington it was pigeonholed "until the proper time" and forgotten.

On January 15, 1846, he wrote the adjutant general again, saying he would like to be transferred to any new regiment being organized for "Western service"; he was not interested in more rank but simply "a more active kind of life." Later he would tell Ellen and his sister Elisabeth that he had requested any expedition "however desperate"—the more hazardous the better—but there are no such letters in the adjutant general's files.

In the meantime the army's administrative wheels were turning. Early in 1846 Colonel Gates received orders to select two first lieutenants for service elsewhere. One was to be sent to Captain Samuel Ringgold's new and dashing "flying artillery." Designed for rapid intervention on the battlefield, it was the talk of the army and service with it was highly coveted—

and it would be in the thick of the fighting. The other assignment was a humdrum one: recruiting duty in Pennsylvania and neighboring states.

Sherman did not have a high opinion of his commanding officer, who he felt spent too much time off-post and neglected his command; whatever Colonel Gates thought of Sherman, he thought more of Lieutenant Randolph Ridgely, who may have been the colonel's favorite, as Sherman claimed. Sherman got the recruiting duty. Ridgely was at Palo Alto, the first battle of the Mexican War, where the "flying artillery" won instant fame; Ringgold was killed shortly thereafter, and young Ridgely assumed command.

Sherman read of these stirring events in Pittsburgh. He spent his days there interviewing recruits and his evenings lamenting his fate and writing letters reflecting his mood. He had seen Ellen again, but their meeting had not gone well. One of the letters he wrote her from Pittsburgh he himself described as "shabby"; he wrote again immediately to apologize for it. Dejection and self-pity were reflected in the letters he sent her almost daily: "Every castle that I build is undermined and upset at the very moment I flatter myself of its completion, but the fact is I'm getting pretty well used to it."

Then one day in June came a letter from the adjutant general's office: he was to report to Captain Christopher Tompkins, commanding Company F, Third Artillery, which was in New York City and about to take ship (Tompkins had specifically asked for Sherman and in the adjutant general's office his request had been matched up with one of Sherman's pleas for reassignment). Lieutenant Sherman had to rush to join Company F before it departed. On July 14, 1846, he was aboard the U.S.S. *Lexington* when it sailed for the Mexican province of Alta California, by way of Cape Horn. The *Lexington*'s hold was filled with war matériel and her passenger list was made up largely of the officers and men of Company F; among them was Sherman's old friend Ord. There were a few other passengers; one of them was Lieutenant Henry Halleck of the Engineers, another acquaintance from academy days. They all talked excitedly about the war they were headed for; unfortunately for these young officers eager to win laurels, Company F was the only one of the Third Artillery's ten companies that would neither see Mexico nor fire a shot in anger.

The voyage lasted six and one-half months. Sherman spent part of this time writing long letters to Ellen, including elaborate illustrated descriptions of their ports of call, Rio de Janeiro and Valparaiso. He himself was fascinated by the life at sea. He filled a fifty-page journal with nautical data and he wrote Ellen more than she wanted to read about sailing vessels and seamanship. He told her he had observed the handling of the *Lexington* so closely that in an emergency he believed he could bring her safely into port. He had only to seize the speaking trumpet and shout, "clew lines, let go halyards, lay aloft . . . "

At Valparaiso the officers of Company F learned that they would not be landing on a hostile shore, for American forces had landed in Alta California and occupied Monterey in July. When the *Lexington* entered Monterey Bay on January 26, 1847, all was quiet; the town appeared not only peaceful but somnolent–silent and without movement. Sherman could make out "straggling mud and adobe houses stuck about without order." Perhaps a dozen were large and comfortable enough to be properly called houses, with the rest simply hovels. A larger two-story building, the Customs House, dominated the port. The town was half empty, for most of the adult males had prudently taken sanctuary among the hills that formed a backdrop to Monterey. Such was the former capital of Mexican California.

Soon Sherman found himself working for Colonel Richard Mason, the new military governor. Mason made him assistant adjutant general for the new Tenth Military District and named Henry Halleck secretary of state; Sherman would concern himself with military affairs and Halleck would be responsible for the civil sector. With these two assistants and a few clerks drawn from the ranks Mason moved into the Customs House and set to work governing California, which proved to be more of a challenge than conquering it.

Today the California of 1847 would be classified as an underdeveloped country. Much of its 150,000 square miles was inaccessible to wheeled traffic. It had the thinnest sprinkling of inhabitants. There had been no census, but probably there were no more than 25,000, counting the Indians, who made up more than half the population. Of about 10,000 whites, the vast bulk were Hispanic Californios, as they called themselves, with probably not a thousand Americans, almost all of them recently arrived. There were a few Europeans, among them the Swiss Johann Sutter, whose name would be famous on two continents by the end of 1848. There were already some few Chinese, for a tenuous commercial relationship existed with China—also with the Hawaiian Islands, then called the Sandwich Islands.

The region's economy had scarcely moved above the subsistence level, with only a trickle of foreign commerce through the Customs House. Import duties were high, so most goods were smuggled in. California had no commercial centers and thus no urban life to speak of—Sherman would find San Francisco a "village" of some 700 souls. There was scarcely any coinage in circulation. Barter was widespread; cowhides, "California bank notes," were a common if cumbersome medium of exchange.

As for viable political institutions, Colonel Mason and his assistants found very little to build on. Only alcaldes, or mayors, seemed common to the whole area, and their powers varied greatly from one town to the next. The colonel began by laying down the fundamental principle that military authority was superior to civil functionaries. Sherman found this step eminently sensible. Mason ordered the full resumption of customs collections, with the proceeds going into army coffers as a "military contribution." He gave the alcaldes their orders; he reviewed their actions and decisions and often overruled them.

With local legal practices and procedures Colonel Mason had little patience. In the case of hangings in which the rope broke or the knot slipped, the event was considered among the population as something of a miracle, and the culprit often released. When two such cases occurred on the same day, the attending priest rushed to Mason to report the double miracle and have the prisoners released. According to Walter Colton, then alcalde of Monterey, Mason answered that "the prisoners were condemned to be hung by the neck until dead, and when this sentence had been executed the knot slipping business might perhaps be looked into."

Sherman developed great admiration for Mason; after working with him for four months he wrote Ellen: "Colonel Mason is an excellent man, well suited for his office, a little severe for civilians, but just and determined." It was the sort of judgment that could well have been made about Sherman himself by some officer serving under him a decade and a half later. In truth, prior to coming to California Sherman had had no experience in civil-military relations, so that the two years under Mason were a kind of apprenticeship; certainly the papers that crossed his desk presented the whole range of problems that came with maintaining control over hostile and potentially violent civilians—problems Sherman would later have to deal with in the Civil War South. And like Mason, Sherman would base his powers on the laws of war.

With not a thousand troops at his disposal, Mason scattered them in small posts, believing that it was important to "keep up a show of troops, however small in numbers." In the Customs House they worried about insurrection; in December Sherman wrote to warn his friend James Hardie that there was fighting to the south in American-held Mazatlán and that "the infection may spread from Lower to Upper California." Thereafter the threat diminished, but the colonel and his assistants soon had their hands full with other concerns.

It is clear from the surviving records of the Tenth Military District that Mason did not regard Lieutenant Sherman as simply an amanuensis. It was the lieutenant who prepared in their final form the endless reports that the colonel sent to Washingon, as well as orders and correspondence; here and there one can recognize in them the vigorous language that was Sherman's signature. Mason did not hesitate to ask his assistants for their views on policy and personnel matters. It was Sherman who endorsed removal of the obstreperous alcalde of Sonoma by force, and then undertook it at the head of a party of armed sailors (he wrote Ellen the alcalde was "a citizen who is kind of rebellious and needs a little military law"). And it was he who recommended Henry Halleck for the position of secretary of state.

Because Sherman was known to be actively involved in what happened at headquarters, as well as approachable and obliging, he received a considerable flow of letters from fellow officers, most of it in the form of queries or pleas. Could he put in a good word for an acting alcalde of San Francisco, who was doing a good job and should be kept on? A Captain Shannon wrote Sherman that he had "fairly fallen in love with San Diego" and wanted to be stationed there permanently. Someone from the Commissary Department wrote from Washington to find out if salt salmon of good quality was obtainable in Oregon. He found it was, and supplied information on what else the area offered by way of subsistence for troops. In April 1849 he received a letter of thanks from Commissary General of

Subsistence George Gibson—and later on an appointment in Gibson's department.

The function of assistant adjutant general carried with it what bureaucrats used to call a "heavy paper load"; the extensive nonofficial correspondence made it heavier still. Six days out of seven the diligent and seemingly indefatigable Lieutenant Sherman sat at a table in the Customs House, working his way through stacks of papers; for a time he would read, and then seize a blank sheet and drive the pen across it with his rapid, slanting script. If Colonel Mason appeared and added another stack it would be accepted with a dutiful nod. But after one such day Sherman took up a sheet of paper to unburden himself to Ellen: "I have been bending over a table till my head aches almost to bursting. A great deal of writing devolves upon me . . . "

Gradually Sherman extended his contacts beyond the military community. He became a boarder at the home of a woman named Doña Angustias de la Guerra Jimeno Casarín, then rented a small house in town. The Jimeno family contained several pretty daughters, a major reason that young American officers took to frequenting it. Two of the girls eventually married gringos, and after Doña Angustias's husband died she became the wife of Ord's brother, James. If Sherman developed any such romantic attachments he kept them to himself (a story involving a "Sherman Rosebush" whose blossoms symbolized his love for a Monterey belle has found credence chiefly among tourists and postcard vendors).

He and the other officers who frequented the "family circle" of Doña Angustias learned to speak Spanish after a fashion. Among themselves the Americans sprinkled their speech and letters with phrases such as "quien sabe" and "cástigo de Dios." Sherman never became proficient: "I can talk about all the essentials," he wrote, "but often get stalled when I venture into the higher plane."

The ties of friendship formed with the Californios were genuine and lasting; yet at the same time, among themselves the American officers spoke disparagingly of them as a people. It was the universal judgment that they were "too ignorant and too unenterprising to attempt to remedy the evils under which they labor." It was Sherman who found the trenchant phrase and applied the mordant humor. It was he who took to calling the country "the Italy of America," with the languor and *dolce far'niente* of the Mediterranean world, and it was he who explained to Ellen why vegetables were scarce in California: "It is difficult to hoe potatoes on horseback and any employment on foot is degrading."

He also sensed that beneath the surface the Californios harbored a certain hostility to the new ways being imposed by their conquerors: "They do not love us, they do not like our ways, our institutions, our ruthlessness, our internal taxes, our laws, all are too complicated for their lazy brains and lazy hands." About the country he blew hot and cold; he was taken by its beauty, yet repelled by its poverty and lack of resources.

But Sherman saw something else in California: the American invasion was working a subtle but profound cultural change. Already in 1847 he wrote, "the country is filling up so rapidly that the native population will soon be lost"; with it, he knew, would go the distinctive lifestyle of Mexican California. He did not for an instant mourn its passing, but he was fascinated by the vision of the new California that would take its place. Even before he sailed from New York the idea was on his mind, for he shared it with Ellen: "Do you not think it will be a great thing to be the premier in such a move, to precede the flow of population thither and become one of the pillars of the land?"

The most galling feature of California of the 1840s was its remoteness. The mails were the unique link with the East, and the mails came when the ships came–irregularly and infrequently, bringing letters that had been posted in New York or Cincinnati nearly a year before. For Sherman, reading a six-month-old newspaper from the East was an exercise in frustration, and the break in correspondence with Ellen–they had been writing each other almost daily–was particularly brutal.

He dashed off a letter to her as soon as he arrived, but had to wait two months for a departing ship to take it on its way; he did not hear from her until April, when he received a letter she had mailed the preceding October. Yet the couple wrote faithfully, and their surviving letters reflect the continuing desire to share a future together in spite of all, and also to make their love and devotion manifest. "You must admit that I have been constant and persevering almost to a fault, almost to pertinacity," Sherman wrote. "I left my friends at Charleston to be near you," he continued, overlooking the fact that it was not love but War Department orders that had taken him from Charleston to Pittsburgh. At one point Ellen, for her part, offered to make a signal sacrifice: because she believed herself too ill to make him the wife he deserved, and because she might pass her disease on to any children they had, she would remain single; if he found a suitable bride in California and brought her back to Lancaster Ellen would give her "a hearty, earnest and most affectionate welcome." "Am I not sensible? Am I not reasonable?" she asked. Hers was an argument that begged for refutation, and Sherman of course obliged. There would be no other woman for him.

Ellen's health remained a major subject in the couple's correspondence; she complained especially of pain and swelling in her neck: "It has not been well a day since you left." She was living the life of a semi-invalid, and so fatigued with writing that she had ordered stationery of smaller dimensions so that she would have to write shorter letters. Sherman was understanding and supportive, even to the point of invoking the Almighty: "I will still hope that He to whose service you have devoted so much of your time, your affections and your love will in a measure return it by guarding your health."

Occasionally Ellen's endless preoccupation with things spiritual provoked him. Hugh Ewing was then at West Point and not doing well (he would not graduate); Ellen feared that her brother's troubles were traceable to the moral climate of the academy, which had somehow undermined his Christian faith. Sherman would have none of it: it was "folly" to talk about the state of Hugh's soul–"that has nothing to do with his want of industry and enterprise . . ." He further advised her: "Don't write to him as you do to me about the sinfulness of war and all that."

In Sherman's letters to Ellen he tells a good bit about himself. Though Jesse Benton Frémont, wife of General John C. Frémont, met him in 1848 and recalled him as rail-thin with a bad cough, he told Ellen his health was good; only once or twice did he mention the asthma that would later plague him. He was certainly depressed in the winter of 1847–48, when he unburdened himself in letters to Ellen. There was the heavy load of his work; there was the isolation, in which he saw something almost punitive–he was marooned in the backwater of the war, "banished from fame." For a career soldier there was the steady, gnawing awareness that battles were being fought and laurels won in which he would never share.

By April 1848 Sherman's situation should have become a bit easier.

He had reestablished contact with Ellen; the pace of business in the Customs House was somewhat less hectic. There was no longer any danger of an insurrection by the Californios–Sherman described them as "quiet as lambs." To John he wrote: "California is fast settling back into its original and deserved obscurity."

Even as he wrote, an event of seismic proportions was unfolding. On a calm Sunday three months earlier a man named James Marshall was examining the water-powered sawmill that Johann Sutter was setting up on his property along the American River. As Marshall told the story–he would retell it for the rest of his life–"My eye was caught by something shining in the ditch." Though Sutter tried to keep Marshall's discovery a secret, vague rumors began to spread. Then on May 12, 1848, a man named Sam Brannan walked through the streets of San Francisco holding a small bottle in the air and shouting "Gold! Gold on the American River!"–and the great gold rush was on.

The news seems to have passed over the United States with the impact of a sonic boom. Something like a hundred thousand men simply dropped whatever they were doing and started for California; others hesitated for a time, then began packing as stories circulated of fabulous finds: a man panning in Hudson's Gulch got 244 ounces of gold from a single panful of earth; at Sierra City someone unearthed a nugget weighing 148 pounds. Then too, Federal officials in California produced a report saying the strike was indeed genuine.

The officials were Colonel Mason and Lieutenant Sherman; Sutter had sent samples to Mason, who called in Sherman. There were no acids available for the "hot acid" test, but Sherman tried the "malleability" test: he bit down on a sample, struck it with a hammer, and pronounced it gold. In June and July the two officers made an official visit to the site of the strike and a second visit later in the fall. Both times the area swarmed with men, tents and lean-tos dotted the landscape, and every stream was crowded with miners washing panloads of earth. The numbers drawn to the gold fields continued to swell; by the fall of 1850 there were probably a quarter-million people in California–ten times the population of 1846.

It was not just men fresh from the East who waded about in cold streams with pans full of dirt. Many of California's farms and crops lay abandoned; her small towns were nearly depopulated. "The aged have called for their crutches," Sherman wrote Commissary General Gibson, "and children have caught the common infection." And such was its virulence that "the lazy Californians are actually *working*."

Nor was the army immune to this "yellow fever." "Our captain has pretty strong symptoms of it," James Ord wrote Sherman from Santa Barbara, "he is borrowing all the works on geology and mineralogy, makes surveying expeditions most every day, and brings home any number of stones." While most officers were drawn by the idea of wresting the gold from the earth by their own labor, and while Sherman tried his hand at panning, he preferred to take a different tack. He was attracted by the possibilities in acquiring gold in less direct and less arduous ways.

It was the profits to be made in hauling and selling supplies that most appealed to him. In his visit to the gold district in the fall of 1848 he and Norman Bestor used their team to "make a little," as they put it, in the belief that "Colonel M." would not mind. Profits from selling items were even higher than from hauling them. Prices for goods at the "diggings" were four times Monterey prices; a thirsty miner with plenty of gold dust would not hesitate to part with an ounce—worth $18 at the U.S. Mint—for an ounce of whiskey. For a time that fall Sherman, another officer named William Warner, and Norman Bestor maintained a store, with Bestor the front man, since he was a civilian. Sherman wrote John that if he could send $10,000 worth of shoes, blankets, and clothing to the California market, he could realize a very handsome profit.

While the officers were tantalized by easy riches, their men simply succumbed to the gold fever en masse. When the soldiers, who earned $7 a month less fines and stoppages, learned that they could earn $15 a day mining gold, the arithmetic was irresistible. The desertion rate rose even higher after word reached California that peace had been made with the Mexican government. Only in wartime could the army execute deserters; now the worst they had to fear was the lash. At one point the United States Army's strength in California appears to have dropped to fewer than fifty enlisted men; Company F of the Third Artillery shrank to five soldiers.

Colonel Mason's initial response was to take strong action to maintain order. But in August 1848 he had the wind taken out of his sails when the war with Mexico ended. The colonel regarded military rule as legitimate when exercised over occupied enemy territory; but as part of the peace agreement California had now become a part of the United States. Congress had failed to provide a government for its new acquisition, for the status of California–destined to be a "free" state–was caught up in the swelling sectional controversy over slavery and would only be resolved as part of the Compromise of 1850. In the meantime Colonel Mason, finding himself in a jurisdictional limbo, felt he could run only a caretaker regime. Good soldier though he was, by November 1848 he had had enough; he asked to be relieved.

California's governmental problems were accompanied by severe economic dislocation, provoked by the massive infusion of gold into the economy. Prices skyrocketed; wages did the same save where they were fixed by law. This was unfortunately the case with the soldiers' pay. Colonel Mason had the men paid in kind–in rations, which they could sell for considerably more than what was due them in money. The officers were not so lucky; only in 1850 did Congress increase their allowances by the minuscule sum of $2 a month. Having already lost his clerk to desertion, Sherman now lost his black body servant to inflation: the man at least had the goodness to inform his master that he could no longer afford his services: Sherman had been paying him $12 a month; now he was accepting another gentleman's offer of $300. Colonel Mason, similarly deprived of his cook, managed to get his meals as part of a curious menage à trois: every morning one could find the military governor of California, the alcalde of Monterey, and the commander of the navy's Pacific Squadron gathered in the same smoky kitchen, grinding coffee and toasting bread.

Over the next few months the officers in Mason's command survived by belt tightening and "moonlighting," taking leave and finding temporary employment in civilian life. Some became silent partners in commercial ventures, such as the store Sherman had an interest in (and which apparently sold army supplies, a technical violation of the Thirty-sixth Article of War). If they had any money they invested it in hopes of a rich return in California's booming economy. Colonel Mason put 500 ounces of gold into a trading venture involving a shipload of goods to be brought from China.

In February 1849 Colonel Mason received orders to return to Washington. His duties were to be assumed by two functionaries: General Persifer Smith became commander of the Division of the Pacific, while General Bennet Riley became military governor. One of Mason's last official acts was to approve Sherman's application for leave; he used it to give his young protégé a resounding if awkwardly worded endorsement: "If there is an officer in the whole army who from his high military bearing and correct deportment and close application to every duty with a zeal and industry that never tire merits an indulgence, it is Lieutenant W. T. Sherman." General Smith declined to grant the leave and appointed Sherman his adjutant, probably because he would need his help in taking over a new command.

This action did not go down at all well with Sherman, who manifested an increasing frustration with his situation during the spring of 1849. He wrote Ellen in March that he was tempted to resign, but it might take two years to get the necessary paperwork done. He wrote her father a month later: "I am still General Smith's adjutant, and fear I'll find it harder to get off than I expected, but I am determined to do something rash, rather than continue here in service much longer. I have had my share and now want to look after my own interests."

On May 15 General Smith authorized a sixty-day leave for Sherman, his place being taken by an officer named Joseph Hooker. This was hardly the "long furlough" he was hoping for. What he meant about looking after his own interests was working during his furlough, primarily as a surveyor—Ord had made $3,000 in six weeks, doing surveying work in the Los Angeles area. Johann Sutter wanted Sherman to divide some of his property into small farms, while another entrepreneur was planning a new city with the ambitious name New York of the Pacific, and wanted Sherman to lay it out. Opportunities abounded; all Sherman needed was the time to take advantage of them.

Then word came that Colonel Mason had been made brigadier general as a reward for his work in California; a corresponding reward for his able second, a gesture the army often made, had in this case been omitted. Sherman now did something rash: on June 12 he wrote a letter resigning his commission and addressed it to General Smith. In it he spoke of the "implied disgrace" in his being overlooked: "Self respect compels me, therefore, to quit the profession." In truth it was not just his amour propre that was injured, but his material interest as well. Now he and Ellen were agreed that their marriage should take place soon, and if he stayed in the army he would need a captaincy; with pay and allowances of $170 a month and the right billet a captain could hope to support a family.

Curiously, there was no reaction to his resignation, not even an acknowledgment. When he returned from furlough he found that General Smith had simply set his letter aside; now he wanted to talk to its author. Smith encouraged Sherman about his promotion prospects and promised him that later in the year he would be sent East to deliver dispatches and make a report in person to the army's commanding general, Winfield Scott. In the interim he would spend the rest of his time in California as one of the general's aides-de-camp, officers who had much free time. Moreover, Sherman recorded in his *Memoirs* that "General Smith encouraged us to go into any business that would enable us to make money." Sherman returned to surveying in his spare time, making several thousand dollars in the process.

Then at the end of the year General Smith's dispatches were ready. On New Year's Day, 1850, Sherman sailed from Monterey, almost three years from the day he had arrived. With his mind on what lay ahead, on the new chapter about to open in his life, he could scarcely have imagined that he would soon be back in California—and no longer wearing a uniform.

# 5

## ENTR'ACTE

TRAVELING THIS TIME by steamship, Sherman took just a month to reach New York, where he found a new political order prevailing. The Whigs had triumphed in the presidential election of 1848; General Zachary Taylor now occupied the White House. Winfield Scott, the army's commanding general, was in New York, so Sherman called on him first. The conqueror of Mexico City invited him to dinner and quizzed him about the situation in California, then sent him on to Washington to report to the president and to Secretary of War George W. Crawford. Sherman had never seen Zachary Taylor before; now he spent an hour with him, once again talking mostly about California. After Winfield Scott's overpowering presence, Sherman found Taylor a man of "pleasant, easy manners." Having accomplished his mission, Lieutenant Sherman was rewarded with a leave of six months.

It was in Washington that he was reunited with Ellen, for Thomas Ewing was now secretary of the interior and living with his family in the Blair House. The young couple talked at length about their forthcoming marriage. They seem to have worked out a compromise of sorts on the issues that divided them, though there is too little written evidence to do more than guess its details. Sherman would leave the army and set out on a civilian career and Ellen would go with him wherever opportunities seemed best–he had made it clear that would not be Ohio. Ellen apparently dropped or at least muted her overt campaign to convert him to Catholicism (she continued to pray for his conversion). Sherman then headed for Mansfield, Ohio, and a reunion with his mother.

The letters he addressed to Ellen during his stay at Mansfield were filled with the details of their impending marriage. On March 27, 1850, he wrote her what things would be like after he assumed "the high trust" of being her guardian and master: "I will not promise to be the kindest-hearted, most loving man in the world, nor will I profess myself a Blue-beard, as there is a wide medium in which the contented, happy world moves in . . . You shall be my adjutant and chief counsellor, and I'll show you how to steer clear of the real and imaginary troubles of this world. Only be contented, happy, and repose your trust in me . . . "

Specifically she should trust him to decide when he should take up another profession. He warned her that it might take as long as two years; on the other hand, if the right opportunity came along he could even quit the service before his leave expired. Ellen apparently accepted this modus vivendi; at least there is no evidence that she objected. Thus was this issue in their relationship papered over.

Even as he was writing to Ellen about his resignation from the army he was working to improve his prospects there. He had now established good relations with Commissary General of Subsistence Gibson and with Adjutant General Jones; he had been a dinner guest of Commanding General Winfield Scott and had had a tête-à-tête with the president; there was probably not another lieutenant in the army who was so widely known in the right places. His friend James Hardie teased him about having "the reputation of a favorite."

Now Sherman sent a deftly crafted letter to Adjutant General Jones, asking for a captaincy: "My peculiarly bad luck in having served the past few eventful years in California, to my great personal sacrifice and loss of professional advancement, must be seen as my apology for this not very modest request." Legislation had been proposed creating four new captaincies in the Subsistence Department, and one of these would suit him best; Commissary General Gibson would probably look favorably on his candidacy.

The shift from line to staff duty would be an eminently sound career

move. In the peacetime army–and Sherman saw decades of peace ahead–
a first lieutenant in the artillery had little chance to win glory or advancement in skirmishes with Indians. He could do no better than advance to a
captaincy in the Commissary Department; such was the conventional wisdom, firmly grounded in fact. A commissary captain was almost assured of
being assigned to a community of some size, where he could find the
amenities conducive to family life–and absolutely necessary to Ellen.

As the wedding date approached Sherman played the typical bride-
groom. He tinkered with the itinerary of their honeymoon trip–Baltimore,
Philadelphia, New York, and on to Ohio, with obligatory stops at Niagara
Falls and West Point; he warned Ellen not to bring too much luggage, "two
trunks at most." He chose two Catholic army officers as groomsmen–Ord
and James A. Hardie, both close friends and fresh from California themselves. Sherman made the customary jokes, telling Ellen he would need
two good men to back him up "in so awful a danger."

It was a sumptuous wedding, and for a time the talk of Washington.
Sherman wore full uniform of the new pattern, with saber and spurs.
Three hundred guests came to Blair House for the affair, including Presi-
dent Taylor, his cabinet, and a constellation of congressional dignitaries.

After the wedding and honeymoon the couple spent most of the summer in Washington. When President Taylor died suddenly in July, Sherman was tapped by Adjutant General Jones to take part in the funeral
procession; with time on his hands he became a spectator in the great
political debates that led to the Compromise of 1850. He witnessed the
sectional clashes in the Senate, heard Webster and Clay speak, and
watched a gaunt, moribund John C. Calhoun sitting slumped in his chair,
listening to his own speech being read by a colleague.

Vice President Millard Fillmore's move to the White House entailed
changes in the administration; Thomas Ewing lost his cabinet position,
but then was appointed by the governor of Ohio to fill a Senate seat. He
moved his family back to Lancaster, and Sherman accompanied them for
the time being.

The time was fast approaching when Lieutenant Sherman would
return to duty, but just where or in what capacity he did not know. General Scott had assured Thomas Ewing that his son-in-law would receive
one of the four new captaincies in the Subsistence Department. It came

through in September, and to Sherman's delight he found himself stationed in St. Louis. He had liked the town from the first for its clean and prosperous look; his attachment would grow stronger with the years.

St. Louis was a departmental headquarters, under the command of Colonel N. S. Clarke, Sixth Infantry. Major Don Carlos Buell was departmental adjutant general and Colonel Thomas Swords was depot quartermaster; all the offices were housed in a building in downtown St. Louis. Of Sherman's official functions—handling and distributing foodstuffs and keeping accounts relating to them—there is scarcely any mention in his correspondence, nor does he refer to this subject in his memoirs. He had an excellent clerk and all records were in good order; it was a slack time for purchasing operations, so he had little to do.

There was only one problem with this excellent billet: it might be of short duration. A senior officer in the Subsistence Department, Major R. B. Lee, had previously run the service in St. Louis; he was now in California, but was to return within the next few months; though he had not stated any preference for his next duty post, it was thought that he would choose to return to St. Louis; if he did, Captain Sherman would of necessity be moving on. Just when Major Lee might return no one seemed to know, not even General Gibson. Sherman, who knew the major, described him as "of an erratic turn."

For his immediate housing needs he moved into the Planter's Hotel, which he had admired in 1843. He and Ellen were destined to begin housekeeping in St. Louis, but not immediately. She was now pregnant, or as he decorously phrased it in a letter to Hugh Ewing, "detained by a cause common to married women." Neither Ellen nor her parents felt she should leave Lancaster until after the baby was born, and her husband reluctantly agreed.

Without the company of his wife, without much demand on his time as commissary, Sherman soon immersed himself in work of a totally different kind: he became the agent and counselor of Thomas Ewing in the considerable interests Ewing possessed or was in the process of acquiring in St. Louis and the surrounding area. Soon Sherman was collecting Ewing's rents, paying his taxes, and buying and selling land for him. Within a short time he made himself at home in real estate operations. He made recommendations on the prices Ewing should ask, based upon a study he had made of sales of property contiguous to the Ewing holdings. His advice was generally that of a prudent man: "My opinion is that unless you contemplate a long investment . . . you should sell when you have a fair offer, as times may change for the worse."

His wrote steadily to Ellen; one of his main goals was to convince her of the charm, the salubrity, and the coming greatness of St. Louis. Knowing her susceptibility to all matters of health, he wrote her on his arrival that "the idea of this being a sickly place is absurd" (Thomas Ewing had unfortunately used that term to describe the city). Sherman insisted that it was "high, dry, with broad streets and the pure air of the plains." It had a sizable proportion of Catholics, whose spiritual needs were supplied by "well educated priests."

The couple had agreed that as soon as Ellen gave birth—was "safely disencumbered," to use Sherman's expression—he would be informed by telegraph. The telegram arrived on January 28, announcing that Ellen had given birth to a daughter. The child was christened Maria in honor of her maternal grandmother, but like other members of the Ewing clan she was destined to go through life answering to another name—Minnie. In March Sherman went to Lancaster and brought his wife and infant daughter back to St. Louis.

The little family settled into a small house on Chouteau Avenue. The household they set up there—the first of a long series—would last a little over a year. Just how well the two adjusted to married life is not easy to say. In the letters the couple had exchanged after their marriage two old problems had already reappeared. He wrote her that now he had received his captaincy, "one of the best appointments in the Army," it made no sense to leave the service; they should consider his fate sealed. When he sent word that he might possibly be sent to Oregon, she balked at the idea of going there, unconvinced by his glowing descriptions of the Northwest or by his assurances that "my study is your welfare."

Ellen was less than charmed with St. Louis. She missed both her family and the home place in Lancaster. The couple did move in a circle of acquaintances, mostly army officers and their wives. Sherman found in St. Louis an old friend from the Monterey days, Major Henry Turner. The Shermans saw a good bit of the Turners; Ellen and Mrs. Turner compared notes on housekeeping and child-rearing; the men reminisced about California. Sherman would later speak of Turner as his closest friend.

If Ellen was afflicted with homesickness, her husband was increasingly preoccupied with money problems. He had been careful in his personal finances, managing to live on a junior officer's pay, give help to family members from time to time, and still save money; in the first few months in Monterey he had put aside $800, and he had come back from California with considerably more than that. He would have had more still, had he not been so ready to make loans to family and friends. Hugh Ewing owed him money, as did a man named McNulty, who had made part payment on some California property and still owed him $2,500. Sherman had in the meantime begun to buy real estate himself; he talked of "a growing permanent investment."

Then things changed for the worse. Living *en famille* with Ellen proved more expensive than he had thought. He was pinched to find money for furniture; in addition to the nurse for Maria-Minnie, Ellen needed two other servants. The financial picture continued to darken. In November 1851, after eight months of family life, he wrote John that he was increasingly dipping into his savings to pay the monthly bills: "My household expenses here are beyond my means—I thought that gradually I might economize, but 'tis impossible." He made the same confession to Thomas Ewing two months later.

The most frustrating aspect to Sherman's predicament was that a solution was at hand, if only he could grasp it. McNulty was to have paid the remaining $2,500 on his note at Mansfield in November 1850; when the time came the man could not be found. He was reported to have been seen here and there in Ohio, then word came that he had gone back to California. Sherman put John on the case, but John was soon in a thicket of difficulties: it was not that simple to collect in Ohio a debt contracted in California. Months passed before Sherman got his money; by then he had been forced to take several economy measures; he sent Ellen and Minnie back to Lancaster for the summer and sold the house; he became a boarder again, and would remain so for the rest of his stay in St. Louis.

Meanwhile Major Lee had pursued his leisurely course back toward Washington. Sherman had a friend in the Commissary Department, Lieutenant A. E. Shiras, and Shiras kept him informed. At first it looked like Sherman might be sent to Santa Fe, New Mexico, then Oregon, then in

April 1851 it appeared that he might be kept on at St. Louis; the suspense continued into 1852 as Major Lee toured South America.

Then Mary Sherman died suddenly in Mansfield. She had just written her son she was in good health and looking forward to a visit from him, so the blow was all the more telling. In a long, anguished letter to Ellen he summed up his mother's life in a phrase that could have served as her epitaph: "She has had hard times."

Finally there was the separation from Ellen, which he was not taking well. He blamed her for it in a letter of August 13, which he knew would cause her pain. The letter has not survived, but he referred to it in a second letter, this one of apology, dashed off the following day. They had been married over two years, and had scarcely been together half of that time, and he confessed that it had been hard on him. Now they were faced with another period of separation, for she was pregnant again with her "confinement" not far off, while he would soon be packing up for another move to an unknown destination.

The tone of the second letter to Ellen was sad and affectionate, but the brief one he sent her father about the same time–a note regarding real estate transactions–was polite but distant and formal. In it he announced that he would not need a building lot after all: "I have extinguished all idea of having a permanent abode and therefore do not desire any place for myself." Sherman's letter to his brother-in-law Tom Ewing Jr. was in a tone he rarely used with family or friends–blunt and angry. He told Tom he was under orders to another post and would have to leave without seeing Ellen or Minnie: "This is too bad and is only due to the intense love you all bear Lancaster. I have good reason to be jealous of a place that virtually robs me of my family and I cannot help feeling sometimes a degree of dislike for that very reason to the name of Lancaster." He informed Tom that when Ellen rejoined him, it would be for good: "As to her being home next summer when you get there I doubt it exceedingly–I think she has been at Lancaster too much since our marriage, and it is time for her to be weaned. Unless New Orleans should be visited by an epidemic she will not likely leave for two or three years."

For Sherman now knew that New Orleans would be his next post. On September 4 Lieutenant Shiras had informed him in a confidential note–marked "Burn this at once"–that General Gibson had made the decision.

In the end Sherman's fate had not depended on the choice Major Lee made; indeed the major still had not made up his mind. Bad practices and a burgeoning scandal had been uncovered in the New Orleans office; the commissary there, Major Waggaman, was being transferred to St. Louis and Sherman would take his place. His orders followed soon after.

On his arrival in New Orleans he took a hotel room and plunged into his new assignment. His first task was to deal with a considerable backlog of correspondence left by his predecessor. That dispatched, he had to change the way of doing business. What had led to Major Waggaman's departure in semi-disgrace was a classic scenario, one as old as capitalism itself: during the Mexican War a Major Seawell had served as acting commissary in New Orleans, and had seen fit to make a large proportion of purchases from the firm of Perry Seawell and Company, this Seawell being a relative of his. When Major Waggaman took over he continued the practice. Then Waggaman's brother, thrown out of work, found a clerk's job at Perry Seawell and Company. By then the firm had almost all of the commissary business.

In Washington commissary officials noticed that while the New Orleans market prices for such foodstuffs as sugar, coffee, and bacon were lower than at New York or Baltimore, the other points of purchase by the Commissary Department, the prices the government paid did not reflect this advantage. Moreover there were problems of quality, for commissary officers at posts in Texas, customarily supplied from New Orleans, were increasingly asking that their provisions be shipped from other depots; finally, other New Orleans merchants complained.

Quizzed on these matters, Major Waggaman admitted freely that he did most of his business with Perry Seawell and Company because they were prompt, reliable, and fair in their prices; he could see no impropriety in doing business with a firm in which his brother was now a partner. Thereupon Waggaman was transferred. Sherman thought this punishment sufficient, for he was convinced from his review of the records that Waggaman had not profited personally.

When Lieutenant Shiras told Sherman he was going to a post with such a troubled past, he assured him that the reasons General Gibson chose him were "most flattering." In truth Gibson was looking for an officer with energy and probity whom he could count on to set things aright without delay, and he knew or at least had been told that Sherman was his man. He made the right choice: in a short time commissary operations in New Orleans were completely overhauled. Sherman shopped along the levee for the best quality and lowest prices. He leased a warehouse and hired a guard for it at an annual cost of $1,080, but wrote General Gibson, "I will warrant to save twice that sum;" he could now buy nonperishables when prices were lowest and store them. Major Waggaman had let Perry Seawell and Company do the packing, but now Sherman's own personnel did this under his eye; he was known to roll up his sleeves and fish through a barrel of salt pork to verify quality.

Captain Sherman soon became well-known on the levee and in commercial and social circles generally. He was after all one of the city's major purchasers of foodstuffs, buying sugar, coffee, or hams, tons at a time. He could not help boasting to his brother John: "As I handle considerable funds I am somewhat of an item here–Dinners with elegant banquets–clubs, etc. are frequent, but when it is found that these will not win favor, they will in a measure cease."

Sherman found time to rent a house on Magazine Street and move in; he sent Ellen an elaborate description of the place, complete with a sketch of the floor plan. He soon received word that he had another daughter, Mary Elisabeth Sherman, to be called Lizzie. Ellen had had an easy time of it and was soon back at letter writing.

Sherman's letters to Ellen were mostly about New Orleans, and one can see in them a systematic effort to sell her on all its advantages. "The city is very healthy," he assured her, "I never saw more healthy people than crowd the thoroughfares." While it was true that there were a few yellow fever cases, they were kept under confinement in hospitals; "We never hear or see of them out doors." The climate was very agreeable: "our weather continues beautiful," he wrote her on December 14, "and I fear for you in the cold miserable climate of Ohio." Culturally, he stressed the French, Catholic influence in the city–so profound that "it will take centuries to change it." In the meantime she would find plenty of churches.

Strangely enough, while he was trying to convince Ellen of the wisdom of his move, he himself was beginning to have doubts about it. He had confessed to her that he had found the city a surprisingly expensive place to live. When she joined him in late December, bringing two infants, a nurse, and Sherman's young sister Frances, his expenses shot up alarmingly. To his brother John he confided that his army pay–adequate to support a family in most places–would not cover the cost of living in New Orleans; his superiors knew this but could do nothing, for his pay was fixed by law. If he stayed he would be obliged to sell what real estate he had acquired over the past two years.

On November 21 Sherman wrote to his friend Henry S. Turner, renewing contact that had been broken when he left St. Louis. The letter does not seem to have survived, but something of its nature may be seen in Turner's response. From it one may deduce that money was not Sherman's only worry and that he was having doubts about a commissary's life. "I infer from your letter that you are not altogether pleased with New Orleans," Turner's letter began. Sherman had confessed to a certain "restlessness," and Turner thought it came from giving up the exciting world of real estate speculation for "tame commissary duties." But Turner felt his friend would be no happier if he went back to the artillery: "The duties of a subaltern in time of peace are wholly incompatible with your intellect or energy."

Then Turner announced that he was turning a page in his own professional life with a "California venture," a bank to be opened in San Francisco as a branch of Lucas, Simonds and Company of St. Louis, in which he was a partner. Finally he came to the essential point of his letter: "Would a fair consideration induce you to embark in the same business with myself?" To Sherman the offer was enticing for several reasons. First of all it would put him in a better financial position–by "fair consideration" Turner meant a partnership in addition to the salary. Then there was the sheer challenge of it, a challenge he felt he could meet; Turner clearly felt he could, for he had written, "I know you possess an extraordinary business capacity." Finally, this would take him back to California.

Sherman had kept in touch with California friends; even Doña Angustias wrote him. As he had kept in touch, so had he preserved a sort of emotional bond to the place. In June 1852 he confided to Hugh, his kindred soul among the Ewings, that he reminisced a good deal: "My fancy will roam back to Monterey and the valley and hills of Carmel," he wrote. He admitted that such daydreaming as he sat in the commissary office was "enough to strip all quiet life here of all interest and charm. I cannot shake off its effect on me."

Not all of his dreaming had been idle. He and Ellen seem to have talked over a number of scenarios for him professionally in those first years of their marriage, and probably among them was a return to California. In one of his letters to her from New Orleans he speculated that the high cost of living there might eventually compel them to leave, in which case they could fall back on what he called his "Pacific plan." More significantly, in his exchange of letters with Turner, the two men discussed the possibility of Sherman exchanging positions with Langdon C. Easton, then in charge of the San Francisco commissary office. This would be an ideal way to gain entrée into commercial circles there; Sherman may even have written Easton.

Once launched, the California project gained momentum. Turner was leaving St. Louis immediately for New York and on to San Francisco, to get the bank opened; he told Sherman to send his thoughts on the offer to Lucas—then decided to change his itinerary and pass by New Orleans, where he and Sherman talked at length. Then, in Washington, Turner used his army connections to prepare the way for Sherman's leave of absence. Sherman won Ellen's consent; he exchanged letters with the principal partner, banker James Lucas, who also came to New Orleans.

By the end of February the deal was done: Sherman would go to San Francisco to manage the bank on a trial basis, at an annual salary of $5,000 and a one-eighth partnership; the army obligingly gave him a leave of six months. He started Ellen, Frances, and the children on the trip back to Lancaster, and that same day he had the contents of their house auctioned off; on March 6 he left New Orleans by ship, dashing off a letter to John before he departed. He would not leave the army with any sense of bitterness; he had gone as far as he could in the service, but it was not enough for a family man. And he might still change his mind: "You may depend on it," he told his brother, "I will not throw away my present position without a strong probability of decided advantage." He sailed to Nicaragua for a crossing of the Isthmus, which he made by small boat and mule; then he took a ship bound for San Francisco to try his fortune in a new profession.

# 6

## RETURN TO CALIFORNIA

SHERMAN'S RETURN WAS not an auspicious one. His ship struck a reef before dawn. He made it ashore in a small boat, then found a schooner headed for San Francisco with a load of lumber. While entering the bay the schooner was caught by current and winds that rolled it over on its side. A passing vessel—the fourth he sailed on that day—carried him ashore.

The vast changes struck him at once. The docks were extensive and new; alongside them was a forest of masts—more ships than the San Francisco of 1848 would have seen in a year. The town that lay beyond was no less surprising. He wrote John: "large brick and granite houses fill the hill where stood the poor, contemptible village." The city's population was now 50,000. Though the alcalde had given way to an American-style mayor, much of the turbulence of the Forty-niner years remained.

Sherman found his bank, Lucas, Turner and Company, already open, and had his first disappointment when he learned from Turner that it was capitalized at only $100,000, about a third of what he thought was adequate. Turner promised to ask Lucas about an increase, but did not seem optimistic. Sherman found his old friend timid and hesitant in his new role, but Turner would soon go back East and Sherman would be his own master.

California's financial institutions were like none he had ever seen.

While a shortage of specie and a plethora of paper bank notes were characteristic of the American marketplace, in California the situation was exactly reversed: the state was awash with gold; on a single good day in the gold fields production could attain a third of a million dollars; in 1850 the value of California's gold production hit $80,000,000, about twice the size of that year's Federal budget.

And this flood of gold percolated through the state's economy with complete freedom; those who produced it, traded in it, accepted it for deposit or held it as collateral were subject to no regulation. The state authorities chartered no banks, nor did they send examiners to inspect the books of financial institutions; they essentially ignored them; yet there was an absolute ban on paper money. There were no usury laws; the prevailing interest rate was three percent–per month. California's bankruptcy legislation was extremely generous to the debtor. The state's banking practices were decidedly different: a merchant detained at a construction site once settled with a creditor by giving him a check for $28,000 written on a shingle with a piece of red chalk; it was honored without question. Lucas, Turner and Company once made a loan with 200 barrels of butter as collateral, and acquired through foreclosure sizable quantities of pick handles, drugs, pickles and other "funny things," as Sherman called them.

In the freewheeling economy of the early 1850s gold was everywhere, most often in the form of fine flakes or "dust." Every bar and hardware store had its set of scales. The dust had drawbacks as a medium, its purity or "fineness" could vary considerably; yet dealing in it could be very lucrative. It could be bought in the fields for $10 an ounce, sold in San Francisco for $14 and at the U.S. Mint in Philadelphia for $18.

Sherman saw very clearly the eccentric nature of the state's economy and its fundamental weaknesses and inequities: "The mines are California, the miners its producers, and they won't pay taxes, bear any of the burdens of the state or acquire any fixed property or houses. The mercantile class simply supply their necessities. We have ranches, farms, etc., but of limited value compared with the gold." And since gold was the life's blood of California's economy, an interruption or even a slackening in its flow registered in every sector. If floods isolated the gold regions for any length of time–or worse yet if a drought shut down mining operations–the whole economy slumped. When three days of heavy rain ended a drought in

April 1856, Sherman estimated they were worth $10 million to the state.

The demand for credit was insatiable—Sherman claimed he could easily place ten times the funds he had out at interest. But the high rates also reflected a heightened element of risk. In a business community that had sprung up almost overnight there were no old established firms with a history of unimpeachable credit. Many in San Francisco's business community were like Sherman, new to what they were doing; almost every borrower was something of an unknown quantity.

Then there was the dearth of good collateral. There were no "gilt-edge" securities, bonds or shares of profitable, well-run railroads, for example, or of old, well-established commercial or industrial firms. The value of state and municipal obligations fluctuated; San Francisco's paper "warrants" were heavily discounted. Real estate, normally the most solid security for a loan, made poor collateral. It was heavily taxed and land titles often disputed. Many of the structures were of wood, and this in a town without fire insurance and periodically swept by large conflagrations.

The director of Lucas, Turner and Company went into banking with the same energy and drive he had displayed in the army. With the assistance of Benjamin Nisbet, chief clerk, he quickly mastered the bank's various operations. He got to know the competition: nineteen banks, nearly half of which would go out of business the following year. Page, Bacon and Company, by far the largest bank, was like Sherman's establishment an extension or branch of a parent firm, Page, Bacon and Company of St. Louis.

Banker Sherman explored other aspects of the city's economic life, noting down figures, asking questions, and making acquaintances about the town; he made brief trips to the surrounding country. By early June he decided the banking venture would be successful if more capital were available. "My business here is the best going, provided we have plenty of money," he wrote John. "Without it I stick to Uncle Sam emphatically." The question of additional capital would have to be settled in St. Louis with Mr. Lucas. Sherman sailed for New York in July, went for a brief visit with his family at Lancaster, and then on to St. Louis. There Lucas agreed to increase the bank's capital by $100,000 and pay for a new bank building. Sherman in turn gave his word he would remain with the bank until 1860.

He must have come away from the meeting in a state of considerable elation. Some of that elation shows up in the letter he immediately drafted to his father-in-law, and along with elation more than a touch of self-congratulation: "In going to California six months ago I knew it was experimental and it has resulted even better than was anticipated . . . I stand in the position of chief of a banking house in the great city of the Pacific—with a cash capital paid in of $200,000, and with the promise of a firm credit in New York for overdrafts. You are acquainted with all my associates and know them to be wealthy and respectable men . . . They assert perfect confidence in me and give me everything I ask."

The second stopover with the Ewings did not go quite as Sherman had planned. In the family discussions Thomas and Maria could not hide their unhappiness at the prospect of their daughter's departure. They were now an aging couple living in a large house that had once rung with the happy sound of their children; during Ellen's stay with her daughters the house had come back to life, and now that would end. Couldn't Minnie stay with them a little longer? Ellen had found it hard to deny her parents. She and her husband had Lizzie and would doubtless have other children (she became pregnant again that September); yet she made her agreement contingent on that of her husband, and he reluctantly gave his assent.

One more step remained to be taken, one that may have moved Sherman more than the temporary separation from his elder daughter. On September 3 he wrote out a letter of resignation from the army and addressed it to the adjutant general, thus leaving a career he had entered seventeen years before on the plain at West Point.

Sherman, Ellen, Lizzie, and the infant's nurse took ship at New York and reached San Francisco without major incident. Once there they settled into a rented house and Sherman plunged back into banking; in his spare time he worked on plans for the new bank building.

By the end of 1854 Sherman and others with their fingers on the city's economic pulse detected signs of a subtle but general decline. Stock companies were failing, and merchants complained of sagging sales. Then, on February 17, 1855, a run began on the bank of Page, Bacon and Company, forcing it under. Six days later the contagion spread to Lucas, Turner and Company. Sherman's tellers paid off hordes of depositors all day, while he reassured big depositors and made the rounds borrowing cash, including some from the army. By three o'clock, an hour from closing time, "our trays looked slim enough," Sherman recalled. "Our clock seemed provokingly slow." Finally four o'clock came and the doors closed. Sherman made the rounds again and was braced for heavy withdrawals the next day, but the run was over.

Dwight L. Clarke, who made a close study of Sherman's banking career, had nothing but praise for his conduct during the famous run, marked as it was by his "coolness, courage, integrity and boldness to improvise"—the same attributes that would translate into "coolness under fire" at Shiloh. Not surprisingly, his management style was military: "In banking, as in the Army, there can be but one master," he explained to Hugh Ewing, "and so long as I am here I must occupy that post." He supervised the affairs of Lucas, Turner and Company closely. In the Commissary Department he had adopted the habit of making all money disbursements himself; in the bank he adopted a similar policy: "I signed all bills of exchange myself and insisted on Nisbet consulting me on loans and discounts."

Sherman made it a point not to leave the bank until each day's balance had been made. Though frequently kept awake most of the night by asthma attacks, he was rarely away from his post. This close supervision Sherman believed essential to success in banking. "If I were Mr. Lucas," he wrote Turner, "I would have but one bank, and that right under my eye."

He had a number of ideas about how the bank could be improved; one of the more successful was the creation of certificates of deposit—quite similar to modern ones—by which Lucas and Turner could attract the savings of people of modest means who rarely frequented banks. But it was not the flair for innovation that most typified Sherman's banking operations, but rather his profound conservatism. It was a quality that he had already recognized in himself. In 1851 he confided to John: "The troubles I have seen others encounter in making false calculations makes me timid and suspicious in money matters." The turbulent economy of California and the run on his bank only strengthened his views; after the run was over he told an army friend: "We weathered it by excessive caution, by scattering risks & by not having *too* many sinking friends to bolster up."

Certain practices that other bankers tolerated in their big depositors Sherman would not put up with—allowing them to overdraw their accounts, for one. John Littell, who knew Sherman in the 1850s and later authored a history of trade and commerce in San Francisco, suggested that his stance on this issue kept him from becoming the leading banker of San Francisco: "The merchants were all willing and anxious to give him the lion's share of their accounts, with a large average of deposits, but they demanded in return that he should occasionally allow them to make considerable overdrafts, for which they would pay him high interest. He demanded, however, more security than they were willing to give, and they gradually withdrew their patronage."

Someone who knew him well said Sherman's view of all credit operations was dominated by a "high, almost quixotic sense of pecuniary honor which characterized his whole life." He never borrowed if he could avoid it; he was scrupulous in repayment and as a banker and creditor he expected the same promptness and exactitude. He wrote one customer, "I find with surprise your note of $100 unpaid. My idea is that under the circumstances you should pay it if you have to steal the money."

Some who dealt with him found him Scrooge-like. Isaac Wistar, a steady customer, took his business elsewhere after Sherman made piddling objection to a loan. When James A. Garfield visited San Francisco in the 1870s he recorded in his diary that unflattering memories of Sherman the banker lingered: "Old citizens tell me they have seen him dunning boarding house women for their rents and forcing little settlements for his bank."

In addition to the challenge of increasingly hard times in the spring of 1856, Sherman found himself at the center of a major political crisis. By then he was a well-known and well-liked member of the city's business community, quite active in civic affairs. Captain Sherman was often called on for advice on military matters, and that spring Governor J. Neely Johnson named him to command the Second Division of the California Militia, embracing San Francisco, with the rank of major general.

Though a new city, the San Francisco of the 1850s already had the ingrained patterns of corruption of an old and settled municipality, among them bribery, peculation, jury tampering, bid rigging, and ballot-box stuffing; and periodically its citizenry became exasperated with the resulting scandals. But whereas popular discontent in an older community would usually produce a reform movement, whose candidates would win office and "turn the rascals out," in San Francisco the public's exasperation produced by a kind of spontaneous combustion the extra-legal Committee of Vigilance, which temporarily displaced the existing government and dispensed summary justice to those it identified as evildoers and troublemakers.

The city was not particularly well-run in 1856; bribery and corruption were widespread, and two killings raised public feeling to a feverish state. A man named Cora killed a policeman, then was let off by the jury; and a newspaper editor named Casey shot a rival journalist with the bizarre name King of William. This latter victim lingered some days, public indignation building all the time. Ultimately it produced an afflux of armed men in the center of town and the creation of another Committee of Vigilance, patterned on the Committee of 1851 and with the same popular mandate: to clean the town out with a vengeance.

The two culprits, Casey and Cora, took refuge in the city jail, and the sheriff raised a posse of fifty or so to protect them. As masses of armed men poured into town to do the bidding of the Committee of Vigilance, the men in the posse felt they needed a leader, and they elected Sherman as their captain. He declined to serve since he already had a military function, and as he put it, he would be "major general or nothing." The mayor then called out three volunteer companies, but only a handful of men appeared. On May 16 Governor J. Neely Johnson came to talk to both sides, but with no result. On the eighteenth the committee forces simply overawed the defenders and took Casey and Cora without a shot being fired. Now everyone waited to see if King of William would survive.

Sherman remained essentially a privileged spectator in those first days of the crisis; he gave advice to the mayor, who seemed overwhelmed by events; he toured the jail and told the sheriff it would be impossible to defend against serious attack; and he had an interview with Governor Johnson. On the twentieth, just after news came that King of William was dead, Sherman finished a long letter to Turner that reveals that the author's antipathy to the press was already pronounced: "Between you and me 'tis well King is out of the way," he confided. "If he had recovered he would have been the veryest tyrant on earth & his paper would have become law. Casey's execution will end the *Sunday Times*; thus two birds are killed by one stone. Is this the operation of providence?"

The Committee of Vigilance tried and condemned Casey and Cora. Their hangings and the funeral of King of William were held simultaneously. Having accomplished these tasks, the Committee set to work cleaning house generally, rounding up various gamblers, swindlers, courthouse hangers-on and the like, then putting them on ships bound for New York, where it was said most of them came from. Sherman could only applaud this work: "So at last we are turning the tables on New York."

In this period he received various overtures from the Vigilance Committee. Several prominent members urged him to resign from his militia command, but he declined to do so. He published a somewhat equivocal "Open Letter to John Nugent" in the *Herald* in which he said he was opposed to the action of the committee, though he knew and respected members of it—and he was in favor of ridding the city of its undesirables.

But as the chief military authority he could not stay neutral. On May 29 Governor Johnson warned him that "measures of extreme severity" might be necessary; on June 3 the governor declared San Francisco in a state of insurrection and directed Sherman to call out the militia. His call went out the same day. The response was very disappointing; the *Bulletin* described it as "the fizzle-call of General(?) Sherman." The press had a field day. Sherman was ridiculed as "A Mighty Man of War taken from the desk of a counting house." Unfortunately he had put in his order the palpably false statement that the call-up had "nothing to do with the exciting issues of the past two weeks"; newspaper editors delighted in printing his order just under the governor's directive to him to suppress "acts of violence and rebellion."

Then there were the questions addressed to him by the press: Was it true that his bank held $35,000 in an account payable to Casey once King of William was dead? Had he solicited his own militia appointment? Had he said he would "put down the Committee or reduce the city to ashes"? Sherman soon had enough. On June 9 he resigned. Unfortunately he decided to publish an explanation for his action, which appeared in the *Alta California*. There he said he had given up his post because his attitude of "moderation and forebearance" had not found favor with his superiors. This angered Governor Johnson, who said Sherman's departure was *his* idea—and made further statements that drew on Sherman the ire of General John Wool, the army's highest ranking officer in California.

In truth Sherman was far less roughly handled in this imbroglio than other targets of the journalists' invective. But these men were accustomed to life in the public eye; they of necessity accepted the slings and arrows that went with it. Sherman was not able to accept them and never would be. He complained that the newspapers "poured out their abuse of me," and he later confided to a friend who was in the newspaper business, "I conceived a terrible mistrust of the press in California."

Sherman did not cut a very impressive figure in this whole affair, in which he appears to have been at the very least irresolute. Isaac Wistar, who took the command of that demoralized jailhouse posse that Sherman turned down, spoke of Major General Sherman's "supineness, inaction, and excessive caution"–but then Wistar had quarreled with Sherman and disliked him. Wistar did add that he thought Sherman had an "apparent dread of personal unpopularity," and here he seems to have been on firm ground. Certainly Sherman's public utterances, and especially the texts he drafted for publication, read like attempts–very inept ones–to appear to the public in favorable light.

Then the act of resigning one's command at a time of armed confrontation–something like retiring from the battle line–was something that Sherman was at some pains to justify. He acknowledged that General Wool was "annoyed" with him, as was Governor Johnson. There was a lesson to be learned here. Henceforth he would leave public and political affairs alone; that was a rule he would follow, with a few unimportant exceptions, for the rest of his life.

A decade earlier, aboard the old *Lexington*, he had written Ellen about a dream or vision he had of seeking and finding his fortune in some new place, of rising with a nascent society to become one of its respected leaders, "a pillar in the land." The vision had come to him on his first trip to California; now, a decade later, he seems to have buried it there.

But there was another facet to Sherman's life in those California years, and an important one. Alongside Sherman the banker and Sherman the civic leader there was now Sherman the husband and father. With the exception of seven months in 1855 when Ellen returned for a visit to her parents, the couple maintained a household in San Francisco for three and a half years. Ellen was particularly diligent about making diary entries for 1854 and 1855, so that it is possible to establish with some surety the pattern of their lives together. On arrival Ellen had immediately established contact with San Francisco's Catholic community. In her journal she meticulously recorded the holy days; on February 11, 1854, she "sent jel-lycake to Bishop," and the entry for March 28 reads, "Archbishop called."

Sherman accepted that it would be Ellen who attended to the reli-gious instruction of their children—and ultimately their education, since she destined them for parochial schools. Sherman's own religion, or lack of it, was a permanent source of worry for his wife. He was probably the subject of her diary entry for March 20, 1854: "Prayed for the conversion, etc.," but quarrels over this subject are not mentioned in her diary. Their children came to regard it as a normal thing that their father should go horseback riding while they and their mother went to Sunday Mass.

The greatest bone of contention between Sherman and Ellen was Cal-ifornia. Ellen arrived predisposed to dislike the place because of its remoteness from Ohio; she was further unnerved by the perilous sea voy-age required to reach it. Ellen poured out her feelings in letters to her par-ents. She would never be content to live in "this miserable city." Everything that happened there fed her discontent.

Sherman for his part unburdened himself to Henry Turner and Hugh Ewing. "She don't appreciate the advantages I have here," he complained to Hugh; "I cannot yet tell what will be done. But I can have the satisfac-tion of knowing that in any alternative the blame will fall on me. There is no doubt I ought to make a home for them here or in Ohio. If here we are all together, but in Ohio I never expect to live & therefore the children would grow up as Minnie now is, an utter stranger to me."

Minnie was constantly on Ellen's mind in those first months. When she had no news about her daughter she imagined terrible things had hap-pened to her and she would be moved to tears; then when her father wrote to report that Minnie was thriving, "fat as a little pig and all the time in motion . . . not yet entirely spoiled," that would be the occasion for fresh tears. "Good cry at night," she confided to her diary on New Year's Day, 1854; "had a cry about Minnie before I went to bed," she wrote the day fol-lowing.

At the approach of each spring, when travel became less hazardous, Ellen talked of returning home for a visit. As early as February 1854, Sher-man wrote Hardie: "I have my share of family cares inasmuch as Mrs. Sherman is now preparing to go home and we don't know what to do with the children." She most likely dropped her plans because of a series of maritime disasters that spring. Then in the spring of 1855 Ellen began packing again, and this time Sherman seems to have encouraged her; with her and the children living with her parents for several months he could save money. Then Ellen decided the hazards were too great to take the children, so she would leave them behind. That is what she did, so for seven months Sherman and the children shared their house with Samuel Bowman (the bank's attorney) and his wife, with Mrs. Bowman running the household.

While Ellen agonized over her absence from "home," her husband agonized over the financial burden of keeping her and the children with him in California. Though nominally well-to-do, he found that maintaining a household in California was taxing his means. The Shermans lived well, occupying three successive houses, each larger than the one before, and Sherman gave serious thought to building still a fourth. Ellen had three female servants and for a while Sherman employed a young man as well. The couple entertained frequently. A good bit of money went for furniture, and an expensive piano graced the parlor; for a time Sherman kept a horse and buggy. While spending lavishly, he complained about the expense and blamed Ellen. But if he spoke to her about it she had a ready answer: he had only to take her back to Ohio, where she would be happy in a hovel. So it was Turner who heard most of Sherman's complaints: Ellen was "utterly deaf to all ideas of economy, not extravagant but not caring for money where the children are concerned." After maintaining a household for a year Sherman confessed to his partner that he was unable to set any money aside and was in debt not only for his house but for his furniture as well.

Ellen's diary reveals much about the couple's daily life, which was marked by persistent health problems for both partners. Ellen was a semi-invalid, probably housebound much of the time; she was still troubled by boils and headaches, and mixed in with these were episodes with colds, cholera morbus, and an unspecified malady for which she was taking tincture of opium. This seemingly endless succession of afflictions tried her husband's patience. Even before they were married he had teasingly suggested that she derived some pleasure from taking medicine, especially "your favorite oil, over the administration of which you linger with a species of pleasure not unlike the wine bibber." In later years he would tell her flatly that her illnesses were imaginary.

Sherman's own health was none too good, with the chief affliction being asthma. Curiously, it had not troubled him much during his first stay in California, but now it settled on him in earnest. He suspected at various times the sea air he was breathing, the moisture in walls, or "carbonic acid" emitted by trees during the night, but he was never able to fix on the cause. For long spells at a time asthma would rob him of needed sleep. He had a particularly bad bout in 1854. He wrote to Turner that summer: "For the past seven months I have been compelled to sit up, more or less each night, breathing the smoke of nitre paper, and know that the climate will sooner or later kill me dead as a herring." Occasionally he was too afflicted to go to the bank, and in that case customers who needed to see him personally would call at his home.

Ellen's diary occasionally affords more intimate glimpses into the couple's life together. They apparently continued to share the same bed through their various illnesses, though the niter paper Sherman burned tended to make Ellen's headaches worse. Frequently she noted that her husband was "up in the night." For April 28, 1854, she recorded, "Cump and I have talk before getting up," and a few days later there is a curious entry: "Cump rubbed me with whiskey." Frequently they had guests for dinner or evening callers; occasionally they invited another couple for an evening of whist. There is also a description of the two spending a quiet evening together: "Cump & I sat upstairs in the evening, Cump reading and nodding and I sewing."

Two things stand out in this pattern of activity. First of all the friends who came to visit the Shermans were in the great majority people connected in some way with the United States Army, headed by General Ethan Allen Hitchcock and his successor, General John E. Wool, and thereafter a succession of colonels, majors and captains; there was Major Hammond, who been with him in Alabama in 1844 and had more recently given his bank a transfusion with army funds during the run; and there was Lieutenant George Stoneman, who would become General Stoneman and command a cavalry corps in one of Sherman's armies. Then, while the couple often received guests, they rarely paid calls or attended events together. Sherman frequently went out by himself. Often it was a meeting he attended, but Ellen's diary also indicates he went to the theater or attended dinners.

In this period it is difficult to gauge the depth of the couple's feelings for each other—missing are the tenders of affection that would have been in their letters to each other. One has the impression that in their regard for each other the partners sometimes worked at cross purposes. Sherman spent more than he should have—and probably more than Ellen asked him to—in a vain effort to make her like California. He had told her father that he would now have the means to keep her in some style and he did so. Though he complained to Turner about her extravagance, the "discussions" recorded in Ellen's diaries did not turn on finances; moreover Sherman asked Turner not to reveal to her that he was not saving money. Then, when he had given up this effort and urged her to take the children back to Ohio as an economy measure, she refused in part because she would not abandon him while he was ill. When she did visit her parents in 1855 he was manifestly happy to have her back. Ellen, for her part, was proud of her husband and fiercely defensive of him. In the spring and summer of 1856, when his name was in the San Francisco newspapers, she began to paste clippings into a scrapbook; and she would continue the practice, documenting in her "archives" the various steps in her husband's subsequent career.

Ellen's diary suggests there were many conversations in which Sherman did the most of the talking. Ellen read very little beyond religious texts and tracts. Her husband would unburden himself to her on various topics, for in her own pronouncements on issues of the day one can often recognize his strong opinions. She in turn was endlessly supportive as his confidante, occasionally his counselor, and—though Sherman would have bridled at the term—his confessor. Later, during the war years, his letters to her would often be the best key to his innermost thoughts.

There was now another element in the Sherman menage and an increasingly important one: their children. They had brought Lizzie with them and Ellen gave birth to two sons in California. She proved to be an exemplary mother, dutiful and caring in the conventional mold. Sherman, on the other hand, played his role of father in curious ways. With the first two daughters, Minnie and Lizzie, he seems to have considered their appearance as phenomena for Ellen to deal with. It was not unusual in that age for fathers to distance themselves from their own progeny, but Sherman did it in his own fashion; for a time he referred to each of his infant daughters not as "she," but as "it." His description of Lizzie in a letter to Hugh is detached and clinical: "The youngest baby is growing rapidly but resembles no body that I ever saw yet, but as time develops her features some distinguishing traits may be seen whereby we may classify her."

But this detachment soon ended; as the girls grew, so did their father's affection for them, though often expressed in unconventional ways. When Minnie was a year old he penned this description of her: "It takes the whole establishment to keep her out of the drawers, boxes, and closets, all of which she inspects even to the last item of straw dust. When spring comes we expect to turn her out to grass in some of the public parks." When Lizzie was two he wrote Hugh: "It is out of all use for me to attempt to sit down at home, for she leads me out into the street and drags me all about, never tiring." Here was the essence of the rapport that would link Sherman and his young daughters: they would turn about him, entreating, cajoling; and he would grumble at the interruption and the distraction, even as he battened on their presence and their attentions.

When William Tecumseh Sherman Jr. was born on June 8, 1854, his father's reaction was altogether different. He had the keenest interest in his first son from the very beginning, broadcasting news of the baby's arrival, misdating one letter in his excitement. The boy had hair "most decidedly red," and weighed nine and three-quarter pounds, "which among those versed in such matters is considered very large." And thereafter one could follow Willie's development in his father's letters, which record the appearance of his first tooth, his first steps, his first efforts to speak. This infant is prodigious in every respect. He is "very large and strong"; by the age of two he is "the admiration of the town," and his father boasts: "He can hold his own with a two year old anywhere." For this boy—at least in his father's description—is a force of nature, with a will as strong as his body. He is "almost beyond control"; he tyrannizes his sister Lizzie, declares his independence of his nurse, "rides roughshod over us all," and gives clear promise that he will "make his way in the world"—all this before he is three years old.

In fact it is only Sherman who describes his son in this fashion. Ellen spoke of him as a gentle, obedient child, ever anxious to please. Letters of others in the Ewing family who knew him give no indication that he was headstrong or unruly.

A second son was born to the couple while they were in California, Thomas Ewing Sherman, born October 12, 1855. Though at birth he was even larger than Willie, his arrival did not elicit the fanfare his elder brother had received. In a letter to Turner a few days afterward Sherman wrote: "I came near forgetting to tell you that Mrs. Sherman presented No. 4 just a week ago . . . "

One can only speculate on whether Willie Sherman the man would have or could have manifested all those extraordinary traits of character that an admiring father saw in the child, for Willie was fated not to reach manhood. As for Tommy, he was destined to grow up in the shadow of his deceased elder brother, and to know a fate scarcely less tragic.

# 7

## LOUISIANA

CLOSING THE BOOKS ON 1856, Sherman conceded that it had been a bad year for Lucas, Turner and Company and for San Francisco. By now hard times were striking essential services: the city fathers could afford to burn only one street light per block and the school system was shutting down for lack of funds. Yet 1857 was beginning well, with the abundant rains essential for increased mining. It would not end well, for gold production would be even lower than before. As for Sherman, he would be gone before the year was half over. The word came from Turner in January: Mr. Lucas had decided to shut down the bank; Sherman was to arrange an early and orderly closing; then, if he accepted a new position being offered him, he would manage a branch to be opened in the country's financial heart—New York City.

Sherman was torn. To leave was to retreat—never pleasant for a soldier—and to abandon the start he had made for himself in San Francisco, where he was "a big chief." Then there was something about California that appealed to him and always would. Yet there were signs everywhere that for the time being the state's economy was in severe depression—he could see this all the more clearly as he prepared to shut down operations.

Closing down a bank was a delicate business: if word got out too soon it could spark a run that he was now far less able to withstand. He began to accumulate cash, calling in sums out on demand and turning down loan applications even when there was good collateral. By April the bank was ready to settle with its depositors and a formal notice was inserted in the newspapers. To Sherman's delight there was no great rush to withdraw; some depositors still had funds in the bank after the April 20 deadline. There were two major problems remaining: one was delinquent debtors and the other was real estate that the bank had acquired, and for which the market was then too depressed to carry out a liquidation. Together these assets involved a considerable sum—roughly $300,000—to be realized later.

The Shermans sailed for New York in May, with Sherman preoccupied by what he was leaving and what lay ahead. He must have watched the coast of California disappear with regret. He had confessed to Turner: "I wish I could promise to stay out here all my life, which I want to do." New York would be a daunting challenge: he worried that he would be "swallowed up in that vast gulf of man-kind where I will be small indeed"; he did not know whether he could compete with the "acute operators of New York."

Ellen was delighted to leave. She had no objection at all with the arrangements Sherman had made for her, after clearing them with his father-in-law: she and the children would remain in Lancaster while he tested the waters in New York. During the leg of the trip from the isthmus to New York the family celebrated Willie's third birthday on board the ill-fated *Central America*, which would go down four months later in one of the great maritime disasters of the century.

Before June was out Sherman had deposited his wife and children in Lancaster, gone on to St. Louis to confer with Lucas and Turner, then headed back to New York, where after a flurry of preparation he opened the bank on July 21, at No. 12 Wall Street. He had the misfortune to launch the venture just as a major financial storm was bearing down on the whole country—the Panic of 1857. The basic causes of the crisis have been identified—shifts in the country's foreign trade patterns, rampant speculation, especially in lands, and so on—though historians still argue about their relative importance. Certainly one factor was the fragility of the country's financial institutions. The number of banks had doubled since the beginning of the decade; there were probably too many of them—sixty in New York City alone—and many were shaky from the start; there was no federal regulation of banking and state laws were often inadequate or simply lacking. The fundamental weakness of these institutions was inadequate reserves for the paper they had out. When the storm struck many were swept away overnight.

Among the first to succumb was the New York branch of the Ohio Life Insurance and Trust Company, which suspended payment four days after Sherman opened for business, and the collapse of that house undermined the parent firm in Cincinnati and two dozen New York banks as well. Many New York bankers were counting on a transfusion of specie to help them survive, since they had gold coming from San Francisco on the *Central America.* Everyone was awaiting its arrival; instead came news that on September 12 the ship had gone down in a hurricane off the Carolinas, taking with her over 400 passengers and several tons of gold. A panic was now spreading throughout the country: on September 25 it was the turn of the Bank of Pennsylvania; then early on the morning of October 7 Sherman was awakened with the news that his parent firm in St. Louis had suspended payment. Things worsened in New York; on October 13 Sherman wrote Turner, "all hell has broken loose"–there was a run on every bank in the city. But for Lucas, Turner and Sherman the greater crisis was in St. Louis; Sherman went down to No. 12 Wall Street, filled a trunk with all the assets he could find, and took the train to St. Louis, where he stayed until the end of the year, sorting the debris of Lucas's empire.

There was money to be realized from bank assets in San Francisco, property to be sold or leased and notes to be collected or, failing that, auctioned off. Sherman headed for San Francisco, where he pursued debtors and sold or leased what property the bank still owned. He put his own house on the market, asking $6,000 for a home that had cost him $10,000; he found no takers so he leased that out as well.

There was another matter outstanding: "trust accounts" held by a number of Sherman's army friends; they stood to lose some $130,000. Though legally the bank could not be made to cover these losses, Sherman's quixotic notions in money matters led him to make reimbursement. With all the funds he could muster for this purpose he still lacked some $15,000; he therefore "stood some of the loss," at the cost of his savings. He would say in later years that none of his army friends had lost a cent.

By July 1858 he had done all he could; he took ship for New York and an uncertain future. While at sea he wrote a friend, "I go to Lancaster, Ohio to start where I began 22 years ago." By the beginning of August he was back, where he came to rest for a month. Yet he could not rest—not with a wife and children to support. James Lucas had offered to help him get established in the wholesale grocery business, but he had no stomach for it. In a mood of depression he had told Turner his experience in California had completely destroyed his self-confidence; he considered himself "utterly disqualified for business."

Thomas Ewing had renewed his offer of employment at the Hocking Valley salt works, and Sherman, then still in California, had accepted: "All I stipulate is that I don't want to live in Lancaster. You can understand what Ellen does not—that a man needs a consciousness of position and influence among his peers . . . At Lancaster I can only be Cump Sherman." But once back in Lancaster he backed away from the agreement. "For that part of Ohio I had no fancy," he recorded in his *Memoirs*.

In the end he chose another course, with Thomas Ewing's blessing. Tom Jr. was making a living for himself in Leavenworth, Kansas, practicing law and taking care of the family's interests in that area. Sherman wrote him, outlining his difficulties. Could Tom use a partner? "I am on the market cheap, what will you bid?" Tom agreed to take him on and he went to Kansas in September. Ellen and the children would follow if things worked out—a familiar scenario by now.

Sherman knew nothing of the law, but could run the office, keep the books, and handle collections. Leavenworth was an army town, in fact little more than an appendage to the fort. In that installation Captain Sherman had automatic entree; there he found a number of army friends. He also had easy entree to the Kansas bar when Judge Lecompton, after a brief chat, admitted him "on general knowledge."

He felt like an impostor and dreaded the day he might have to make a court appearance on behalf of a client. On one or two occasions he had to argue a case when his partners were not available, and invariably he floundered and lost. Army friends chuckled over the story that he once based his legal argument on a statute that had been repealed thirty years before; in his *Memoirs* he related his brief legal career as a comic interlude characterized by "little paying business and plenty of lawyers." He had other modest sources of income: he recorded that he occasionally "picked up a stray 50c" as a notary public, and then his army connection brought him extra work and income; he ran an auction of surplus horses and mules and surveyed a road and contracted for repairs to it; there he cleared $50— the largest sum he collected during his entire legal career.

Tom had a house large enough for two families, so Ellen and the children came out early in November—minus Minnie, who sent a note saying she could not abandon her grandparents. The Shermans would live on a shoestring: their only steady income was a few dollars a month from two rental houses Ellen owned in Leavenworth and two "tenements" Sherman still possessed in St. Louis. He had enjoyed a salary of $5,000 a year in San Francisco, but was now trying to maintain his family six months on $650, including the purchase of furniture and Ellen's travel expenses from Lancaster. The Leavenworth household did not last long. Tom decided to part with his house and rent another, while Ellen was once again pregnant, ill, and increasingly homesick. Sherman was running out of money—he estimated he and his two partners were not clearing more than $25 apiece from one month to the next.

In March Ellen and the children returned to Lancaster; Sherman moved out to Indian Springs, Kansas, where he homesteaded for Thomas Ewing, who wanted a farm prepared for a relative, with buildings erected and fruit trees planted. Those next months out on the prairie, "the Pampas of Kansas," he called it, were the nadir of Sherman's fortunes. He had abandoned the career for which he had been prepared and he could not return to it. He had not been a success in the civilian world—at least in the lasting, material terms that a family man must have; moreover he had now dropped several rungs on the social ladder, taking on work that gentlemen didn't do. And now time was running out. He was approaching forty, then the threshold of old age; his capacities would wane, his cares increase. He had been humorously if cruelly ironic when describing his lack of success in life, calling himself "the Jonah of banking," for example; but now he was pitiless in his self-appraisal: "I look upon myself as a dead cock in the pit, not worthy of further notice."

From Ellen, who usually sustained him, came letters painful to read. Pregnant and ill, she wanted to know "what will we have to pay fuel, home, rent, doctors, shoes and hat bills next winter if you are making nothing now?" She flung in his face his apparent indifference to her preg-

nancies and the birth of their children: "Have you ever thought or wondered what the child might be? Have you no preference as to the name it
might bear?" She was now convinced that her pregnancy had only interrupted the progress of a fatal disease whose ravages would soon resume.
After she succumbed she wanted him to remarry so that their offspring
would have "a good Catholic mother."

Sherman's diary in those spring months of 1859 tells just how far the
officer and banker had fallen. He laid out building lots and fixed up his
one-room dwelling: most of April he worked on a barn and a corn crib; on
May 3 he planted sweet potatoes and tomatoes, and on the fifth "went to
buy a cow." He spent whole days hauling and plowing, days that started
early: "6 A.M.–Yoked up," ran one entry. And the former banker who had
handled hundreds of thousands of dollars still kept meticulous accounts:
"July 7–paid $1 for ox whip."

He had heard on the army grapevine that there were personnel problems in the Paymaster's Department, where an officer was in trouble over
his accounts; there could be a vacancy there–and this was the only department in the army that could fill vacancies with men taken directly from
civilian life. Sherman wrote to Don Carlos Buell, who had been one year
behind him at West Point and was then well-placed in the office of the secretary of war. He told Buell that if "political names or money bonds" were
a factor, he could "command these ad infinitum."

Buell replied that while there might indeed be a vacancy, it would
probably be filled by the secretary of war on the basis of political considerations. The secretary, John B. Floyd of Virginia, would appoint someone
from the South. But Buell knew of another job that had just come to his
attention and he would be glad to recommend him for it. The state of
Louisiana was setting up a school organized and run along military lines,
and its Board of Supervisors was looking for a superintendent; Buell oblig-
ingly sent along a brief description of the position.

Sherman sat down and addressed to Louisiana's Governor R. B.
Wickliffe a four-page letter of application that is a curious mixture of confidence and timidity. He began by what was almost an apology for putting
forward his "pretentions" for a position "so high and honorable." He then
outlined his education and career, saying he had solicited no endorsements from "personal and partial friends," but suggested the Governor

could get a better judgment from military men in Louisiana: his West Point classmate Paul Hébert, Colonel Braxton Bragg, and Colonel A. C. Myers; what is more, Sherman could "refer with confidence to every officer of the Army from General Scott to the youngest captain." He ended by requesting that in the event he was not chosen, his application should be kept confidential, "because I do not wish to seem an unsuccessful applicant for anything."

Here a curious chain of circumstances came into play, helping open doors for him: Buell had received advance notice of the position because he knew the president of the new school's Board of Supervisors, George Mason Graham; not only had Buell served with him in Mexico, he had married the widow of Graham's half-brother—who was none other than Richard B. Mason, Sherman's old commanding officer in California. Were this not enough, Graham's sister taught at the Academy of the Visitation, where Ellen Ewing Sherman had been one of her favorite pupils.

The selection of the teachers for the new institution occupied the board for most of July; there were to be five faculty members, one of whom would also serve as superintendent. For the five positions the board received eighty-one applications; though the filing deadline was July 31, on the sixteenth General Graham had already offered the superintendency to one of the state's native sons, P. G. T. Beauregard, who turned it down. Other names were considered but set aside.

Then Sherman's name came up. General Graham favored him because of Buell's very strong letter of recommendation, but one of the board members violently denounced Sherman as "the son-in-law of that blackhearted Abolitionist Tom Ewing." The interjections of another board member named Michael Ryan—"that wild Irishman," Graham called him—created further tumult and confusion. The board became so tangled in procedure that it accidentally elected to the superintendency one of the other applicants, a young man named Francis W. Smith, just twenty years old.

Graham saw his chance: "In getting that undone, I succeeded in getting Sherman elected." Contrary to popular belief, neither Beauregard nor Bragg played any role in the selection of Sherman. Beauregard favored another candidate. Bragg only heard of the appointment later, but then wrote Graham to applaud his choice, calling Sherman "one of the ablest men I know, moral, open, candid, sincere—ardent in all he undertakes . . . "

The position as superintendent and professor of engineering at an annual salary of $3,500 was duly accepted by "Major Sherman," for so he was addressed in the letter of appointment. Graham and Sherman agreed they should at least visit some other military schools before they opened their own. Graham went to the Virginia Military Institute; Sherman looked over a military school recently opened in Kentucky, his visit being delayed until Ellen had given birth to "Number 5"–Eleanor Mary, to be called Elly. Sherman also wrote George McClellan (USMA, 1846) for his views, since he had recently made a fact-finding tour of Europe. He also wrote to West Point for information; Major Delafield, still superintendent, promptly replied.

As Sherman and Graham shared their information and ideas, first by correspondence and later in conversation, it became evident that they had differing opinions on a number of points. Graham, who had attended the academy but had not graduated, regarded West Point as the model to be followed as closely as possible. Sherman was more open-minded: differences in climate between Louisiana and upstate New York might make it difficult to follow the West Point model too closely; summer camp would not be practical. Uniforms could be changed with advantage from cadet gray to dark blue and the tight-fitting coatee replaced by the Army's frock coat (Sherman consulted both McClellan and Buell on this point).

Sherman left for Louisiana in October, passing by way of St. Louis. As he went down the Mississippi his mood seems to have been somewhere between hope and resignation, hoping that ahead lay an upward turn in his fortunes; if not, the Hocking Valley salt wells were always there. On the way he wrote a long letter to Ellen: "Should my health utterly fail me or abolition drive me and all moderate men from the South, then we can retreat down Hocking and exist until time puts us away under ground. This is not poetically expressed, but is the basis of my present plans."

The rising sectional antagonisms did not make the autumn of 1859 a propitious time for a man from Ohio to take employment in Louisiana. Sherman wrote John, who was then in Congress, reminding him of the delicacy of the situation in Louisiana and asking him to "take the highest stand possible" consistent with his party's beliefs. Later he wrote him, "I would like to see your position yet more moderate." Sherman's own views, when limited to the matter of slavery, could hardly give offense in

Louisiana: "The relations between master and slave cannot be changed without utter ruin to immense numbers, and it is not sure that the Negro would be benefitted thereby." Then too, Sherman's easy sociability could be counted on to help him when the sensitive issue came up in conversation, and as he pointed out to Hugh, "I understand the tone and temper of the Southern gentleman."

When he met George Mason Graham the two men took an immediate liking to each other. "Major Sherman" was already known in Alexandria, thanks to a flattering article Graham had written for the *Louisiana Democrat.* The town proved to be disappointingly small; he went out to Pineville for a first look at the seminary, where he found a huge three-story building of vaguely medieval style with crenellated walls and towers, sitting by itself in a 400-acre pine forest. The only people about were carpenters putting finishing touches on the structure. The academy building had absorbed virtually all of the available funds, some $135,000; in its seventy-two rooms there was not a stick of furniture.

The area surrounding the seminary was sparsely populated and the whole region was very remote: the mail came only three times a week; to reach New Orleans, the nearest city of any size, took thirty-five hours of travel on stage and steamboat. There was only one house near the school, which Professor Antony Vallas had already rented—Sherman understood now why Graham had strongly suggested he come down alone, bringing his family later. There were plans to erect two houses for Vallas and Sherman, they being the only married faculty members, but for the time being Ellen and the children would have to remain in Lancaster.

This was just as well, for at its best life in the piney woods of Louisiana would not be easy for Ellen Sherman. There was no Catholic church nearby, no shops or stores at hand, and no pharmacists or physicians except the surgeon at the seminary. Alexandria would offer little to a woman with Ellen Ewing's needs and tastes. Then there would be the servant problem. Ellen could not live without them; if she brought them with her Sherman predicted they would quit on arrival. She would then have "to wait on herself or 'buy a nigger.'" Convinced as he was that blacks would work only under compulsion, he suggested Ellen buy a couple of slaves, both as an investment and as a solution to the servant problem. She indignantly refused. "This is going to be a trial for Ellen," he wrote

Tom Ewing Jr., "far, far harder than San Francisco or Leavenworth."

Once he had taken his bearings, Sherman concentrated on the launching of the seminary. The Board of Supervisors was a largely honorific body, a civic distinction for the planters and professional men named to it. One of the fourteen members had never attended a meeting, others put in appearances rarely. Often only a rump could be assembled; it could pass only nonbinding resolutions. The driving force behind the institution was General Graham, who was kept busy promoting it in the legislature and defending its military character against an antimilitary faction on the board led by Dr. S. A. Smith. Instead of a West Point of the South Dr. Smith favored a replica of his alma mater, the University of Virginia. The institution they created was an awkward compromise with a cumbersome name: the Louisiana State Seminary of Learning and Military Academy.

Though the seminary was scheduled to open with the new year, much of the practical preparation had not been done or even seriously thought of. What was needed was a man with initiative, energy, and industry, and in Major Sherman the board had found that man. With General Graham and a couple of the faculty members he framed the regulations of the new school, borrowing heavily from the Virginia Military Institute (the original draft, in Sherman's hand, is preserved in the Louisiana State University Library). Curiously, Sherman and his associates did not retain the VMI court-martial system; justice within the walls of the seminary would be dispensed by the superintendent.

No one had thought of advertising, so Sherman drew up an announcement of the seminary's opening for distribution to the state's newspapers. Then he took the lead in working out the first semester's program of study. For books and apparatus he made a shopping trip to New Orleans; finding little of what he needed there, he placed orders with New York firms, despite some grumbling on the board. Later, at the cost of more grumbling, he placed the order for cadet uniforms there; a New York City tailor found his way to Alexandria to take measurements.

The seminary could accommodate 160 cadets, and first projections were that the school would be filled to capacity. But that estimate soon changed; it was only by dint of aggressive recruiting and lowering standards and age limits that enrollments reached fifty-nine at the end of the year. Then on the appointed day, January 2, 1860, the number of cadets reporting was only eighteen: "Not punctual, according to Southern style," was the superintendent's comment. Other cadets came trickling in as the semester got under way.

The faculty was spread very thin. Sherman was professor of engineering, architecture, and drawing; Francis W. Smith was responsible for mathematics, "natural and experimental philosophy," and artillery tactics, and in addition he was commandant of cadets; Professor Vallas held forth on chemistry, geology, mineralogy, and—in theory at least—infantry tactics. David French Boyd taught English, Latin, and Greek; Berte de Saint-Ange was to teach French and Spanish. Since Vallas and Saint-Ange were foreigners and Francis Smith fresh out of school himself, Sherman counted on Boyd for his "social qualities." The two men would become fast friends and remain so until Sherman's death.

Just as Sherman was beginning to feel at home, another door opened to him. A banking job in London, only a glimmering possibility a few months before, now took on form and substance. It was the idea of a group of entrepreneurs that Hugh Ewing was associated with, and now their spokesman came to Louisiana and laid on the table a dazzling offer: a two-year contract at $7,500 per year, more than twice what the seminary would pay. Ellen and the Ewings thought it was an opportunity not to be missed. But after a quick trip north and a renegotiation of his salary at the seminary, he decided to stay.

Sherman had now made his choice, and he wrote Ellen about the life they could finally have together in the tranquillity of Louisiana. But Louisiana and the entire South were anything but tranquil as 1859 ended and 1860 began. For months repercussions of the John Brown raid continued; now, closer to home, John Sherman was beginning to make himself known in the South. He had foolishly endorsed a book he had not read that urged the abolition of slavery. Hinton R. Helper's *Impending Crisis of the South* created a firestorm in the South and made John Sherman's name anathema.

The *New Orleans Bulletin* called attention to the "singular fact" that John Sherman's brother was head of the state's military academy, but fortunately no hue and cry developed in the Louisiana press. In an effort at damage control Sherman passed John's letter of explanation to General Graham, who undertook to show it around to various influential people.

Sherman himself was curious to see if anyone would bring up the subject of his brother in his presence. While no one took him to task about John, he received subtle invitations to give assurance of his own trustworthiness. One of the board members remarked to him that in such parlous times Louisiana needed to have her schools "in the hands of friends." Sherman plainly stated his view: he was not opposed to slavery but he was unalterably opposed to disunion.

Within the seminary Sherman had some problems with his faculty, especially with Professors Vallas and Saint-Ange, whom he described as "foreigners with their peculiarities." He wrote Buell: "They want to share in the government and think I have too much power." But there was another source of resistance that would give him a great deal of worry—the corps of cadets. He had foreseen the basic problem well before the school opened and pointed it out to General Graham: at West Point the cadet held in effect a highly coveted appointment and a four-year scholarship, something that he—or at least his parents—would value too much to risk losing. But in Louisiana the parents would be paying for the young man's education; if they found the burden of work too heavy for their son, what would keep them from simply withdrawing their offspring and sending him somewhere else?

Sherman had at least a partial answer: make the program less onerous and more pleasing to the cadets. "I am convinced that we can make certain drills, guards, and military parades and exercise so manifestly advantageous to the cadets that their own sense, judgment, and fancy will take the place of compulsion, and the course of studies, being more practical and useful, will be preferred by the cadet and parent to the old routine of grammar and everlasting lexicon." He argued successfully against a West Point–style summer encampment. Finally he questioned whether these cadets could be held to the strict military rule of West Point. But he acknowledged that the decision was not his to make; his role was to execute. Soon he was to find that role a challenging one.

Anyone who had been at the seminary when it opened could have predicted that the school would have its full share of disciplinary problems. Sherman had confided to Ellen that he expected to have his patience put to the test "by a parcel of unruly boys." Despite official assurances that "a youth's time is so regulated that dissolute and expensive habits cannot be contracted," the cadets had plenty of leisure; marooned in the middle of the piney woods, living cheek by jowl in the upper recesses of the vast, barnlike seminary, with virtually nothing provided by way of harmless outlets for their energies, even the best-behaved youths would have been tempted to mischief.

The cadets were not the best-behaved youths. The planter class that provided most of the complement seems to have had its full share of spoiled, headstrong sons, accustomed to having their way and ordering servants about from their earliest childhood. To the parents of such youths the military-style strictness of the seminary must have seemed just the corrective their sons needed, so they made their problems Sherman's problems.

Sherman was not well-equipped to maintain discipline. The board had never given formal approval of his regulations; it declined to give him the disciplinary powers of a military commander, nor would it maintain discipline itself. He extemporized each time a case arose, and some of his decisions were controversial at the very least. In January Cadets Hyams and Haworth exchanged angry words, then blows; Haworth broke off the engagement to run to his room for a Bowie knife but was intercepted before he could use it. Sherman expelled him on the spot, on the grounds that "no possible provocation can justify such an act." Then he decided that Hyams had started the scuffle and sent him home too. Hyams *père* came to the school accompanied by a local judge. There was an angry scene: the elder Hyams told Sherman his military regime would soon end; Sherman shot back: "One hundred young men in this building under a civic government would tear down the building and make study impossible."

In June a rash of misbehavior erupted that seems laughable in its triviality, but that Sherman pursued with relentless determination (Boyd called him "a perfect detective"). Several of the cadets had organized a nocturnal band called the Midnight Marauders; their pranks were meticulously catalogued by the superintendent as "the Mose Chicken Case," "the Bucket Case," "the Affair at the Spring" (the ducking of a slave), and so on. Sherman eventually learned the identity of two chief malefactors, Cadets Campbell and Ringgold, who admitted their roles, and he intended to send them away. But not content with their confessions he wanted to have the testimony of eyewitnesses, "beginning with Cadet Stafford, then Cadet Hillen, etc."

Here he ran into a stone wall. He lectured the assembled cadets about the dangers of "concealing real wrongs & outrages because it looks like 'tattling.'" To no avail. Cadets Stafford and Hillen departed rather than give testimony; Cadet Liddell, reputed head of the Marauders, left on his own initiative—apparently in protest to the tattling policy. After an angry exchange with Liddell's father Sherman decided the young man's unauthorized departure could not be passed over in silence. He wrote the father, "It must be published to the legislature and to the world that he deserted." Mr. Liddell made strenuous objection.

Ultimately this dispute and others of similar nature seem to have been set aside, to be referred to the Board of Supervisors, which would serve as court of appeal for Sherman's decisions. In the interim it was General Graham who increasingly took up the role of mediator between angry parents and the seminary's superintendent. Some of the board members felt Sherman's punishments were too severe; Graham himself acknowledged to at least one parent that he was "not satisfied" with the severe punishments meted out by Major Sherman. Admittedly the job of keeping order would have been a challenge to anyone; but judging from fragmentary evidence Sherman does not seem to have handled it particularly well. His responses appear to have been abrupt and heavy-handed, made without reflection or consultation—and without a clear idea of the reaction they would provoke. This is not to say that in general he failed to run the seminary effectively—not a class was missed, not a drill was canceled.

Then too, the Board of Supervisors would have had an excellent opportunity to get rid of him when he was offered the London banking job; instead they gave him a raise. Just how popular he was with the cadets is difficult to say, since the evidence comes mainly from Sherman himself or from Boyd, a loyal friend whose testimony was given some thirty years later. The cadets invented no derogatory nicknames for him, nor was he the victim of their pranks, as Professor Vallas was. He had no classes—engineering was for advanced students—but he occasionally taught classes for ill or absent colleagues; thus he taught Vallas's Spanish class, using what he called "the true greaser pronunciation." On Friday evenings he would talk to the cadets about some episode in American history; able raconteur that he was, he had large and enthusiastic audiences. According to Boyd his talks about California in the era of the gold rush enjoyed a packed house.

And thanks largely to Sherman's efforts there were the beginnings of organized social life at the seminary. He arranged for periodic dances, or "hops," attended by daughters of local planters. There was an elaborate ball that enabled the cadets to show off their dress uniforms. The Fourth of July was celebrated at the seminary with some ceremony. Cadet Cornelius made a declamation in the florid style of the era. His topic was apparently chosen by himself; it would not have been one selected by the superintendent: "The Future and Utter Extinction of our Happy Union because of the Overweaning Pride of National Power."

Yet the subject was eminently appropriate, for the nation was entering a grave crisis. When Sherman returned to the seminary after a trip north in the late summer of 1860 the coming presidential election dominated all conversations. Even the cadets were affected; Sherman described them as "nervous." He had been thinking about bringing his family down to occupy the house that had been built for him, but in the aftermath of Lincoln's election he changed his mind. All eyes were now on South Carolina, where there were increasing calls for Secession. Boyd was with Sherman when the news came that South Carolina had seceded: "Sherman burst out crying and began in his nervous way pacing the floor and deprecating the step which he feared might bring destruction to the whole country. For an hour or more this went on."

One by one other Southern states followed South Carolina. On January 26 it was the turn of Louisiana. Bragg sent Sherman word of the Louisiana secession convention's action, saying "separation is inevitable. Nothing can prevent it now. Why should there be any strife over it?" Louisiana left the Union, but Sherman did not leave Louisiana. He stayed on a full month, and his reason was at once compelling and embarrassing: he needed to collect the rest of his salary, especially $500 that had to be authorized by the legislature.

He sought to justify his staying on to Thomas Ewing—and probably to himself: "I owe no allegiance to Louisiana. I am working like any man for my hire." Ewing and John Sherman both feared he might be in danger, but he replied, "there is no personal danger that I would weigh in the scale with the mortification of hanging about loose and unemployed," and he told Ewing frankly his situation: "I cannot afford to leave here unless I know I can do something right away." The plain fact was that he had been unable to save anything.

He would have to find other work, and soon. Yet there was something fitful and contradictory in his efforts: while he told his father-in-law that he saw no chance of work in Ohio, he wrote Tom Jr. that he would seek employment there or St. Louis; he told Ellen, "If you want me to come away you must get me something to do," then in the next line admitted that it was a ridiculous request; then in a letter written six weeks later he again urged her to "make some exertions" on his behalf. On the eve of Louisiana's secession he proposed that she and the children join him for a month or so, "then you could return and I could start for California or wherever else I think the best chance lies for occupation." To Hugh he wrote that he was looking for work in "some obscure place," and the salt wells would do. At once frantic and undecided, he was turning in circles, drawn to one possibility, then to another. "I will be able to forego my preferences," he confessed to Ellen, "if I can control myself and settle down."

Sherman's letters in this period show other emotions. There are flashes of anger, most often directed at Ellen. He lectured her on her wifely duties: "'Tis your part to follow me, without imposing conditions that cripple my action." She had the temerity to rent a house in Lancaster for herself and the children, when he had had one built for her at the seminary: "Maybe you intend to secede and set up on your own hook." And in a letter to Boyd he reveals an unaccustomed remorse. "The truth is that I have socially been too isolated from my children." But he intends to make amends: "I will settle down," he pledges, "with my little Minnie, and Willy, and the rascal Tom . . . "

Sherman may have been remorseful for another reason; Ellen was having difficulty maintaining discipline over their children. In truth all were reasonably well-behaved but one—Tommy. He would hit his sisters and Willie too; at one point when Tommy struck his elder brother Ellen told Willie to give him a proper thrashing, but Tommy fought so furiously that Willie, whose heart was not in it, abandoned the effort. Ellen herself then took his place.

Handling the day-to-day business of the seminary still took much of Sherman's time, though now he took little pleasure in performing his various tasks. General Graham was gone, tired of quarrels over the institution he had fathered. Dr. Smith had taken his place, and Sherman's own field of competence was reduced. The mails were slow and irregular, a foretaste of "the anarchy and confusion to come." To his disgust letters now came addressed to "The Military Academy of the Independent State of Louisiana." The cadets were increasingly difficult to control. As the new year began he confided to Ellen, "it takes me all I can do to suppress disorder and irregularity. Yesterday I had a cadet to threaten me with a loaded pistol because I found a whiskey jug in his room."

The Board of Supervisors had increased Sherman's salary $500 a year by designating the seminary an arsenal and paying him as its director. Now he was earning this salary in a most humiliating way, for Governor Moore ordered the seizure of stocks of arms in various federal facilities and sent the weapons to Sherman for storage. Cases of rifles and other ordnance filled whole rooms and drove him from his quarters in the seminary building. He was forced to settle nearby; every evening he did his correspondence among crates and cases in the front parlor of the house that was to have been his family's home.

Finally, on February 19, it was time to go. He had already submitted his letter of resignation and it had been accepted with genuine regret. The Board of Supervisors had given formal expression of their sentiment in a letter of appreciation. Now he said goodbye to the assembled cadets and his colleagues in an emotional scene. For once in his life words seem to have failed him. Finally he could only point to his heart and say, "You are all here," then he was gone.

He went to New Orleans where he stayed a day or two before heading on to Lancaster and a reunion with his family; during his stay he visited with his friend Bragg, who thought there would be no war. But he was there on February 22, when the city held a grand military parade and illumination; he wrote in his diary that evening: "Glorious rejoicing at the downfall of our country." On the twenty-third he wrote a long letter to his friend Boyd. Not knowing if he would ever see him again he ended his letter with a somber "Goodbye." The next morning at seven-thirty he climbed on board the train and began the long voyage north. He was leaving the Deep South with a portmanteau; he would return with an army.

# 8

## WHAT MANNER OF MAN

SHERMAN'S STATE OF MIND as he departed New Orleans can well be imagined. His destination, Lancaster, had become a dreary way station between failed ventures. And now, in addition to the crisis in his personal life, there was the unprecedented crisis that gripped the nation–the whole country was rushing to Armageddon, and the prospect oppressed him even more than his own predicament. But America's tragedy would be Sherman's salvation. The man who boarded the northbound train that February morning was headed for a profound change of fortune, one such as he could never have imagined: within ninety days he would wear a brigadier general's stars; a little more than a year ahead lay Shiloh and national acclaim.

Such meteoric ascents are endlessly intriguing to the historian, who looks for their causes both in the times and milieu in which the event occurred, and in the traits and qualities of the individual involved, "the circumstances plus the man." And here, on the eve of his metamorphosis, is the propitious moment to take a closer look at this man. At forty-one years of age he is fully formed and tempered. The intellect has jelled, the patterns of thought and behavior are fixed, though crises to come will soon reveal more of their nature; fixed also is the physical constitution that will sustain him through another thirty years of intensely active existence.

The other passengers on the train out of New Orleans would have noticed Sherman. Physically he had a certain presence and impact—"very striking," David F. Boyd described it. There was his hair, red-brown and often remarked on. He was outsized, standing nearly six feet, yet neither ill-proportioned nor encumbered by heavy extremities; on the contrary, these were finely formed. He wore a size nine shoe, and photographs show slender hands with long tapering fingers. One observer described his physique as "narrow and almost effeminate."

By 1861 his skin was aging prematurely, most notably on his hands and face. There were already pronounced crow's feet at the corners of his eyes; these would gradually spread over the ensuing years to corrugate his cheeks. At war's end Ellen found that her forty-six-year-old husband had the face of a sixty-year-old.

Putting the man's frame and features together, an acquaintance offered this word portrait: "tall and lank, not very erect, with hair like thatch which he rubs up with his hands, a rusty beard trimmed close, a wrinkled face, sharp, prominent red nose, small, bright eyes, coarse red hands."

People who met Sherman in this period of his life sometimes felt there was something about him that was typically or distinctively American. "A very remarkable man such as could not be grown out of America," said one observer, "the concentrated quintessence of Yankeedom." Perhaps the tall, lank body and the bladelike nose evoked an image then emerging as a national symbol; with appropriate garb and beard Sherman would have made a very passable Uncle Sam.

His clothing did little to enhance his appearance; in his cadet days he was already noted for a certain carelessness in this regard. Even when he reached high rank during the war there was no pretension to elegance, much less the foppishness that some general officers indulged in. At best his attitude toward such things seemed to be one of indifference; on campaign it descended to plain negligence. He would wear his coat flapping open and his vest buttoned only at the bottom, in a fashion followed by other Western generals; moreover his uniform was often wrinkled and soiled. An officer who served under him remarked, "he seems to have an aversion to new clothes."

But even when he was a penniless lieutenant in California he was ordering clothes from John J. Fraser, a highly regarded tailor of New York

City; by 1860 his measurements were with Brooks Brothers, who supplied his uniforms during the war. His coat might be wrinkled and soiled, but it was a gentleman's coat, made by a gentleman's tailor.

Then anyone sitting near Sherman in the railway car would have noticed a restlessness about the man. Most likely he descended at every stop the train made, there to puff nervously on a cigar as he paced up and down in his vigorous, distinctive gait. One observer said the man "jerked himself along."

Even when seated, Sherman knew no real repose; so strong was the impulse to movement in his waking hours that one might wonder whether his sleep was not equally fitful and agitated. The energy, the tension were always there, at times bridled and held in check, at other times erupting spontaneously in an almost spastic way. Here is how a wartime associate described Sherman listening to a concert: his eyes would be "dancing in every direction and on everything. He is never quiet. His fingers nervously twitch his whiskers—his coat buttons—play a tattoo on his table or chair, or run through his hair. One moment his legs are crossed, and the next both are on the floor. He sits a moment, then paces the floor."

This "restlessness of a Hotspur" explains a great deal about the Sherman of the prewar years. Here one can see in clearer light the indefatigable commissary officer who shopped the New Orleans docks at break of day, when he might have sat comfortably in his quarters and let others do the work for him, as his predecessor had done; credible also is the banker who spent the better part of a day tracking down $18.05 missing from an account, and the seminary superintendent who rose from his paperwork every thirty minutes throughout the day to make a tour of the grounds and satisfy himself that the slaves were working as they should.

Sherman was in a boastful mood but hardly wide of the mark when he wrote John in 1852: "They regard me in Washington as prompt, energetic and vigorous." He was strongly motivated, as he explained to Ellen: "I always have and do feel a strong desire to advance in anything I do." But it was also his boundless energies, directed toward useful ends, however paltry or inconsequential, that had made him an enviable reputation in the peacetime army and quickly distinguished him in civilian pursuits.

This life of constant, restless activity, which would continue virtually without interruption to a week or so before Sherman's death, was made possible by a robust body and generally good health. Some thirty years ago a pathologist and Civil War scholar named Paul Steiner reviewed Sherman's medical history and concluded that the man's body remained fundamentally sound for seven decades. Despite morbid thoughts and false presentiments of death that came with periods of depression, Sherman himself seems to have concluded that he had a strong constitution. He even came to believe that the asthma attacks that assailed him from time to time had in fact strengthened his lungs; they were a form of exercise, forcing him to breathe more vigorously. To his youngest son, "Cumpy" (Philemon Tecumseh Sherman), he quoted what probably became a favorite adage of his in old age: "Asthma is a sure sign of long life."

Toward the end of that life he told Cumpy that during an illness in 1850 he had consulted an army doctor who prescribed a regimen that he seems to have adopted as his own. He was to take no medicine; moreover he was to "build up the strength of the body by healthy food and rich wines and whiskey in moderation." He was to "take all possible exercise, especially in the open air." The doctor apparently made no strictures regarding tobacco; Sherman was a cigar smoker for practically his entire adult life (during the war he told a friend that his "ration" was six per day, but he often exceeded it).

While the story of Sherman's various wounds, injuries, and other bodily afflictions is only a series of footnotes to his history, the intellect lodged within that body was a remarkable one, worthy of examination for its qualities, peculiarities, and contradictions. He seemed to have a memory of infinite capacity; what he learned, he almost never forgot; at the close of the war he could probably call 5,000 officers by name. And data drawn from this bank he could manipulate with marvelous speed and facility: at age sixty-five he dashed off an itinerary for a trip west John Sherman wanted to make; while John looked on in amazement he filled three pages with names of people and places, "without change or erasure of a single word."

In temperament he would be classified as mercurial, with discernible "highs" and "lows." The former were characterized by soaring self-confidence, cocksure pronouncements, and boasting; in the "lows" Sherman was inclined to bitterness, self-pity, and even feelings of worthlessness—thus his merciless description of himself in 1859 as "a dead cock in the pit."

Those who had occasion to be in Sherman's presence deduced his thought from his speech, which was logical enough, but often misleading. When he was excited, when he warmed to a subject, his speech became an almost manic chatter, and this rapid, chaotic delivery suggested to some a similar disorder in thought; one contemporary compared Sherman's mind to "a splendid piece of machinery with all of the screws a little loose." General Joseph Hooker, who knew him in California and later served under him, said of him "if he is not flighty I never saw a flighty man."

David French Boyd reached this conclusion about his friend: "He could not reason—that is, his mind leaped so quick from idea to idea that he seemed to take no account of the time over which it passed." Boyd suggested that Sherman reached conclusions the way he believed women did: instantly and intuitively. Paul Steiner judged that it was only Sherman's rapid-fire speech that gave this impression—and also got him into trouble from time to time—while what he wrote was always "sensible," but Steiner worked from published materials, including the general's personal correspondence, brought out under family auspices and edited to remove bizarre or unflattering passages.

Sherman himself acknowledged that he was fast and loose with the pen: "I write hastily, as hastily as I think and speak, and I know full well that I often write and speak things that should have remained unsaid." Of his haste there is no doubt; otherwise he could not have maintained the staggering volume of his daily correspondence.

In his writing as in his speech he tended toward extravagance of expression. He used hyperbole for emphasis the same way someone else might interject an oath or make an emphatic gesture of the hand. His simple interjection into a conversation could stop it cold. When in that first summer of the war he heard several congressmen assuring one another that the population of the city of Washington was loyal, he stunned them with this remark: "The sentiment of the people of Washington is such that they would cut the throats of our wounded on the sidewalks with table knives if our arms should meet a disaster in this neighborhood." His tendency to exaggerate may have found expression in his horrific projections of the casualties to be expected in the war, as well as his surprising estimates of the conflict's duration.

The mature Sherman often affirmed that he said just what he thought, and that it was this frankness that had created problems for him; some students of his life have also endorsed this view. But there were more than a few situations in which it is clear that what he said or wrote departed significantly from what he thought. From time to time he clearly used flattery on his superiors; he did so without any particular skill, but often with considerable success. Or he might justify an action with three different and mutually contradictory explanations, each tailored to the susceptibilities of the recipient. His *Memoirs* would be notable for their deft "retouching" of the past.

Sherman told Boyd that he had schooled himself to make up his mind quickly when he was a banker: "He often said 'yes' or 'no' when he didn't know if he was right or wrong, only he believed he was right, but that it was the *time* to decide, and that he would rather promptly decide a thing wrong at the *right* time than to be undecided and put it off. So it was with him ever after in campaign and battle; he was ever prompt and decided." This last phrase was generally but not universally true; there would be occasions in the war when Sherman would be neither prompt nor decided.

It is also true that there were few subjects or issues on which he had trouble making up his mind. One of them, oddly enough, was California. At one moment he saw it as America's Eldorado, at another an accursed place with a deadly climate and no future: "California is now and always has been a riddle to me," he wrote after two residences there, "I don't like it yet I do."

But generally there was a ring of certainty, even finality, in decisions and judgments that increased as Sherman attained maturity and then fame. Cadet Sherman often sounded cocksure in his letters, and before taking up his duties in Louisiana Superintendent Sherman informed General Graham: "I generally have strong opinions on a subject of importance." In fact a Sherman opinion on anything was likely to be a strong one; he was rarely lukewarm about anything. Perhaps this was another subtle form of exaggeration, of hyperbole on Sherman's part. Still, one has the impression that the images projected to his mind's eye were like those of an old-fashioned television screen when the contrast knob has been turned too far; all was either black or white.

He was aware that his vigorous language and free use of his tongue could get him into trouble—certainly as early as 1856, when the San Francisco press gave him a battering. But these were traits he could not suppress, and when combined they got him into endless difficulties. Newspaper reporters were drawn to him as to a magnet; not so much from malice, as he claimed to believe, but because Sherman made good copy on any subject.

As with the man's words, so with his actions. Sometimes they too struck others as controversial, inappropriate, or simply baffling. All San Francisco bankers worried about overdrawn accounts, but only Sherman warned his depositors with announcements in the newspapers: "Overdrafts will not be permitted." His scheme as superintendent to placard about the countryside the "desertion" of a sixteen-year-old youth of prominent family because he had gone home without permission was at the least very ill-considered, and other actions he took as superintendent made it harder for the seminary's friends to defend the school, and led to closer supervision by the Board of Supervisors.

The man had what one observer called "his crotchets and his prejudices"; expressed in various ways, they inevitably caused their little ripples. Later in his career, when he controlled the lives and fortunes of hundreds of thousands, the consequences of his decisions could be profound. During the war there must have been times when one of his orders or decisions caused the members of his headquarters staff to exchange covert looks—for example, his directive of October 28, 1863, authorizing his chief subordinates to impress for "forced service" in their depleted ranks "any citizen whatever," providing food and equipment to those forced into uniform, but no compensation, "in the nature of a *posse comitatus* called out by a United States Marshal."

To admit error was harder for Sherman than for most men—and he did so very rarely. As a contemporary remarked, he never admitted a mistake and never made it a second time. In his later years, when he enjoyed high position and wide popularity, he could indignantly deny such things or explain them away. But the Sherman of the antebellum years seems to have been particularly sensitive and vulnerable to criticism; his spirits rose and fell not with his material fortunes as much as with his image: in 1854 he was flying high, expansive and ebullient, the young banker who was going to set San Francisco on its ear—and something of a braggart in his letters back home: "I have at my absolute control hundreds of thousands of dollars," and so on. This posturing and boasting stands in sufficiently striking contrast to his depression and self-pity in time of misfortune to raise the question of its psychological implications—but at this point in his life there is not sufficient documentation to supply any meaningful answers.

If Sherman were asked to describe or define himself he would not have done so in psychological terms, but in geographical ones. He saw himself first and foremost as a man of the West, which he saluted with such terms as "the Great West" and "the Mighty Northwest." He told John, who was planning to propose him for a political appointment on the eve of the war, "don't represent me as a Republican but as an American, one who believes that in a short time the inhabitants of the Mississippi and its tributaries will *command* this continent." The great river he called the country's spinal column; he firmly believed that his favorite town, St. Louis, would soon become the nation's second city in wealth and importance. As a banker he resented the financial domination of the Easterners; as a Civil War general he would regard the South as the enemy and the East as the competition. "Let the mighty armies of the East and Centre keep pace with us," he wrote a fellow general, "and then they may taunt us if we fall behind."

Curiously, though he was proud of his Western affiliation, he showed little attachment to Ohio. As early as 1844 he told Hugh Ewing he would never live there and made the same remark to Ellen several times, and in the end he held to his pledge. The reason he gave Hugh was this: "I do not like the people in it as well as some other states." He added that he felt young men always did better if they moved on to some new place. As for Lancaster, he summed up the summer he spent there in 1838 as "seeing nothing and doing nothing." When he was there for a few days in 1860 the daily entries he made in his diary consisted of a single word: "Lancaster."

Some have argued that this early prejudice can be traced to the looming, intimidating presence of his foster father and father-in-law; but long afterward, when Thomas Ewing had disappeared from the scene and Sherman was one of the three or four most famous men in the country, his view of his birthplace had not changed. His granddaughter, Elly Sherman Fitch, recalled that "Grandfather, after 2 or 3 days seeing old friends, was bored stiff in Lancaster. He wanted to get going, to see people and places, so Lancaster was a constant source of friction in the family."

In his political orientation Sherman must be placed firmly on the right. He was a Whig by inclination, but not affiliated. He claimed he voted only once, for Democrat James Buchanan in the elections of 1856, and significantly that was when he was a civilian. Yet Sherman probably showed more interest in politics than most of his colleagues in uniform. He read the political news voraciously, being addicted to the very newspapers he would so roundly denounce. In Louisiana he subscribed to the *National Intelligencer* and the *St. Louis Republican* and stayed abreast of the Louisiana press as well. He was very conversant—and outspoken—on the issues. Though he boasted he had never read a political platform and was speaking purely as a detached observer, once he warmed to his subject he was anything but detached.

Like most conservatives he was a warm friend of law and order. This is how twenty-three-year-old Lieutenant Sherman described Charleston, South Carolina: "One of the neatest cities in this country, policed to perfection and guarded by soldiers enlisted by the state, who enforce order with the bayonet . . . never are you bothered by crowds about the doors of inns or brothels, nor roused from sleep by their yells."

Sherman's idea of the well-ordered society was hardly original; indeed he acknowledged Daniel Webster had given him the idea that the American governmental system resembled the solar system: "Each little planet or asteroid, if kept to its place, remains in perfect harmony, having an influence upon the whole proportionate to its mass but no more." And to continue the analogy, Sherman was increasingly concerned that the harmony and balance of the system were in danger. Here the "system" was the government established by the Constitution of 1787, and Sherman's concern heightened as the Whigs declined and the Democrats rose: the people no longer confided the direction of national affairs to the leadership of sober, responsible, principled men of substance; using their increased franchise the masses were now giving power to irresponsible demagogues; the common good was lost sight of and centrifugal forces increased as factions and sections sought to impose their will.

Since Andrew Jackson and his Democratic successors seized the helm, the ship of state had veered off course and was headed for the rocks and shoals of anarchy. This sort of doomsaying has been frequent in American history; there have been solemn warnings that some of our best-known presidents would take us over the precipice–Jefferson, Jackson, and Franklin Delano Roosevelt among others. But in 1860 there was a precipice before us and we went over it. When secession and war came Sherman was not surprised; he would incorporate the catastrophe into a larger scenario of apocalypse that, like many of his notions, bore his own peculiar stamp. He saw the Southern rebellion as but the opening round of a long struggle that would end either in chaos or in the return to dominance by the "intelligent classes" and a government whose powers had been so reinforced that no one would dare rise up against it again. This coming political reaction he sometimes referred to as "the Revolution."

In one respect Sherman did not fit the conservative mold of his day. Though his parents had been Episcopalians who changed to the Presbyterian Church when they moved to Lancaster, Sherman was a member of no church. Nor did he like to discuss his religious views, but he did drop clues here and there suggesting that he at least leaned toward Deism. Despite Ellen's pleadings, he declined to pray for his safe voyage to California in 1846: "The world is governed by universal laws," he wrote her, "and I do not expect any of them to be set aside for my especial benefit."

Sherman's place, then, in the American social pyramid was with the "principled men of substance." His sense of position and status was instilled in the Lancaster boy of "good family," confirmed by association with others of the same status at West Point, and further reinforced by years as an officer and gentleman in the highly stratified world of the United States Army. He understood early that in his world ability and ambition were better rewarded when supplemented by "connections." Through his father-in-law and later his brother he had those connections and he used them and profited from them, beginning with his appointment to West Point.

Biographers have probably made too much of occasional references in youthful letters in which Sherman says he will no longer accept the help of others, notably of Thomas Ewing. Not only does he continue to do so, but from time to time he flaunts his connections; he does not hesitate to solicit his father-in-law's aid in furthering the careers of others: his brother Lampson, for example, and his good army friend Richard Garnett of Virginia, who would meet his death in Pickett's charge. And these debts Sherman would later repay in the same coin, fostering the Civil War careers of Ewing's sons, three of whom would become Union generals.

Here and there one can catch glimpses of a young man who not only knows he is a member of a privileged class but is also something of a snob. For some reason in 1843–44 he made unusually long and reflective diary entries; talking to himself, so to speak, he also tells a good bit about himself. He noted the attractive women he saw in his travels through the South, sometimes lingering on their attributes. He particularly noticed the "strapping" country girls who lived around Bellefonte, Alabama. Young and unattached, he might have permitted himself a passing dalliance, but there is no evidence that he did; in fact these young women were not that attractive to him, for he considered there was something important lacking in them: they had not had "the opportunities for cultivating their minds and cultivating the graces of motions and body"–and they were not for him.

Sherman's travel accounts for this period also indicate the desire of a young man in his twenties to avoid close contact with fellow citizens of the meaner sort. In his trip down the Mississippi in 1843 he was clearly uncomfortable when he could not get a cabin on the steamboat and was thrown in with a "promiscuous crowd" of deck passengers. "Thank God I don't know a person on board," he noted with relief. He entered a brief description of the crowd on the boat that was witty but hardly kind. His view of such contacts with the masses had not changed seven years later when he was figuring out how he might make a trip East: "Six days and nights, jammed nine to a small coach, mulattos, loafers, women and children, no."

The reference to mulattos suggests that the youthful Sherman considered himself the member not only of a privileged class but also of a superior race. His long stay in the South may have shaped his attitudes in this respect, for the word "nigger" entered his vocabulary early and stayed there the rest of his life. He held blacks to be inferior to whites, opposed putting them in uniform, and when obliged to accept them as soldiers saw to it that they were not used in combat. Nor did the very creditable record of black troops in the war lead him to change his mind. "A nigger as such is a most excellent fellow," he wrote a friend in 1865, "but he is not fit to marry, to associate, or to vote with me and mine."

Such were his views of his country, its people, and the problems and prospects in its future. What did he see and seek for himself? If one searches among the random remarks and musings he made in this period concerning his goals and purposes in life, an impression gradually emerges that status and position, especially as they were perceived by others, were of central importance to this man. Wealth was not a goal he sought or cared particularly about attaining; his career certainly indicates no great consuming effort in that direction. When Lucas, Turner and Company was foundering he did not hesitate to liquidate his own assets so that friends who had invested with him did not suffer loss. "You must know by this time," he wrote his friend Henry Turner in 1855, "that I don't love money but that ambition is wrapped up in my plans." Immediately following this passage he said that he had been offered the Democratic Party's nomination for city treasurer of San Francisco, a position that could start him up the political ladder, and the first of many such offers he would decline. Yet Sherman maintained a steady involvement in civic matters in San Francisco; he was widely known and almost universally liked in the business community, ever ready to pitch in, to be of public service, to boost the City by the Bay. He would have made the perfect Rotarian.

When he wrote that he was "somebody" in San Francisco and a "big chief," he was saying that in a sense he had arrived. When he wrote Tom Ewing Jr. from Louisiana, telling him he would laugh up his sleeve to see Cump Sherman "hobnobbing" with the governor and talking familiarly with the state's legislators, he was saying the same thing. He acknowledged that the Louisiana position had attracted him in part because it was a "good social position," one that would give the holder wide entree. But to what end? Would he have been satisfied with simply "the esteem of men" or did he covet something more? At this stage in his life it would be impossible to tell, and Sherman himself most likely did not know. These questions are in any event idle ones, for the great storm was approaching; both he and his country would emerge from it transformed.

# 9

## BULL RUN

Among students of warfare General Sherman has long had a reputation as prophet and seer, notably in anticipating the nature of the
great conflict in which he made his reputation. Lloyd Lewis subtitled his
biography of the general "Fighting Prophet," while Gamaliel Bradford, in
the Sherman essay of his "Union Portraits" series, was explicit on this
point: "Sherman saw and foresaw everything." And for this prescience
Bradford said the general paid a heavy price; others, lacking his foresight,
"decried him and almost displaced him as a sheer, unbalanced lunatic."

Yet Sherman was not particularly perceptive in sensing the approach
of the great storm, perhaps because of the turbulence in his own life.
Though he spent several months in "Bleeding Kansas" in 1858–59, and
apparently crossed paths with John Brown, in his diary and letters he had
relatively little to say about this overture to the great conflict. It was only
in Louisiana that he gave the crisis his full attention. Then most people
thought the war would either confirm the solidity of the Union or divide
the country in two. Sherman saw other, more complicated scenarios,
including a whole series of successor-states or "new combinations" as he
called them.

While the popular view, both North and South, was that the war
would be a short one, with the enemy's resolve fading almost as quickly as

the smoke from the first battle, Sherman saw it as a very long struggle (the same prediction he had made for the Mexican War). One of his estimates in the spring of 1861 put the duration of the war at thirty years; in other references he suggested it might take a half-century, then six months before Appomattox he spoke again of thirty more years of fighting. The war would be the hecatomb of American manhood: the complete reduction of the South would "involve the destruction of all able-bodied men of this generation and go pretty deep into the next." He spoke of casualties running 300,000 a year for an indeterminate period.

These chilling estimates may have been another expression of his penchant for hyperbole, put forward primarily for their shock value. But it is likely that they were linked to the nature of the conflict as he saw it. For him the stakes in the war were higher and the questions it would settle more profound than his contemporaries realized: "These side issues of niggers, state rights, conciliation, outrages, cruelty, barbarity, bankruptcy, subjugations, etc., are all idle and nonsensical."

In the sundering of the Union he saw a more fundamental issue: secession was the final, fateful challenge to that venerable and near-perfect polity bequeathed by the Founding Fathers. In the demagogic oratory he had heard in Louisiana, the exultant crowds he had seen on the streets of New Orleans, and the strident bellicosity of the Southern press, he recognized many manifestations of the old enemy that Hamilton and others had inveighed against—the furious masses running their destructive anarchical course.

The conquest and reincorporation of the rebellious South was only the first phase of the "Revolution," beyond which lay the reduction of factionalism, the diminution of states' rights, restriction of the franchise, and bringing the mobs of the cities to heel. These things done, the ship of state could be brought back on course, with representatives of "the intelligent classes" at its helm.

For a deep-dyed conservative, for a soldier with the profession of arms ground into his composition, it was a war not to be missed. Yet when it came Sherman was in no hurry to join the fray. Jobless though he was, he declined to offer his sword in those anxious days when a military showdown was looming at Fort Sumter; more than that, he would also fail to answer several subsequent calls for his services. A few months earlier he would have leaped at the chance to get back into uniform; now he was content to let the war start without him.

This change in attitude was fairly abrupt. As late as December 1860 he was interested in using the leverage John would have with the incoming administration to get back into uniform. But his goal was a position in either of two "staff" departments, that of the inspector general or the adjutant general. His guiding motive here was probably security. These were safe, comfortable jobs in war or peace, eminently suitable for a family man. If war should come he would not be called on to wreck his career with one bad day on the battlefield, leading green troops. He speculated—correctly—that those generals who conducted the initial campaigns would suffer just that fate; it was undoubtedly for this reason that when war did come, he preferred to bide his time.

He may also have been disheartened by what was transpiring in Charleston harbor. He had been following the fate of the Federal troops there with special interest, first because he had served there, and second because the garrison at Fort Sumter was now commanded by Robert Anderson, a friend of long standing who had been his company commander in Florida twenty years before. Anderson had desperate need of supplies and reinforcements, but with South Carolina gripped by secession fever the Buchanan administration feared doing anything that might appear provocative—and so it did next to nothing. Sherman was outraged, denouncing the administration as "pusillanimous."

In January he wrote John to say that he would not consider entering the army again except with a high commission. Two weeks later he was more specific; he felt "no inclination to take part in this civil strife" unless he was confident about the military dispositions, and he could have that confidence only if he were "high enough to have a word in council." Actually, for the moment what he wanted more was a civil post in Federal service, the sub-treasurership in St. Louis. It was a political appointment, and now that John was prominent in the party coming to power, Sherman thought it might be within reach.

The sub-treasurership—or some other civilian employment—may well have been what was foremost in Sherman's mind as he traveled north in those last days of February 1861. He reached St. Louis on the twenty-seventh and stopped there, where he made contact with his friend Henry Turner.

The two men no doubt talked about the chances for the treasury post. St. Louis was the political fiefdom of the powerful Blair family, now headed by Montgomery Blair, postmaster general in the incoming administration, and his brother Frank, who was a member of the Missouri delegation in the House of Representatives. If John pushed his brother's candidacy for the treasury position he would have to contend with the Blairs.

But there were other employment possibilities in St. Louis, notably running the Fifth Street Railway, a streetcar line that was already built and operating. Sherman would need to be elected to the position, but since Mr. Lucas and an associate were majority stockholders, that would present no problem; Sherman gave his tentative agreement, then headed for a reunion with his family at Lancaster, where he arrived March 2. He did not linger, for there was a message from John, urging him to come to Washington to explore possibilities there.

He reached the capital on March 6 and John took him to the White House. Abraham Lincoln had been in office forty-eight hours; his new residence was swarming with friends and well-wishers, office-seekers, and hangers-on. The two brothers found their way through the crowd to the president, and according to Sherman a singular conversation ensued:

John: "Mr. President, this is my brother, Colonel Sherman, who is just up from Louisiana; he may give you some information you want."

Lincoln: "Ah! How are they getting along down there?"

Sherman: "They think they are getting along swimmingly—they are preparing for war."

Lincoln: "Oh, well! I guess we'll manage to keep house."

There were probably some other words exchanged, for Sherman later claimed Lincoln told him, in a dismissive tone, he would not have much need of "military men." In any event Sherman came away from the meeting profoundly discouraged. He told John that he and his fellow politicians were putting the country into "a hell of a fix." He had been put off by the boisterous, clamorous atmosphere of the White House and the casual, countrified manner of its chief occupant. So far as he could see the new administration had no greater sense of impending catastrophe than the one it had replaced. He told John that Lincoln and his advisers were men sleeping on a volcano.

On March 8 he was on his way back to Lancaster; by the eighteenth he had installed his family and belongings in a rented house in St. Louis. He went to work for the Fifth Street Railway, which named him president on April 2. His salary was $2,000 a year but he soon found his family expenses running $300 a month, with little possibility of reducing them. Even using the interest on money that came from the sale of some of Ellen's property, it would not be easy to make ends meet.

He threw himself into his work. He found that with some tinkering with the line's operations he could reduce labor costs substantially; he cut the number of employees by a third. When there was no work in the office and nothing else to occupy his time, he would go out and position himself on a street corner, watch in hand, making sure that the drivers kept their intervals.

He had learned that the sub-treasurer's place had gone to another applicant. He held the Blairs responsible, "a selfish, unscrupulous set of ——;" thereafter he considered them enemies. When Montgomery Blair wrote to ask him if he would accept the position as the army's chief clerk, with promotion to assistant secretary of war as soon as Congress met, he wired back, "I cannot accept" and followed it with a brief letter in which he wished the new administration well in "its almost impossible task of governing this distracted and anarchical people."

He was in his office on April 13 when the news came by telegraph that Fort Sumter had fallen. He was quick to endorse Lincoln's call for 75,000 volunteers but thought the president should have made it 300,000. Napoleon said it took three years to make a soldier; these men, called for three months, would be home again before they had even learned how to clean their muskets. He sent a stream of advice on military matters to John and did not hesitate to offer suggestions on high strategy: If General Scott meditated a move against Richmond he would be well-advised to wait until the Confederate Congress was in session, then the psychological impact would be greater. He offered his frank opinions on the Union military leadership: John could have confidence in McClellan, who was "the best man"; McDowell, however, had seen too much of his service "in a smooth office chair." General Wool was not to be trusted: "He failed me at a critical moment in California."

Sherman was a spectator to the struggle over secession in Missouri; though nominally a slave state, with considerable agitation and strife it stayed in the Union, thanks in part to the efforts of the Blairs. Sherman visited the local arsenal and noted with satisfaction that it was ready to withstand an assault by "the people." He and Willie were caught in a fusillade that began when a belligerent drunk clashed with a skittish contingent of militia; though several bystanders were killed neither Sherman nor Willie was hurt, and the father noticed with pleasure that under fire his son was "cool as a cucumber."

Now, as the country was rushing to arms, overtures and feelers about returning to the service continued to arrive. Militia organizations sought his expertise as they prepared to answer the president's call; John, who was beginning to lose patience with his brother, was convinced that he could have had the command of the Ohio militia and the rank of major general if he had made the slightest show of interest; Tom Ewing Jr. wrote from Washington, asking him what would induce him to return to the army and promising to go see General Scott on his behalf; James B. Fry, a friend from West Point days, tried to convince him that the mass of new volunteers would need the guidance of every academy graduate possible; Ellen wrote John behind her husband's back, imploring him to find her husband a suitable command: "I am convinced that he will never be happy out of the Army."

Sherman himself took up the pen on May 8 and wrote a brief letter offering his services to Secretary of War Simon Cameron.

He explained that he had not simply volunteered because, rightly or wrongly, with his background he felt himself "unwilling to take a simple soldier's place": at the same time he had traveled about so much that the other volunteers would not know him and thus not elect him as one of their officers. "Should my services be needed," he concluded, "the records of the War Department will enable you to designate the station in which I can render best service."

Congress was creating several new regiments in the regular army; he still had his army connections in Washington, and by May 20 he had heard that he was slated to receive the colonelcy of one of the new regiments. He reminded John that the new colonels would rank in order of their appointment: "If you are behind the curtain try to get mine as well up as possible." On June 6 he got the summons to Washington. There was a frantic period of repacking. Sherman sublet the house and left for Washington; Ellen, the children, and the furniture took the road back to Lancaster.

When Sherman reached Washington he found that he would indeed be a colonel, commanding officer of the Thirteenth U.S. Infantry, a unit that for the moment existed only on paper (he would be listed as its colonel for two years, though he would never actually lead the regiment). At first he thought he might be charged with recruiting his men, but in the end this was done for him. He did have a voice in the selection of the regimental officers; after considerable maneuvering he got a captain's commission for his brother-in-law, Charles Ewing. Schuyler Hamilton, who had briefly worked for Sherman in his bank, was now Winfield Scott's aide, and probably instrumental in getting the new colonel a temporary assignment on Scott's staff.

In the spring of 1861 the nation's most distinguished soldier was seventy-five years old. Physically he was virtually an invalid; overweight and tormented by gout, the commanding general was directing the nation's war effort from his parlor, where he spent much of the time with his foot immersed in a tub of icewater. In mind and temperament he seemed as keen and irascible as ever. A young naval officer whose duties took him to Scott's house likened his visits to "paying a call on a sick bear."

Scott well-remembered the young officer who had brought him dispatches from California a dozen years before. He had him to dinner and gave him the tasks of an inspector, visiting and reporting on units and installations about the capital. For that purpose Sherman spent $12 for a "reconnoitering glass" and passed much of his time on horseback, familiarizing himself with the terrain around Washington. The duty with Scott lasted only ten days. At the end of June he received new orders to take command of a brigade of volunteers that was manning one of the forts guarding the capital.

Fort Corcoran and its outworks had been thrown up along the south bank of the Potomac, across from Georgetown; it occupied eight acres of a country estate called Rosslyn, which subsequently gave its name to a section of Arlington. The fort was the temporary home of several militia regiments, five of which belonged to Sherman's brigade: the Thirteenth, Twenty-ninth, Sixty-ninth, and Seventy-ninth New York regiments, and the Second Wisconsin. They well-illustrate the frantic and tragicomic effort that had been made that spring to prepare the nation for the greatest war in its history. The men were nominally infantrymen, with at least a passing acquaintance with their rifles; but they knew nothing at all about the massive cannon they would serve in event of an attack.

Nor were the regiments very well-equipped for campaigning as infantry. The Second Wisconsin went to war in gray uniforms practically indistinguishable from those of Confederate units they would shortly meet in battle. The Sixty-ninth and Seventy-ninth were from New York City. They had distinct cultural identities, the Sixty-ninth being emphatically Irish, with a number of "New Ireland" militants in its ranks. Among its officers was the activist Thomas Francis Meagher, whom Sherman would have cause to remember. Had been condemned to three years of banishment in Tasmania by a British court. The men of the Seventy-ninth were predominately Scots; their colonel was the brother of Secretary of War Simon Cameron. These last two regiments were frequently called on for parades and other public functions and they were equipped accordingly. The Sixty-ninth brought a fife and drum corps and a very large Green banner. The Seventy-ninth "Highlanders" were distinguished by the Cameron tartan, highly visible in the trousers of their undress uniform and the kilts of their parade dress; the latter, according to the regimental historian, "attracted a good deal of attention."

Sherman's new assignment gave him a great deal of trouble. First of all the fort was still under construction, much of the labor being provided by the soldiers, who had neither the skill nor the taste for the work. Then the army's presence had resulted in various claims by the local inhabitants, especially the owner of Rosslyn, the estate on which the fort had been built. Ever scrupulous with accounts and with other people's money, Sherman reduced the owner's claims to a fraction of what was sought, adding this comment: "His man is selling to our mess vegetables from the garden charged for as destroyed and milk from the cows charged for as lost."

But it was the 3,400 officers and men of his brigade that gave Colonel Sherman the greatest concern. In the regular army companies and regiments had needed little attention from higher authority; usually a word to a company or regimental commander was all that was necessary for things to be put right. But in these units no one knew what to do. It was a bit like the first days at the seminary in Louisiana; he could not go fifty feet before someone called for Colonel Sherman.

"Each of these regiments has its peculiar troubles," he explained to Ellen, "all of which are referred to me." The colonel of the Second Wisconsin was Dr. S. Park Coon, whom Sherman described as "a good-hearted gentleman who knew no more of the military art than a child," while the lieutenant colonel had been to West Point and was quite able. Sherman had to promote Dr. Coon to a largely ornamental staff position so his lieutenant colonel could command the regiment. In the Sixty-ninth New York there was a potentially explosive situation. It had immediately answered the president's call for ninety-day volunteers back in April; now those ninety days were about up, and while many were willing to stay to see a battle, others were ready to go home. The War Department examined their case and had a surprise for them all: their ninety days began only when they were formally mustered into federal service; they would have to serve until August.

Worrisome as these problems were, beyond them loomed a far greater concern: What might happen if these green troops had to fight? This question preoccupied the army's leadership from General Scott on down. But pressures to act were building: two Confederate armies now hovered in northern Virginia, a constant menace to the nation's capital. The politicians and the press were calling for action and the public was becoming exasperated by the delay—and unless the army marched soon, it would melt away as militia obligations ran out. Against his better judgment General Scott sent the order to advance to General Irvin McDowell, who in turn started a cascade of orders down through the divisions, brigades, and regiments of his hastily constructed and untried army.

Just before the army marched Sherman composed a brief letter to Ellen, a "last letter," such as many other men were writing. It was devoid of the usual flourishes and witticisms. "Whatever befalls me," he wrote, "I know that you appreciate what good qualities I possess and will make charitable allowances for defects, and that under your care our children will grow up on the safe side." The letter ended: "Goodbye and believe me always most affectionately yours."

On the afternoon of July 16 General McDowell put his army in motion; long columns of men and vehicles began to move westward along the dusty roads of Virginia. The immediate goal of McDowell's column was the little town of Centreville just over twenty miles away, where his five divisions would concentrate to form an army of 30,000.

It took McDowell's army two and a half days to cover the twenty miles to Centreville. The heat that July was intense, there were frequent stops to fill canteens, and the progress along the dusty roads was impeded part of the way by trees that the Confederates had felled across them. But there were other reasons: confusion over the order of march or the route to be taken, traffic jams caused by slow-moving herds of cattle, and supply columns that could not find their destinations.

The men could not be hurried; they might be roused by reveille at 3 A.M., but they could not be ready to take the road until nearly three hours later. They went at the pace of tourists, which in a sense they were. They might linger on their way through a village despite all exhortation, or desert the column en masse to fill their canteens at a spring or chase chickens in a barnyard. They committed depredations, slipping out of the column and rejoining it with a ham or a pail of milk; once or twice houses mysteriously caught fire during their passage. General McDowell sent daily reports back to Washington, and his problems with his soldiers began to appear in his dispatches: on July 18 he wrote, "I am distressed to have to report excesses by our troops."

Shepherding his regiments through the countryside of Virginia, Sherman was in a position to see those excesses firsthand. He had his two aides police the column as much as they could, calling out, "Colonel Sherman says you must keep in the rank," "you must close up," "you must not chase the pigs and chickens," and so on. The soldiers would shout back, "who are you, anyway?" And as they passed up and down the column the aides were followed by jeers and catcalls.

The Third Brigade and its commander had their baptism of fire on the eighteenth, in a brief preliminary engagement at Blackburn's Ford; it was toward the end of the fighting there that they were hurried forward to relieve another brigade. Sherman was troubled by the fact that they were encountering far too many stragglers; for their part the men were eager to see the enemy, some climbing trees in hopes of getting a glimpse of them. Once in position they came under fire from artillery they could not see; the bombardment lasted thirty minutes and produced minimal casualties.

Sherman tried to calm and reassure his men. A man in the Seventy-ninth Highlanders recalled that their commander told them it was a waste of time to duck, since by the time they heard the shell it had already passed. "Hardly had the words left his lips when a big shot or shell came crashing through the trees and but a few feet above him; down went *his* head, close to the pommel of the saddle, and when he raised it again it was to confront a line of grinning faces. 'Well, boys,' said he, a broad smile softening his rather hard features, 'you may dodge the big ones.'"

He wrote Ellen after the affair was over: "I am uneasy at the fact that the volunteers do pretty much as they please, and on the slightest pretext bang away. Danger from this desultory firing is greater than from the enemy as they are always so close, whilst the latter keep a respectful distance." To his brother John he confided: "The volunteers test my patience . . . 'Twill take time to make soldiers of them." They were already taking up much of his time: "I am pretty well up all night and sleeping a little by day."

On the twenty-first the army pushed westward again, toward almost certain battle, and Sherman could only speculate on how his men might behave: "With my regulars I would have no doubts, but these volunteers are subject to stampedes." The First Battle of Bull Run or Manassas began early on that morning and lasted through the day. Sherman had his brigade on the road well before dawn, and firing began almost with first light. General McDowell had set in motion a battle plan that would strike General Beauregard's forces from both the east and the north (tacticians would give the general high marks for his conception and low ones for its execution). General Beauregard was in the meantime being reinforced from the army of General Joseph E. Johnston, so that the Confederates had a slightly larger number in the battle than McDowell.

The final and decisive act of the struggle took place in the afternoon, and the stage was a prominence known as Henry House Hill, from the home that stood atop it. The Confederates held it and the Union forces assailed it. Sherman had shown his qualities of energy and enterprise that morning, locating a ford in the stream that would give the battle its name and pushing his brigade across it to pursue the enemy. That afternoon his brigade took part in the attacks on Henry House Hill; one after the other his regiments mounted to the assault and were driven back by the cannon and rifle fire of the resolute defenders—Confederate General Stonewall

Jackson had an important role in holding the hill, and that afternoon he would acquire his fame and his nom de guerre.

Like all scenes of mass strife and confusion, the struggle for Henry House Hill is still the subject of controversy among historians; one sees a critical shift in the tide of battle where another does not. But all agree that sometime between three and four o'clock that afternoon something occurred to decide the issue in favor of the Confederates. In his official report Sherman wrote that about half past three "began the scene of confusion and disorder that characterized the remainder of the day. Up to that time all had kept their places, and seemed perfectly cool, and used to the shell and shot that fell, comparatively harmless, around us." But now a change occurred: "Men fell away from their ranks, talking, and in great confusion."

Had the conviction spread over this mass of struggling men like some dark cloud that the day was irretrievably lost? Or had they simply had all they could take, as Sherman seems to imply? Some of the Wisconsin soldiers apparently felt their gray uniforms were drawing friendly fire and they should retire; their departure may have given the example to other regiments in Sherman's brigade. There was another observer of this scene, Captain Daniel P. Woodbury of the Union Army, who had another explanation. He estimated that by 4 P.M. there were on the battlefield 12,000 Union soldiers who had completely lost touch with their own regiments. "They could no longer be handled as troops, for the officers and men were not together. Men and officers mingled together promiscuously; and it is worthy of remark that this disorganization did not result from defeat or fear, for up to that point we had been uniformly successful. The discipline which keeps every man in his place had not been acquired." Captain Woodbury was an engineer; he analyzed the debacle at Bull Run almost as if it were a phenomenon in mechanics. But he probably came as close to the truth as anyone: subjected to inhabitual stresses and strains, without its constituent parts securely fastened in place, the machine came to pieces.

Movement was imparted to the mass, at first slowly, then with growing speed and urgency; the retreat became a rout. General McDowell gave thought to halting to take up new positions at Fairfax Court House and discussed the idea with several of his subordinates. But he failed to consult the rank and file of his army, who had decided they would not stop until they crossed the Potomac. They streamed back into the capital in the rain, bedraggled and exhausted, carrying their leaders with them. Sherman and his men had participated in this hurried exodus, but he managed to maintain some degree of control over his men; at a certain point he formed them into a defensive square and faced down the pursuing Confederate cavalry; his men had fought comparatively well on Henry House Hill, taking considerably more casualties than other Union brigades. No laurels would be handed out in McDowell's army, but Sherman's performance had been noted by his superiors.

But back in Washington he became embroiled in another difficulty. The evening of the twenty-second a number of men from Sherman's command took shelter in a barn not far from Fort Corcoran, and Sherman found them there. There are two separate accounts of this confrontation by men who were there (Sherman made no mention of the incident), and they both agree that it was the stern, "official" Sherman who demanded to know what they were doing there. They told him, adding that they had few blankets, no tents, and no food. Then, according to the more charitable and plausible of the two narrators, Sherman said, "Well, you had better go down into the woods and build brush huts; I want my horses in here" (the other version has him order the men to go out and stand in the rain). The more charitable narrator continues: "We were in no condition to remonstrate–audibly–but had our opinion of an officer who would turn men out of shelter for the purpose of giving it to dumb brutes."

By the twenty-third some semblance of order had returned to the camps and installations about Washington; President Lincoln and Secretary of State William H. Seward made a series of visits to try and boost the soldiers' morale; sometime that afternoon they showed up at Fort Corcoran. In his *Memoirs* Sherman described this visit in some detail. He greeted his guests and joined them in their carriage; they drove about, stopping from time to time so the president could talk with the soldiers who crowded about his carriage. At this point Sherman relates that an officer had pushed his way up to the president and announced that he had a grievance–Colonel Sherman had threatened to shoot him. The officer had wanted to make a trip to New York on business, Sherman had prohibited it; the exchange became heated, with Sherman saying that if the officer tried to leave without permission he would shoot him "like a dog."

"Threatened to shoot you?" the president asked, and the officer said yes. Then in a loud stage whisper Lincoln said: "Well, if I were you, and he threatened to shoot, I would not trust him, for I believe he would do it." The assembled soldiers burst out laughing, while the officer turned and disappeared into the crowd.

This anecdote is well-known, having been passed on by most of Sherman's biographers. However, other eyewitness accounts of Lincoln's visit do not corroborate Sherman's story; nor did Sherman relate the affair in a letter to Ellen, though it was his custom to pass on such happenings to her, particularly when they showed him in a favorable light. It is quite possible that it did happen as Sherman related it; it is also possible that, writing his memoirs a dozen years afterward, the general substituted it for another, far more unpleasant occurrence that day.

The morale in the brigade was low; among some the feeling against their commanding officer was running high. The Sixty-ninth had lost both its colonel and its lieutenant colonel, and was now commanded by a captain, who in hot haste laid out the regiment's specific grievance in his official report: Sherman had "peremptorily denied" them the wagons that should have accompanied them on campaign, causing them to be "greatly fatigued and harassed." In the Seventy-ninth the men talked about how their brigade commander had mysteriously disappeared during the battle (Sherman later acknowledged to John that at some point, after the fighting was over, he had become separated from his brigade). The officers and men of the Seventy-ninth were thinking about simply marching off and leaving the brigade; a delegation of officers from the regiment paid a call on the secretary of war and laid out their complaints against their brigade commander. The secretary promised to take action.

And now, to fan the fires of resentment, the story of the barn episode was making the rounds, becoming more embellished with each telling. William Thompson Lusk, one of those who had been in the barn, said that as Lincoln's carriage drove by, the men of the Seventy-ninth swarmed about, jeered at their brigade commander, and "reminded him who it was that first basely deserted us on the battlefield . . ." The historian of the Seventy-ninth, who put his narrative together after consulting other veterans of the unit, gives a different version: one of the Highlanders complained to Lincoln, "Mr. President, we don't think Colonel Sherman has treated us very well," and went on to relate how they had been turned out in the rain the night before. Lincoln heard the man out, then, half-turning toward Sherman, who was sitting "like a statue," he addressed the throng: "Well, boys, I have a great deal of respect for Colonel Sherman, and if he turned you out of the barn I have no doubt it was for some good purpose; I presume he thought you would feel better if you went to work and tried to forget your troubles." The words had their effect on the men; the president made a bow, waved his hand, and the carriage rolled off to the next regiment.

On the twenty-fourth Sherman wrote Ellen a brief, somber letter, the kind a man would write just after he has passed a very bad day. Of Lincoln's visit he said only, "the President and Mr. Seward visited me." Though this was his first letter since the battle, he said little about that either, beyond the fact that he had come out unharmed, only "to experience the mortification of the rout, confusion and now abandonment by whole regiments" (the Seventy-ninth marched off that day, transferred to another brigade on orders from General Scott). "Well, I am sufficiently disgraced now," he wrote; then he struck what would become a familiar chord: "I suppose soon I can sneak into some quiet corner."

Four days later, when he wrote Ellen a ten-page letter, he was clearly in better spirits. His judgments were more measured and his tone more moderate. He commented on his first exposure to combat: he had remained clear-headed, noting his own reactions: "There for the first time I saw the carnage of battle—men lying in every conceivable shape and mangled in a horrible way—but this did not make a particle of impression on me—but horses running about riderless with blood streaming from their nostrils—lying on the ground hitched to guns, gnawing their sides in death."

In retrospect he could understand why his men could not prevail against the storm of shot and shell the enemy directed at them: "I do think it was impossible to stand long in that fire." But what followed had shocked him: the men fell back in disorder, insisting loudly that they had not been supported, that they could not tell friend from foe. "I had read of retreats before—have seen the noise and confusion of masses of men at fires and shipwrecks, but nothing like this. It was disgraceful as words can portray." The comportment of the men as they streamed back through

Virginia had been another revelation. The soldiers were predatory and destructive: they took what they wanted as they passed through the countryside, despoiling and even burning houses: "No curse could be greater than invasion by a volunteer army. No Goths or Vandals ever had less respect for the lives & property of friends and foes . . . "

Over the next few weeks problems with the volunteers continued to preoccupy military authorities in the Federal capital, with Sherman having his full share. He wrote Ellen that he had not slept out of his clothes a single night since before Bull Run, adding, "the volunteers will not allow sleep by day." Some regiments insisted their time was up but could not convince the War Department; others, like the Sixty-ninth New York, felt they were due a transfer to New York City after their recent fatigues. Here and there the upshot was mutiny.

In the army traditionally this was a rare occurrence, something launched by desperate men, marked by violence, and punished with the greatest severity. In Washington that summer it was by no means rare, and its form and outcome were not in conformity with tradition. Now the mutiny usually took the form of a sit-down strike. The men would remain in camp but ignore any orders given them. The case of the Thirteenth New York was typical. Sherman tried to keep the lid on, saying they would "probably" go home in a few days, or that he was "almost positive" they would be discharged. But on the morning of August 13, the date they felt their enlistment was up, a member of the unit wrote home, "Hell reigns in the 13th today." "The 13th New York have refused duty," Sherman wired General McClellan's headquarters, "they simply refuse to form ranks, to go on details or obey any orders whatsoever. Appeals to them are treated with ridicule."

The customary usages of war permitted draconian punishments for such actions. Sherman was in favor of a severe response: in the case of the Thirteenth New York he said "a terrible example" was necessary. But here and in other cases the mutiny was usually ended by a show of force—having the mutineers ringed with a body of regulars or artillery trained on them—then with a combination of orders and threats, most of the men could be compelled to obey; the recalcitrants were then marched off to confinement. In the case of the Seventy-ninth Highlanders twenty-one "ringleaders" got six months' confinement at Fort Jefferson in the Dry Tortugas, and twenty-seven members of the Thirteenth New York joined them there.

The first three months of his active service were a revelation to Sherman. He had speculated on the difficulties of restoring the Union and he had certainly not underestimated them. But he had counted on an increase in the regular army to meet the crisis; in Washington he found that for essentially political reasons the war would have to be fought with armies of volunteers. The recent disastrous campaign had demonstrated all the limitations of such forces, and beyond that it had revealed a government with little capacity and little grasp of the challenge before it.

He was not impressed with Lincoln in his first contacts with him. He was even less impressed with Secretary of War Simon Cameron, a political placeman and sometime journalist whose corrupt ways would drive him from his cabinet post ten months after he took office. Sherman was appalled at the hesitation and timidity the government demonstrated in those first months. Then too, in Washington he could observe firsthand how the machinery of government functioned and also malfunctioned. What he saw gravely troubled him; in the first months of the war he paired his observations with conservative notions he had long held, and eventually he developed a distinctive view of the war and how it ought to be fought.

The most obvious of the nation's mistakes in making war was reliance on the volunteer soldier. While the army's leaders, McDowell foremost among them, were saddled with responsibility for the debacle at Bull Run, McDowell's plan of battle was good and his subordinates fully capable of executing it. "The difficulty is with the masses—our men are not good soldiers. They brag but don't perform . . ." Neither they nor the population from which they were drawn could comprehend the gravity of the threat and the enormous effort they would have to make to overcome it.

Journalists sounded no alarm, but continued to flatter and titillate the public that bought their papers. Politicians, anxious to perpetuate their careers, pursued the same reckless policy. Even if Lincoln and his collaborators saw the danger, Sherman found them "perfectly powerless in this emergency." "No one has the courage to tell the truth," Sherman complained to Ellen. "Public opinion is a more terrible tyrant than Napoleon." And so the people were condemned to learn the hard way: "Our people won't realize the magnitude of the opposition until we are invaded several times à la Bull Run."

Even then he had his doubts whether the young manhood of the republic could be induced to accept the regimentation and the obedience required of a good soldier. "I doubt if our Democratic form of government admits of that organization and discipline without which an army is a mob." Here was the kernel of the problem: "Our adversaries have the weakness of slavery in their midst to offset our democracy, and it's beyond human wisdom to say which is the greater evil."

# 10

## KENTUCKY

THOUGH SHERMAN WAS troubled by what he had seen in the Bull Run campaign, he could take pride in his own performance. General McDowell and Sherman's divisional commander, General Daniel Tyler, both mentioned him favorably in their reports. Then his name was placed first on the list of new brigadier generals of volunteers created that summer, ahead of Ambrose Burnside, George H. Thomas, and Ulysses S. Grant.

His promotion meant that he would now command the new volunteer units being raised by the states, and of necessity say goodbye to the regulars; but he explained to Ellen, "this will still keep me where I want in a modest position until time and circumstance show us the daylight." For a first generation of military leaders would suffer reverses because of their raw troops, who would "mar the plans and blast the fair fame of anybody." And the generals would be sacrificed, "since they, the people, can do no wrong. If defeat arrives then it is mismanagement, masked batteries and such nonsense."

That August there was little time for speculation on such matters, for the mutinies were at their height; Sherman was called to the White House for a conference on the problem with Lincoln, General Scott, and others. There he exchanged a few words with his old company commander, Robert Anderson. Then on the seventeenth Anderson asked him to come see him at Willard's Hotel; Sherman found him there in the company of a number of political figures from Kentucky and Tennessee, including future Vice President Andrew Johnson. Anderson was still showing the effects of his ordeal at Fort Sumter, but it was something else entirely that he wanted to talk about.

Anderson had just been ordered to Kentucky, where there was a crisis, and would leave before the end of the month to take over the Department of the Cumberland; Kentucky was in a precarious situation, and Anderson, a native Kentuckian and hero/martyr of Fort Sumter, seemed the ideal man to keep the state within the Union. His mission was to assist the pro-Union forces in asserting effective control over the state. He had been told he could take with him three brigadier generals of his own choosing, and he had chosen George H. Thomas, Ambrose Burnside, and Sherman—with Sherman as his second-in-command.

Anderson and Sherman then met with President Lincoln several times to talk about their mission, and in the course of the meetings Sherman made an unusual request of Lincoln. He wanted the President's assurance that he would serve only as Anderson's second, and not have to take command himself. Lincoln gladly acquiesced, adding with a chuckle that it was refreshing to meet an officer who was not looking for advancement. According to Sherman's recollection at some point the president expressed his concern about assigning George H. Thomas to such a crucial theater, since the man was a Virginian and several other Southern officers had already "played false." Sherman joined Anderson in assuring the president that Thomas could be counted upon.

Sherman seemed well-pleased with the new assignment. It would take him away from Washington, where each day seemed to bring a new mutiny. Besides, he recalled in his *Memoirs*, "I always wanted to go West." He confided to Ellen, "I think Anderson wanted me because he knows I seek not personal fame or glory, and that I will heartily second his plans and leave him the fame." And again he sounded the note of prudence, even wariness: "Not 'til I see daylight ahead do I want to lead."

The stakes were high in this new assignment, for if Kentucky joined the Confederacy the South would have its sizable resources and a highly defensible frontier along the Ohio. Lincoln, though new to matters of strategy, nevertheless understood the importance of the contest in Kentucky; if the state left the Union, he said, her departure would be "nearly the same as to lose the whole game." Sherman, whose notions of strategy were advanced, if sometimes original, agreed: that part of the country was its "great centre"; if it could not be held, he explained to Ellen, "there is a danger our old government may disintegrate and new combinations be formed."

In May Kentucky had issued a proclamation of neutrality that might have come from a foreign country. Preposterous as it was, the state's neutrality had been rather scrupulously observed by both the Federal and Confederate governments, for each feared doing anything provocative that might drive the Kentuckians into the other camp. Then during the summer the tide of opinion began to shift in favor of the Union; elections in August gave the Unionists a clear majority in the legislature. They could now formally put their state in the Union camp, but they were concerned about the military consequences when they did; Anderson and Sherman were to provide the force to ensure success.

By late summer what the diplomats call incidents were frequent along Kentucky's borders; the inhabitants of Columbus displayed a large Confederate flag in their town; this so angered the captain of a Union gunboat that he sent a party ashore to tear it down. A major confrontation was shaping up in the western part of the state; there both Northern and Southern governments were anxious to establish themselves at a vital "interchange" in the nation's river highway system, where the Tennessee merged with the Ohio, which then flowed into the Mississippi.

By September 1 Anderson and Sherman were at Cincinnati, where they paused briefly. Sherman had written Ellen that time would not permit him to pass by Lancaster, so she, Willie, and Lizzie came to Cincinnati for a brief visit. She could not bring little Rachel, born the previous July. While they were in Cincinnati news came that Kentucky's neutrality had been flagrantly violated by both sides. On September 3 Confederate General Leonidas Polk had occupied Columbus, Kentucky; three days later Ulysses S. Grant seized Paducah.

Anderson headed for Louisville, where he fixed his headquarters, while Sherman made visits to the governors of Indiana, Illinois, and Missouri, seeking troops for the coming battles; everywhere he found state authorities generous in offering regiments—provided the Federal government could equip them. He had rejoined Anderson at Louisville when the welcome news came from Frankfort that Kentucky's legislature had voted resolutions demanding the removal of Southern forces from her soil. At the same time the state called on Anderson to take charge of her defense. The Unionists thus moved Kentucky into the column of loyal states; Southern sympathizers could do no more than create a shadowy rival government that maintained a fitful existence in areas controlled by the Confederate Army.

These were not extensive. By mid-September the Confederates had made three lodgements in southern Kentucky, all of which would figure prominently in the dispatches of Robert Anderson and his second-in-command and eventual successor. There was the force on the Mississippi at Columbus, and at the other end of the state a small army that General Felix Zollicoffer took into Kentucky to seize the Cumberland Gap, virtually the only easy passageway in that mountainous region connecting Kentucky with her neighbor to the south, Confederate Tennessee. Finally, General Simon Bolivar Buckner advanced on Bowling Green and occupied it on September 9. The Confederate hold on Southern Kentucky was further secured by two hastily constructed strongpoints, Fort Henry on the Tennessee River and Fort Donelson on the Cumberland; all these forces were under the Confederate area commander, Albert Sidney Johnston. The Southern troops sent to Kentucky were for the most part poorly armed and imperfectly trained; at no time in 1861 did their number exceed 27,000–a figure worth remembering in light of subsequent events that autumn.

Sherman had hardly reached Louisville after his round-robin tour of state capitals when Anderson ordered him to set up a protective force for the increasingly important Louisville and Nashville Railroad, which ran southwest toward Bowling Green; about thirty miles out the line crossed streams and gorges by means of a series of trestles; these and a strong position at Muldraugh's Hill had to be occupied and held, for they were clearly within the operational reach of the Confederate cavalry–mounted units from Buckner's army had been reported on Green River, fifty-odd miles to the south.

Sherman started out by rail with a mixed force of volunteers and home guards; after considerable delay he got a force on Muldraugh's Hill, but found it far from satisfactory as a defensive position. The supply line

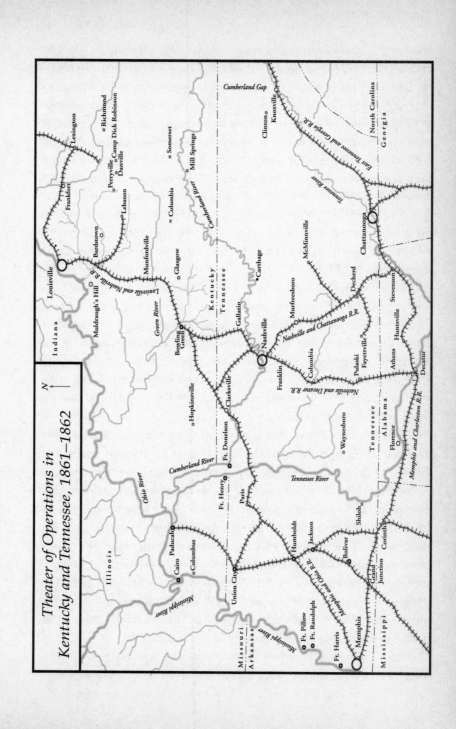

*Theater of Operations in Kentucky and Tennessee, 1861–1862*

was tenuous; the enemy could easily bypass them, then isolate them by burning the trestles and cutting the telegraph line behind them. To unsettle Sherman further, vague rumors of enemy activity came in, difficult or impossible to verify. On the twenty-sixth he wrote a hasty, one-page letter to Ellen in which he said–not without some exaggeration–"we are cut off from everything." Signs of overwork and stress were reappearing. He had not had time to write her in a week, "indeed I have hardly had time to eat, sleep and change clothes." He outlined to her the grimmest scenarios: "I have no doubt that the railroad and telegraph will be cut off behind us . . . The people of Kentucky will not rally in my judgment but turn on us who came to save them from the Despot of the South." And he ended with another tentative "goodbye": "Should this be my last, remember me as you know I want to be."

He wrote again a week later. His rail communications had been restored, but he did not seem cheered by it. Now he was bothered by a shortage of wagons; the people were "nearly all unfriendly" and just waiting for the chance to turn against them. And that chance was coming, for the enemy was gathering at Green River; Sherman had with him about 5,000 men; the enemy was forming with "three times our number." If he were to do anything he would need four times the men he had–but no reinforcements were coming.

He had other worries. In the surrounding countryside he could find "no good position for defense, whilst the ground is admirably adapted to the use of our enemies." He was convinced that among the "country people" of the vicinity there were many spies. He had tightened security, using checkpoints and requiring passes for those traveling on the railway and on the roads; a thousand passes a day were being issued–each one requiring his signature. This letter too ended with quasi-tragic flourish: "For myself I care little but for those entrusted to my care in a desperate cause I am maybe oversensitive." In a letter to Ellen three days later there are flashes of humor–"Where is Hugh that he is not wiping out these southerners?"– but also the recurrence of a troubling refrain: "I don't think I ever felt so much a desire to hide myself in some obscure place . . . "

Only hours after he wrote this last letter he was gone from Muldraugh's Hill, summoned to Louisville by Robert Anderson. Sherman may have had some idea why Anderson had called him to Louisville; in his

*Memoirs* he wrote that he could tell from the tone of the man's letters that "worry and harassment at Louisville were exhausting his strength and health." Anderson had in fact asked to be relieved and recommended Sherman as his successor. The War Department approved his request, ordering him to turn his command over to Sherman and report to Washington.

Sherman had been concerned about Muldraugh's Hill; now he had all of Kentucky to worry about. Ellen had written him that she would like him to address her a cheerful letter for a change, but when he replied, just after taking over from Anderson, he told her there was nothing to be cheerful about. "Personally I am comfortable enough, but the forebodings of danger against which I have struggled keep me far from being easy." Among the forebodings was the possibility that he might fail and suffer in reputation: "Tell Willy that I am very anxious to leave him a name of which he will not be ashamed."

Whatever befell him, he knew he could count on Ellen's support. But to some friends and associates he felt obliged to issue a sort of disclaimer. To one he wrote "I am forced into the command of this department, against my will, and it would take 300,000 men to fill half the calls for troops." This extravagant figure was not the first that Sherman had produced when speaking of manpower needs, nor would it be the last.

His first orders to subordinates in the field were eminently sound ones, urging them to execute a series of sorties and demonstrations, essentially to disconcert the enemy and gain time for the new commander of the Department of the Cumberland to set his house in order: He wanted General Thomas to make a show of force toward London—"I can't give you minute directions but the effect will be good." If Buckner showed signs of advancing from Bowling Green, General William Ward was to threaten his flank so as to make him hesitate.

Sherman did not inherit a well-organized command. There was a severe shortage of trained personnel; the adjutant general's office had made it known that "no inspector general or ordnance officer can possibly be detailed." Competent staff were also lacking in the field. General Thomas, posted at Camp Dick Robinson, south of Lexington, had to manage without a quartermaster for a month. Nor were the army's financial needs well-served; Sherman was caught short and had to sign a note to cover headquarters expenses; General Thomas was obliged to obtain a bank loan to supply his men. Sherman set to work with a brisk efficiency that impressed subordinates; at headquarters he took up much of the slack himself.

In this army of amateurs no one seemed to know the rules, the correct procedure or policy, so Sherman spent considerable time drafting and issuing general orders and circulars that should have gone out earlier, in the process spending much of the day chained to a desk and smoking too many cigars. He put out a circular telling colonels they could not fire their civilian teamsters—that was the job of the quartermaster. He spelled out the duties of the topographical engineer and authored an elaborate set of regulations on the medical service; among other things he advised stretcher bearers that the wounded could be transported in more comfort if "the leading bearer steps off with the left foot and the rear bearer with the right."

Yet the chief deficiency was not in know-how but in materiel. General Anderson had best expressed the situation when he wrote a member of Lincoln's cabinet, "we need everything." The leaders of the Army of the Cumberland lacked maps of Kentucky; its rank and file lacked weapons; everyone lacked winter clothing and tents. Rifles were perhaps the most critical item that fall; with stocks of regulation arms soon exhausted, the quartermasters issued hastily purchased foreign weapons of inferior quality or unsuited to the ammunition available; in their desperation they even cleaned out the racks of the local gun shops; still there were any number of regiments clamoring for arms.

For that other basic element of an army—men—there was at once a plethora and a dearth. Experienced officers were a rarity; Sherman pleaded for more regular officers, but in vain. In the Bull Run fiasco he had encountered officers at regimental level who knew nothing of the profession, but in Kentucky such men wore stars. Brigadier General Jeremiah T. Boyle, with no military service at all behind him, confided to Sherman: "I shall rely on my practical common sense rather than any military knowledge."

The state had put off the decision on raising troops for the same reason she had delayed her decision on secession; then she confided the task to a five-man civilian board that proceeded with great caution. Thus the governors of the states lying to the north of the Ohio River were quicker to raise troops for the Union cause in Kentucky than were the Kentuckians themselves; they understood full well the peril facing them. As Governor Oliver P. Morton of Indiana put it that September, "if we lose Kentucky now, God help us."

Of raw manpower there was plenty; the real dearth was in trained and equipped soldiers; the army Sherman now commanded was gradually filling with volunteer regiments of Hoosiers and Buckeyes, but in the interim he had to supplement the volunteer units with groups such as Kentucky's Home Guards, who soon acquired the unflattering name "Fireside Rangers." Like the Minutemen of the Revolutionary War they came as they were, often bringing their own weapons; they were available for short periods of time only, as little as ten days. Subordinates in the field complained that the Home Guards "disappeared at the approach of danger." Sherman had already had a bad experience with them at Muldraugh's Hill. Then there were the units produced by the spontaneous organizing efforts of local patriots, which Sherman did not encourage: in Maysville the local squirrel hunters had constituted themselves into a company of mounted riflemen; all they needed from the authorities were horses, Sharp's rifles, and "a book giving instruction in the tactics or an old soldier to instruct us." Taken all together, this heterogeneous army struck one observer as "a rather motley crew."

There were endless delays in training volunteer regiments. For a time an outbreak of measles shut down drills and exercises at Camp Nevin. No meaningful training could be done until the men were armed and equipped; as Sherman soon discovered, the volunteers felt strongly on the matter. Without arms they might simply refuse to leave their home state, as was the case with some Indiana regiments. Their governor agreed with them: "It is mortifying to the soldiers and to the state to go unarmed, not having had a chance to learn the manual of arms." Then sometimes the colonel of a regiment, sometimes the men themselves, would refuse to accept obsolete or foreign arms. Sherman could do nothing but lay out his wares and hope for takers. "I have no arms but such as Colonel Steele refused," he wired Governor Morton, "if Colonel Grose will take Belgian muskets, repaired and improved, I would be glad to have his regiment."

Sufficient arms, then, were the precondition for sufficient troops. Sherman wrote his father-in-law, "we only hope to stave off the explosion 'til arms reach us." One of his first acts on assuming command of the department was to wire McClellan: "Do you have anything for us and what?" Two days later he sent Lincoln an equally peremptory telegram on the same subject. In his quest for weapons he launched an effort, entirely unsuccessful, to take back some 15,000 rifles that had earlier been distributed to Fireside Rangers. Fresh from the East, where materiel had not been such a problem, Sherman no doubt felt his command was being neglected. The suspicion would become all too common among commanders in the great valley of the Mississippi that the armies in the West received what the armies the East could spare.

Pressing as Sherman's problems seemed, they were not unique to the army he commanded. The same volume of the *Official Records* that reproduces his pessimistic dispatches also contains similar letters from Confederate military authorities in Kentucky, notably Generals Simon Bolivar Buckner, Felix Zollicoffer, Gideon Pillow, and Leonidas Polk. Their messages to Richmond were filled with the same concerns as were in Sherman's telegrams to Washington: warnings that the Kentuckians were at best lukewarm to the cause, complaints about the scarcity of arms and supplies, and misgivings about how their army of ill-trained volunteers would perform against an adversary whose strength and intentions they could only guess at. On both sides such considerations inclined the military leaders in Kentucky to caution. The Kentucky "campaign" was resolving itself into a series of marches and countermarches and tentative sorties and probings, with occasional minor collisions–this despite calls from both Washington and Richmond for more decisive action.

The new commander of the Department of the Cumberland had been in office barely a week when he received an important visitor, Secretary of War Simon Cameron; in Cameron's party was Adjutant General Lorenzo Thomas and, according to Sherman's memoirs, "six or seven gentlemen who turned out to be newspaper reporters." Cameron had been to St. Louis to look into problems that had developed in General Frémont's command in Missouri and was stopping briefly in Louisville to make train connections. Sherman prevailed on him to stay over until the next morning, so they could talk at length about the situation in the Department of the Cumberland, which he said was bad, "as bad as could be"–an assertion that seemed to surprise Cameron.

The whole party went to the Galt House, the hotel where Sherman had rooms, and once Cameron was comfortably installed and the preliminaries out of the way the secretary said, "Now, General Sherman, tell us of your troubles." The general hesitated, since Cameron's whole suite was present, but the secretary assured him, with a touch of irritation, that those present were "all friends." Sherman then went over and locked the door "to prevent intrusion," though some must have seen the gesture as bizarre. The commander of the Department of the Cumberland then laid out a catastrophic state of affairs: he insisted that desperately needed men and arms were being channeled to other theaters—"We got scarcely anything." The young, active element in Kentucky's youth was flocking to the rebel camp, the population in general was indifferent, and only the old and infirm were strongly attached to the Union cause. As a result Sherman was facing military disaster: "Our forces at Nolin and [Camp] Dick Robinson were powerless for invasion, and only tempting to such a General as we believed Sidney Johnston to be; . . . if Johnston chose, he could march to Louisville any day." Cameron was astounded; he protested that the members of the Kentucky delegation in Congress had assured him there was ample manpower in the state, all they needed was arms and money.

In the conversation that followed Sherman's exact needs in men came up. Before they left Washington he and Robert Anderson had made plans for a force of 60,000. Now Sherman cited this figure, but added that it would be adequate for defensive purposes only; to clear southern Kentucky of rebels and push into Tennessee he would need more than three times that number: "two hundred thousand before we were done." At this point, Sherman recalled, Cameron threw up his hands and said, "Great God! Where are they to come from?" The conversation then swung to the possibilities of raising more troops, with Sherman suggesting the "northwestern states" would be the best source. The conference broke up in a most friendly spirit, and Sherman came away encouraged: "I thought I had roused Mr. Cameron to a realization of the great war that was before us, and was in fact upon us."

Cameron had indeed seemed stirred. On the spot he arranged by telegraph for detachments en route to Missouri to be rerouted to the Department of the Cumberland. He wired Lincoln: "Matters are in a much worse condition than I expected to find them." To his assistant he wired terse and unequivocal instructions to send men and guns to Sherman. But the secretary and those in his party were also struck by Sherman's overwrought state; they spoke of it among themselves and to others. This ripple of concern spread; in Washington a rumor ran through the War Department that Sherman was "touched in the head." President Lincoln later told John Sherman of receiving messages from prominent political and military figures expressing concern over Sherman's "extreme depression of spirits and physical exhaustion."

While sparing no pains to make the weaknesses of his command known in Washington, Sherman went to considerable lengths to hide them from the enemy. Here he seems to have imitated the efforts of his predecessor. Like Anderson, he generally tried to conceal his own movements from the public. He was convinced that the enemy's spies were everywhere, among the "country people" and among the better class with whom he had some contact. He did not trust the telegraph and sometimes used a prearranged code to transmit orders (one Louisville telegraph operator had been arrested for espionage). And here he inevitably collided with the press and was made to feel its barbs.

During his stay in Washington and Fort Corcoran he seems to have escaped any serious press attacks, even though New York newspapers, the *Herald* especially, devoted considerable space to the activities of the Sixty-ninth and Seventy-ninth New York Regiments. But his troubles with those regiments caught up with him in Kentucky when he came in for stinging criticism—not in newspapers but in a hastily published book, *The Last Days of the 69th in Virginia*. Its author, Thomas Francis Meagher of the Sixty-ninth New York militia, had brought out a brief volume describing the unit's exploits and misadventures in the Bull Run campaign in which Sherman appeared as the villain of the piece, "a rude and envenomed martinet" who exhibited the "sourest malignity" toward the Sixty-ninth, whose members repaid his malevolence with a cordial hatred. The book was being bought and read in New York and Washington, and to Sherman's further mortification it was Thomas Ewing who had sent him a copy.

One of Sherman's first acts was to arrange a modus vivendi with George Prentice, who ran the *Louisville Journal* and its evening counterpart the *Tribune*, by which the papers would publish only the war news that the military authorities found unobjectionable. The competing

*Courier* had been suppressed by Federal authorities in September as a "treasonable sheet." Journalists were banned from Sherman's headquarters and the general made some effort to keep them out of that part of Kentucky. His provost guard checked for them on incoming trains, and the reporters who covered Sherman's army spoke of "running the blockade" to get into Louisville.

At least one journalist, Henry Villard, agreed with the general's policy. He acknowledged that the papers could do a lot of mischief through their complete disregard of "military interests" in what they published. "If I were commanding general," said Villard, "I would not tolerate any of the tribe within my lines." In truth the briefest scrutiny of Civil War newspapers reveals their penchant for publishing a wide variety of items useful to the enemy: official reports and other documents spirited out of various headquarters, postmortems of past battles and prognostications by officers happy to see their names in print, notices on the movements of troops, estimates of the size of garrisons, and detailed descriptions of fortifications being erected.

The *Cincinnati Daily Commercial* committed all these sins. It was the most influential newspaper in the region, with a readership of nearly a hundred thousand by 1860, and it would be Sherman's bête noire. The *Commercial*'s regular "Letter from Louisville" frequently contained material on the general himself. It related what it said were his views and intentions; he was frequently the subject of unflattering anecdotes or vignettes. From these it is obvious that the reporters were encountering the "official" Sherman: "He is cold, distant, and of a brusqueness of manner (to use a mild term), that is absolutely repulsive." There was the experience of the reporter who made it into the general's presence at Camp Lebanon: "Arrest his attention and he stops suddenly, looks you straight in the eye, and demands your business. Explain in as few words as possible; he hates circumlocution and loves bluntness of speech. Be Mark Antony in the camp, not Cicero in the forum, with him. Are you through? On he goes in an irregular double-quick, stopping now and then to jerk out his words in a short, quick way: 'Why don't Ohio send troops? I advised her governor ten days ago that this was coming. It has come and no troops.'"

The *Commercial*'s Louisville correspondent did discover that underneath the "official" Sherman there was a highly competent soldier: "Seriously, Gen. Sherman is a very superior man–I think, from close study, the superior of any military man in the West–clear headed and strong headed–loving his profession, his fame and his men . . . But he has one unworthy crotchet in his head–he affects a gruffness that is not at all natural to him; a more genial man does not exist, aside from business."

The most punishing attack on Sherman while he was in Louisville came not from the *Commercial* but from official Washington via the *New York Tribune*. A reporter from the *Tribune* had been one of the party accompanying Simon Cameron. He had obtained a copy of a report on the Galt House meeting that Adjutant General Thomas had drafted on Cameron's instructions. On October 30 the *Tribune* published it, setting off a considerable stir. The version Thomas presented began with Cameron's earlier meeting with Governor Morton of Indiana, which was cordial and optimistic, throwing into relief Indiana's success in mobilizing for the war effort; then in Louisville, by contrast, Thomas reported Sherman had presented "a gloomy picture of affairs." On being asked the number of troops required, "he promptly replied 200,000." There was no mention of the 60,000 that Sherman had said would be adequate for defensive purposes. Sherman probably saw the *Tribune* article in the *Commercial*, which reproduced it in its issue of November 2; bad as Thomas's version of the meeting made him appear, the ironic commentary that accompanied it must have stung even more: "the General only wanted 200,000 men!"

Sherman's reaction might have been more violent had publication not come at a particularly critical time. He had been watching as best he could what was happening on his front with the Confederates at Bowling Green. Now he was sure the rebels would attack his force at Nolin Creek. On October 23 he wrote Ellen, "I know and feel that beyond Green River has assembled a large force." By November 1 he had more confirmation: "Rumors and reports pour in on me," he wrote Ellen, "of the overwhelming force collected in front across Green River."

The situation seemed to grow more ominous each day; at one point the attack seemed so imminent that he went down to share the fate of his troops; when the enemy failed to appear he went back to Louisville. By the beginning of November he clearly wanted out. As early as October 23 he confided to Ellen that the solution would be to have Halleck sent out to join him, for Halleck, now a major general, would in effect take over. On

November 3 he wired McClellan, who had just succeeded Winfield Scott: "Please order Halleck here. Also Generals Buell, Stevens and some officers of experience. The importance of this department is beyond all estimate."

The next day Sherman wired again with an additional argument for replacing him: "The publication of Adj. Genl. Thomas's report impairs my influence. I insist upon being relieved to your army, my old brigade." McClellan did not answer this request for four days; he did tell Sherman to send daily reports of the evolving situation, these he would follow on the map.

Sherman sent the adjutant general a warning and a reproof: "Do not conclude, as before, that I exaggerate the facts. They are as stated, and the future looks as dark as possible." And since he had not heard from McClellan on his relief, he added: "It would be better if some more sanguine mind were here, as I am forced to order according to my convictions." To Robert Anderson he confessed that he was "fearful of the worst consequences," and—more revealing—that he would be "overwhelmed with ignominy."

On the morning of November 8 an urgent message arrived for Thomas Ewing at his home in Lancaster. Ewing was away, so Ellen opened it and read it with sinking heart. It was from Frederick Prime, Sherman's aide: "Send Mrs. Sherman and youngest boy down to relieve General Sherman's mind from the pressure of business—no occasion for alarm."

She decided to go immediately; her brother Philemon agreed to accompany her (respectable women then traveled with male escorts) and she took Willie and Tommy as well. It took them fourteen hours by train to reach Louisville; Philemon recalled they found Sherman "in a great, barnlike room with blazing lights, with a lounge at one end, on which he tried from time to time to catch snatches of sleep, and messengers rushing in at all hours bringing details of disaster or threat."

The day after her arrival Ellen put down her impressions in a letter to John Sherman: "Knowing insanity to be in the family and having seen Cump on the verge of it once in California, I assure you I was tortured by fears, which have been *only in part relieved* since I got here." She went on: "I have not been here long enough to judge well of his state of mind. He wrote me that he felt almost crazy, and I find that he has had little or no sleep for some time. The servant boy who waits on him at table told me this morning when waiting on me alone that General Sherman seldom took a meal lately—that sometimes he would eat nothing all day, when he would carry in to him at his office some dinner towards evening." Officers at headquarters told Ellen they had been quite concerned about him. "He however pays no attention to them or to anyone and scarcely answers a question unless it be on the all engrossing subject. He thinks the whole country is gone irrevocably & ruin and desolation are at hand."

McClellan had by this time agreed to Sherman's request to be relieved: Buell would be arriving to take his place, but the telegram did not indicate where Sherman would be sent. In any event he would soon be moving on and leaving his worries behind; by the fifteenth Ellen felt sufficiently reassured to head back to Lancaster.

Sherman in the meantime had received more indications of the enemy's strength and dispositions: Buckner had "not far short of 45,000 men"; there was "daily and constantly increased evidence of a vast force on our front," moreover the Confederates were collecting large quantities of wagons, preparing for a move. The threat along Green River was taking more definite form.

There was no threat. For the time being General Buckner was quite content to keep his army of some 12,000 where it was. Sherman had taken the bits and snippets of information and misinformation he had received and from them pieced together the worst possible picture.

The damage done was not just to Sherman's nerves. At the time George Thomas was poised to move toward Cumberland Gap, an initiative being strongly pushed from Washington; it would open the door to East Tennessee, where Union sentiment was strong and mass insurrection against the Confederate regime possible. But Sherman told Thomas to stop his advance and withdraw toward Louisville. Thomas wired back: "I will give orders at once for a retrograde move, but I am sure the enemy are not moving between us, my information indicates they are moving south."

It was at this juncture that Buell arrived and the atmosphere of acute crisis dissolved. The new commander of the Department of the Cumberland saw things differently. In Sherman's view Buell was not taking the situation seriously enough, being "imbued with the same spirit that prevails in Washington." As for the powerful Confederate offensive, he wrote John that "they have delayed for some purpose of their own."

It should be pointed out that Sherman was not alone in seeing phantom armies in Kentucky that fall. Even as he was bracing for an imaginary attack, four generals in Confederate-held Columbus, Kentucky–Leonidas Polk, Frank Cheatham, John Porter McCown, and Gideon Pillow–were doing the same thing, warning Richmond that the enemy was making "preparations on a gigantic scale" to assault Columbus; in succeeding days urgent telegrams warned of the "overwhelming numbers" and the "immense force" threatening them–but this threat did not materialize and indeed it did not exist. Columbus remained in Confederate hands for several more months.

Sherman stayed on a week to help Buell settle in and familiarize himself with his command. When his orders came through he found that he was not being summoned back to the East, as he had requested; he headed instead to St. Louis, to report to Henry Halleck, commander of the Department of the Missouri. Halleck gave him a cordial welcome and sent him off on an inspection tour of garrisons west of St. Louis, authorizing him to assume command there if the situation warranted it. There, at Sedalia, Missouri, Sherman again saw a looming danger and took command, issuing orders for concentrating the troops–orders that Halleck found unwarranted and countermanded. He called Sherman back to St. Louis, and there he told his friend to take twenty days' leave, sending him off with reassuring remarks about the need to leave a good workhorse idle in the barn lot from time to time. In the meantime Ellen, without news and full of concern, had come to St. Louis–without escort–to see her husband. She had shared her worries with Halleck while waiting for Sherman's return. Now she took him back to Lancaster.

Despite his soothing words, Halleck was considerably concerned now. He wired McClellan that Sherman had been "stampeded," and could stampede others: "Gen. S. physical and mental system is so completely broken by labor and care as to render him for the present completely unfit for duty." Halleck hoped that with a few weeks' rest he might pull himself together.

The Ewings all rallied to Sherman's side; Thomas Ewing made his considerable weight felt in Washington. Ellen, fiercely loyal to her husband, insisted that he had been laboring heroically in an impossible situation, "poorly and tardily sustained by the authorities in Washington, an armed foe to his front, traitors surrounding him on every side" (Ellen believed that McClellan and some other Union leaders were secret allies of Jefferson Davis).

Sherman's spirits seemed to improve in Lancaster. More than anything else he probably felt that sense of relief he had acknowledged after his resignation in the midst of the Committee of Vigilance crisis: "I'm out of it." Halleck had told him to talk politics and read newspapers for two weeks. He was reading a paper on the morning of December 9 when his eye fell on an article from the *New York Times* about affairs in Missouri. It said that in Sedalia, General Franz Sigel had replaced General Sherman, "whose disorders have removed him, perhaps permanently, from his command." Sherman was shaken. Ellen said "it seemed to affect him more than anything that has hitherto appeared." She dashed off a letter to John, alerting him to what had occurred.

There was worse to come. On December 11 Sherman found in the *Commercial* an article entitled "Gen. William T. Sherman Insane." It related that at times when commanding in Kentucky he was "stark mad." He had telegraphed three times in one day asking permission to evacuate Kentucky and retreat into Indiana; later, at Sedalia, he had issued orders so preposterous that no one would obey them. This same article appeared simultaneously in the *Missouri Democrat*. On the twelfth the *Cincinnati Gazette* carried a "correction" to the report by its rival, the *Commercial*: it was not until Sherman's arrival in Sedalia that his insanity became "clearly developed"; in Kentucky he had merely been having hallucinations. Just how this insanity story got started is difficult to say; years later Murat Halstead of the *Commercial* claimed responsibility, but it may have originated in the "leaking" of a garbled version of Halleck's telegram to McClellan about Sherman being "stampeded."

Philemon Ewing was in Lancaster at the time, and his letters to his father provide the best information on Sherman's reaction. The article in the *Commercial* apparently hit him like a body blow. Philemon reported that when he read it he was "distressed almost to death." Once launched, the story was repeated endlessly whenever Sherman's name was mentioned; indeed it was destined to take on a life of its own. In attenuated form the insanity charge would cling to the man the rest of his days; after his death his children would be called on from time to time to supply explanations, and historians and biographers in turn would have to address the "insanity" issue.

On the precise question of Sherman's state of mind in November and December 1861, the most extensive evidence is found in the letters of Ellen, Philemon, and Sherman himself. Two days after the "Sherman insane" article, Philemon wrote that Sherman had largely recovered and was "as sound in mind as ever he was in his life." In Sherman's own letter to Thomas Ewing there was the reassuring indication that he was not thinking exclusively of himself: "Among the keenest feelings of my life is that arising from a consciousness that you will be mortified beyond measure at the disgrace which has fallen on me by the announcement that I am insane."

It is clear from Sherman's letters written during October and November 1861 that he was rational and functional throughout, and that the measures he was taking in Louisville were sound ones in the face of an imminent attack. John Sherman, who saw him briefly in mid-November, found him lucid and in control, though very anxious to be relieved. And in recent times researchers into this episode have reached the same conclusion: Dr. Paul Steiner, whose study of Sherman has been cited earlier, went over the relevant documentary evidence and prepared a "Neuropsychiatric Record" offering this conclusion: "Today the diagnosis would be that of a mild 'anxiety state.'"

This "breakdown" in Kentucky was not unique; there was at least one similar episode. It occurred some six years before, when Sherman was in California–this is probably the same episode Ellen referred to when she said she had seen her husband on the verge of insanity. In the two crises there are clear parallels in both the causes and the symptoms. Unfortunately the only direct evidence in the 1856 incident is a long, rambling, and anguished letter Sherman addressed to Henry Turner on January 5 of that year. It indicates that the root of the crisis was essentially the same as in Kentucky: apprehension over his inability to deal with the risks and problems facing him–in 1856 these were the troubles of his bank.

In California, as in Kentucky, overwork and lack of sleep contributed to his state of mind: when he wrote Turner on January 5, 1856, he had slept only three hours of the preceding twenty-four. Here also he urged his prompt replacement—adding in this case the argument that he felt he had but a short time to live. In both instances the basic fear is of a total and spectacular failure. In the case of the 1856 crisis Turner, some 2,000 miles away, could do nothing; fortunately Sherman's next letter was more optimistic and the one following that contained an apology for having so alarmed his old friend over his "depression," as he called it.

The episode in Kentucky, well-documented as it is, presents the first opportunity to look further into the Sherman psyche and try to assemble the man's quirks and traits of character into some meaningful pattern of behavior; what is more, it invites a more general inquiry into Sherman's wartime state of mind and its possible effect on the way he waged his campaigns. But with very few exceptions historians have declined this invitation; what is more they have shown little interest in enlisting the aid of psychologists and others who make human behavior their study. As the author of a "reappraisal" of Sherman put it some years ago, "those historians who try to apply psychiatric findings to this great General do not help us in our understanding of him or his contribution. He was not manic-depressive. He was Sherman, unique unto himself."

The craft of history has moved on since those lines were written; thus it is not surprising that a man with Sherman's eccentricities should be the subject of a recent article in the *Psychohistory Review,* or that the article's author should conclude that the general was "a victim of a personality disorder, most likely some form of manic depression."

Admittedly a posthumous diagnosis of mental illness based largely on epistolary evidence more than a century old is at best tentative; still, the line of inquiry is a valid one and should be pursued. In Sherman's case there is another—and more likely—diagnosis: narcissistic personality disorder. Certain of the general's behavior patterns suggestive of bipolar disorder—now the preferred term for manic depression—are also characteristic of narcissism. What is more, in cases of bipolar disorder the outcome without treatment—inevitably the case in Sherman's day—is usually deterioration. The general, on the other hand, was destined to flourish and remain vigorous and active; though he would have other periods of depression, these did not adversely affect the career he pursued for another two decades, and apparently none of these subsequent "lows" came anywhere near the veritable abyss of despair in which he found himself at the end of 1861. As Steiner pointed out, for another quarter-century his mental and physical health would be basically good.

According to current theory the narcissist carries in his psyche a wound, a "narcissistic injury" that was inflicted at an early age and will never be effaced. As a consequence the development of the self, the "true self," is adversely affected, so that it is characterized by emptiness, worthlessness, despair, and depression. The narcissist finds a remedy of sorts in deploying a "defensive self" or "false self," which is the opposite of the true one: its hallmarks are thus notions of exaggerated self-esteem, vaulting confidence, infallibility, and omnipotence. This second self is often described by psychologists as a sort of balloon, which the narcissist strives to keep inflated so that it hides the real self from view. The air required to keep the balloon inflated, the "narcissistic supply," can be furnished in a variety of ways: with a stream of achievements, even if they are in reality trivial; with honors, even if they are hollow or undeserved, and with compliments, even if the recipient has to fish for them.

There is no simple litmus test to indicate the presence of the narcissistic personality disorder. There is, moreover, what could be called a healthy, "normal" narcissism—many of the hard-chargers and movers and shakers of our world seem to have more than a touch of it. Then as a personality disorder narcissism seems to have degrees and varying forms, being recognized by several broad criteria that have various manifestations. Some of these described in the professional literature are especially worthy of note for their relevance to Sherman: "Typically the narcissist is a restless person . . . " "It's not uncommon for him to be a workaholic at a job he does well." "Here we find the man who never admits to failure or shortcomings of any sort, rationalizing his errors and missteps by assigning blame to others, or to circumstances beyond his control."

The cocksure young officer and the workaholic commissary and banker have already appeared in these pages. Then even as a cadet Sherman was not shy about discussing his accomplishments, while he did not accept criticism particularly well. The youthful Sherman does not often speak of his errors or acknowledge mistakes—the mature Sherman almost never. Then the impulsions of narcissism may help explain that inner conflict in Sherman the banker and in Sherman the soldier in those first months of the war. In both instances he revealed a driving ambition, a clear desire to distinguish himself, to stand out. But as a banker he was cautious and conservative almost to a fault; as a soldier he was excessively preoccupied—at least initially—with the consequences of failure. He sounds like a man anxious to increase his narcissistic supply, but even more anxious not to tear a hole in his balloon.

The narcissist's relations with others, however cordial, are fundamentally exploitive. He chooses friends and associates whose devotion, gratitude, and confidence in him will reflect—the psychologist's term is "mirror"—the false self he projects. It has been argued, for example that General Douglas MacArthur chose for his staff men who would mirror his grandiose self-image. A study of the composition and role of Sherman's staff might afford some interesting comparisons in this connection.

One frequently cited hallmark of the narcissist is a lack of "relatedness," a defective sense of empathy. Sherman's associates, especially those who were on campaign with him, occasionally remarked on his indifference when the emotions of others were stirred. "The things that affected others did not affect him," noted a comrade-in-arms. In battles to come he would express grief when a friend fell, but disappointment when his men lay in windrows after an unsuccessful assault.

This coin has another side: a man who cannot empathize cannot accurately gauge the response his words or actions will provoke in others; here is a possible explanation for Sherman's pattern of blunders and miscues in public relations. With the best of intentions, he could sometimes commit appalling gaffes, as in this letter to the widow of Confederate General Lloyd Tilghman: Tilghman had been killed in May 1863; now, three months later, Sherman wrote Mrs. Tilghman by flag of truce to inform her that her son had also been killed. Commiserating with her, he said it was too bad that war did not take just the lives of the "bad men," who had started the conflict, while sparing "the brave and generous and mistaken, such as I know your husband and son were."

"The narcissist is unequivocal in his judgments of others and given to 'black or white rhetoric,'" yet is himself "sensitive to slights and criticism . . . " "In response to criticism, defeat or disappointment, there is either a cool indifference or marked feelings of rage, inferiority, shame, humiliation or emptiness."

In the Kentucky episode, Sherman's initial reaction was one of shame.

What concerns him most is not the military or the political consequences of the catastrophe he sees ahead, but its personal impact on him, the unbearable "ignominy" that will settle on him like a black cloud–thus those anguished words he addressed to Ellen at the beginning of November 1861: "The idea of going down in history with a fame such as now threatens me nearly drives me crazy, indeed I may be so now."

Thus initially he bent in silence under the weight of criticism; only later he would find his voice and speak out on his own behalf; ahead lay more threats to his "fame," as he called it, but these he would meet with growing confidence.

# 11

## SHILOH

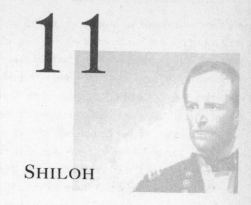

Halleck now sent Sherman to a backwater of the war, the training camp at Benton Barracks, outside St. Louis. There he sequestered himself in a bleak and cheerless semi-exile. He had ceased keeping his diary in October and would say little about this period in his memoirs. His correspondence was largely confined to those close to him; his surviving letters to Ellen and John provide the chief source of information for his cloistered existence.

Though he had any number of friends in nearby St. Louis, it is clear that he had few social contacts in this period, and by choice. He rarely ventured out; as he put it he "kept very close." When his old friend Henry Turner came out to call on him, he confessed to Ellen he could not make himself go out to the gate and see him: "I suppose he too will drop me."

For once he could not lose himself in his work. He wrote very little about his daily activities. The supervision of Benton Barracks and the training of troops were not subjects of great interest to him; if anything he found his work disheartening, for Benton Barracks was "but a specimen of the same disorders which prevail elsewhere, a mass of men partly organized and badly disciplined with their thousand and one wants."

Sherman could not take his mind away from his personal calamity. In his letters he rarely wrote of anything else. Now, looking back, he could see the enormity of his blunder: a general who asks to be relieved because he cannot cope is hardly entitled to a second chance—"it's past hope." He was indelibly marked now—in every camp and cantonment he would be known for the "mad freak" that had wrecked his career.

Were this not enough, another dark cloud had appeared: a congressional committee was slated to investigate the Bull Run defeat. He wrote to John, telling him that he had become separated from his brigade on the day of the battle; he had not mentioned the episode in his official report, and he feared that if it were revealed now the consequences could be serious: "My motive might be misconstrued and in connection with my leaving the Kentucky command at a critical moment may overwhelm me." He wanted John to give him notice if the committee's investigation should "compromise" him.

What were his options if this happened? One possibility he was considering was suicide. He admitted to John that he had been tempted to take his own life, but had held back because of his children (John took the reference seriously, writing back "this is neither manhood nor courage"). Conversely, Sherman wrote Ellen that when he thought of Willie and Lizzie he sometimes felt he should throw himself into the Mississippi.

Lizzie was in this period a frail and delicate child, subject among other things to "fainting fits." Her frailty endeared her to Sherman in some special way, just as the manly precocity he saw in Willie gave the boy a privileged place in his heart. When Ellen wrote of their daughter's health problems early in the year, Sherman made a strange confession: "I almost wish she would quietly slumber in eternal rest and escape the sad events in store for us." More than once he evoked a morbid vision in which Lizzie and he died simultaneously and were entombed together: "How I would like to lay us down together and sleep the long sleep of eternity."

Disturbing as these passages in her husband's letters were to Ellen, she found other passages that must have cheered her. In the depths of his despair Sherman spoke of God, who would "in His mercy," watch over her and the children. And he addressed her in terms that must have made her blush with pleasure: "My dearest wife, who has been true & noble and generous & comforting always." He described his love for her as "the almost only remaining chain of love and affection that should bind me to earth." A little later, when things looked less somber, he assured her he would not forget her support and her faith in him through the dark days; she had endeared herself to him "more than ever."

And as he was frank with her, he sought to be frank with himself. He could now admit that he had committed serious errors in Kentucky and had in fact "signally failed" there; he had grossly overestimated the strength of the enemy. He had other shortcomings: he had not had the "faith which would inspire success" and perhaps he would never have it. To John he wrote with a terrible finality: "I do not think that I can again be trusted with command."

Yet even as he wrote such things he was working to salvage what he could of his career. Thomas Ewing was still urging a libel suit against the *Commercial*. But Sherman refused, because the trial would involve "the searching of records and stirring up matters which I cannot properly explain," perhaps his being separated from his command at Bull Run. He was now hoping that he could at least remain in the army in some other capacity. Buell had once told him he would make an excellent quarter-master general, so perhaps he could advance in this direction—if they would not let him handle troops, they might trust him to handle stores. And he asked John to sound out McClellan concerning his move into one of the army's "disbursing departments."

His hopes of returning to campaigning were not totally extinguished. He confided to Ellen that he would make an attempt to get back "into the field"; he held little chance of success, since he felt McClellan would not help him, nor would Secretary Cameron or Adjutant General Thomas. But if he were to appeal to Lincoln?

In the end it was Ellen who made that appeal in a letter that she told Sherman about only afterward. She wrote the president that her husband had been the victim of a conspiracy and pleaded, "Will you not defend him from the enemies who have combined against him?" Now in the second half of January she accompanied her father to Washington, where on the twenty-ninth they were received by Lincoln. That evening she wrote her husband to assure him that the president had "the highest and most generous feelings" toward him. John Sherman and Thomas Ewing could be counted on to defend his interests in Washington; Simon Cameron was gone, and the new secretary of war, Edwin M. Stanton, was a particular friend of her father's. "A little time will wear away this slander," she promised him, "and then you will stand higher than ever."

The crisis in Sherman's career had a profound effect on his wife. In struggling to aid her embattled husband she became directly involved in the war for the first time. She had not been a reader of newspapers, much less a participant in political discussions; on the eve of the war she was profoundly ill-informed. She had had but two interests, her religion and her family; now she discovered a third—the Cause. She became active in charitable activities, in helping organize "sanitary fairs" and benefits for soldiers.

But in her mind the political, the military, the familial, and the spiritual were all intertwined. The Confederacy was a manifestation of the anti-Christ, an instrumentality of the Prince of Darkness; she sometimes referred to Jefferson Davis as "His Satanic Majesty." She saw the most sinister conspiracies—she had assured Lincoln that just such a conspiracy had cost her husband his command in Kentucky—and she accepted the most baroque explanations for the failures of the Union armies.

As time passed Sherman's spirits improved. Just why is not clear; certainly into 1862 he had little before him by way of bright prospects. But at Benton Barracks he was in that "obscure place" he spoke of when his spirits were low, and now he found it confining. He began to bestir himself (here is another possible indication of narcissism, for the narcissist's depression rarely endures for extended periods). He would not aspire to command, but there were certainly subordinate positions in the field in which he could give good service. While Ellen placed her hopes in Lincoln's favorable regard, Sherman believed his best hope lay in the friendship of Henry Halleck. Halleck's star was then in the ascendant; he had replaced the ill-fated Frémont in Missouri, brought order and success to the Federal cause, and fixed the state solidly in the Union camp. His Department of the Missouri was soon expanded into the Department of the Mississippi, giving him the direction of the Western theater of operations.

As early as the second half of January Sherman had sounded Halleck out about a return in some capacity to service in the field; his friend told him he was still too exhausted for the rigors of such service—and for the moment very valuable where he was; but later he felt "arrangements" could be made. Sherman would be obliged to remain at Benton Barracks for a few more weeks, but now, as he waited for Halleck to make good on his promise, he was again following the course of the war with keen interest and a critical eye.

In the first weeks of 1862 the struggle in the West completely changed its face. On January 19 at the Battle of Mill Springs, Kentucky General George H. Thomas defeated the Confederate force holding Cumberland Gap. Confederate General Zollicoffer was killed; what was left of his army withdrew toward Murfreesboro, Tennessee. On February 2 Ulysses S. Grant led an expedition against Forts Henry and Donelson, which commanded the Tennessee and the Cumberland rivers respectively; the forts were located at a point near the Kentucky-Tennessee line where the two streams were only a dozen miles apart. Using a combination of gunboats and infantry, Grant took Fort Henry on February 6 and Donelson ten days later.

Sherman was quick to appreciate the significance of Grant's enterprise: "This is by far the most important event of this sad war." In fact the Confederacy's "line" in Kentucky had been breached and the railroads that supplied her troops there placed in jeopardy, as Federal gunboats now ranged her inland waterways as far as Mississippi and Alabama, destroying railway bridges. Then on February 16 General Buell took possession of Bowling Green, which the Confederates had hastily evacuated; Buell pushed on south and a few days later crossed into Tennessee, taking possession of Nashville on February 25. By then Columbus, Kentucky, was becoming untenable, so the Southerners pulled out at the beginning of March. The Confederate military presence in Kentucky, which had seemed so formidable to Sherman, collapsed like a house of cards.

New riverborne expeditions were being planned for the further destruction of the Confederacy's rail communications, including one that would move up the Cumberland from Fort Donelson. McClellan proposed Buell to lead it, but Halleck responded: "General Sherman ranks General Buell and he is entitled to a command in that direction," adding by way of reassurance that Sherman's "health" was greatly improved. The appointment of Sherman may have been settled on between the two men, but Secretary of War Edwin Stanton was monitoring these exchanges; when he saw Sherman's name he called for the records on the Kentucky campaign, selected a number of Sherman's letters and reports, and sent them to Lincoln, with a covering letter asking the president to read them and reflect on them: "I recommend his immediate removal and the appointment of General Shields or some other competent commander to take charge of this expedition, believing that it will prove disastrous under the charge of General Sherman."

In the end the expedition up the Cumberland was changed to one up the Tennessee under General Charles F. Smith, who would replace Grant temporarily (Halleck having become peeved with Grant, partly because he had been neglecting his paperwork). Halleck sent Sherman to Paducah where he was put to work directing river traffic and assembling troops for service at the front; at the same time he was putting together the division Halleck promised him he could take into the field. Halleck confided that he was not satisfied with the progress of the war on the rivers and would be making changes, adding, "you will not be forgotten in this." Still, Sherman brought up the subject when the occasion presented itself. On March 6 he reported to Halleck that he had dispatched another contingent up the Tennessee, adding, "Where am I to go?" Halleck was as good as his word: that same day he ordered Sherman to head up the Tennessee with his division loaded into seventeen steamboats; the flotilla steamed up to Fort Henry, where it joined General Smith and three other divisions placed under Smith's command.

Sherman did get to lead an expedition, the advance element of a force General Smith would take up the Tennessee to the vicinity of Corinth, Mississippi. But Smith himself was now incapacitated; he had scraped his leg climbing into a boat, and the wound had become infected. Sherman had hardly arrived when Smith told him to strike and break the Memphis and Charleston Railroad at Burnsville, Mississippi, where its repair shops were located; to the east of the town was another important objective, a railroad bridge over Bear Creek. On the way to his objective Sherman revealed what was to be one of his crotchets: he told his brigadiers and colonels that while their object was not to engage the enemy but to break his communications, if any of the boats were fired on from the shore a force should be landed "and the enemy punished." Such firing never failed to infuriate him.

The mission was a washout—literally. As the cavalrymen he put ashore headed through unfamiliar country in the failing light, torrential rains began to fall. Soon gullies and ravines filled with rushing water; under the deluge the Tennessee began to rise at an alarming rate, six inches an hour. Several times men and horses came near to drowning, and ultimately the force had to turn back. Then he tried a second foray, this time from a base he set up at Pittsburg Landing. Since by now his force was under observation his second sortie was also under cover of night, this time in the direction of Corinth; he hoped to convert his advance into a railway and telegraph wrecking mission if everything went well. It did not; in the middle of the night his head of column collided with a sizable force of Confederate cavalry. Since the alarm had been sounded in the enemy camp he and his men turned back. Reading Sherman's straightforward report of this largely abortive expedition, one has the impression of a man trying very hard to attract favorable notice, even too hard, given the handicaps under which he was laboring: few of his colleagues would have cared to lead green troops on night operations in the rough and uncharted country that bordered the Tennessee.

Sherman had disembarked his division at Pittsburg Landing on March 16; it was one of the few spots along the river's west bank where a sizable force might be installed and was twenty-odd miles from Corinth. His force was soon joined by other divisions that took their assigned positions at the landing. General Grant had been reinstated as commander of the expedition; he arrived at Savannah, a short distance downstream from Pittsburg Landing, on March 17. By the beginning of April he had some 40,000 men, the most of them at Pittsburg Landing. He could expect another 20,000 under General Buell, who had been ordered to join him. In and around Corinth, General Albert Sidney Johnston had collected approximately 44,000 Confederate troops. The stage was thus set for the first great battle of the war.

The Union encampment at Pittsburg Landing was a place of considerable noise and confusion. It was filled with young men who had never seen a battle and who did not yet understand or appreciate the notions of order and discipline that the army was struggling to inculcate in them. There were constant comings and goings as parties of soldiers went off hunting or marauding or simply visiting in other regiments.

There was periodic gunfire along the picket lines, where contact was made from time to time with a string of Confederate pickets set to keep an eye on the Yankees; there was also frequent firing within the Union lines, as the men tested their weapons to see if the frequent rains had dampened their powder or fired them at flocks of geese passing overhead. (Sherman tried to curb the practice by having the cartridge boxes inspected and fining a soldier ten cents for each cartridge missing.)

The evening of April 5 was hardly different in this respect from preceding days; perhaps there had been more firing than usual, and here and there the soldiers noticed a curious phenomenon—unusual numbers of rabbits and squirrels appeared in the encampments, as if some vast game drive were under way to the southwest. But when the Northern soldiers bedded down that night they had no idea that a large Confederate force was within two miles of their camps and would be poised to assault them at first light. Whether the leaders of the Union host were any more enlightened than their men that evening would soon be the subject of long and passionate public debate.

April 6 was a Sunday. Sometime after five o'clock that morning the first firing was perceptible in front of the more advanced Union positions; it gradually swelled in intensity and grew closer to the encampments; eventually the thunder carried all the way to Savannah, nine miles downstream, announcing to Ulysses S. Grant as he took his breakfast that his army was engaged in a great battle.

The Battle of Shiloh was a drama in two acts, played out in as many days. The stage was a triangular piece of land, bounded by the Tennessee River and several smaller streams. This triangle was roughly an equilateral one, with each side perhaps six miles in length. Save for the armies it had the sparsest of populations, a sprinkling of small farms, with cleared places here and there; most of the battlefield was wooded and with considerable underbrush, thus ill-suited for mounted troops. Several features of this rustic landscape would soon become famous: a shallow pond, a peach orchard then in bloom, and the crude log structure Sherman had chosen for his headquarters—Shiloh Methodist Meeting House.

If the course of the battle were schematized on a television monitor, with the Tennessee River side of the triangle running across the top of the screen, then the first engagements would appear as points of light on the lower portion of the screen, marking the places where advancing Confederate troops first encountered the nearest Northern divisions, those of Sherman and General Benjamin Prentiss. The points would become elon-

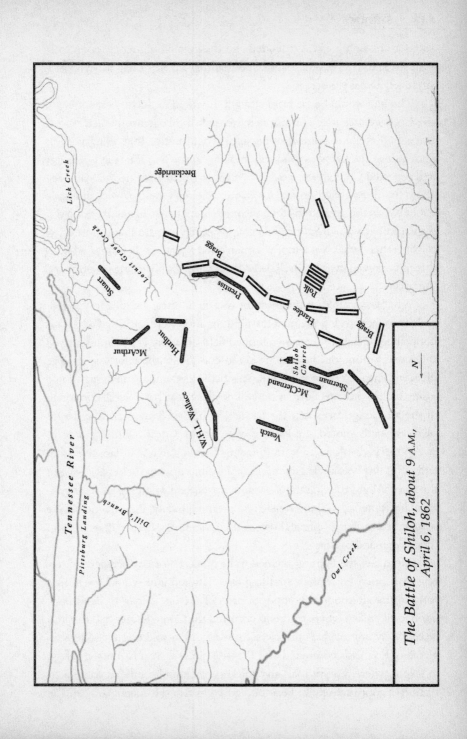

The Battle of Shiloh, *about* 9 A.M., *April 6, 1862*

gated as the fighting was extended and they would ultimately connect in a single line of light—the battle line—slanting up from left to right as it stretched across the screen.

The line would be far from straight. It would undulate perceptibly as the Confederates made an advance here, as the Federals made a riposte there; and at some point the spectator would notice that along with its undulations the line was gradually moving to the left, the result of heavy and repeated Confederate assaults. At times the line seemed to stabilize— about ten-thirty that morning, for example, when Grant arrived, and again about two in the afternoon—then the movement began again. By noon this displacement was already significant, especially on the lower portion of the line; there what was left of Sherman's and Prentiss's divisions had been driven back well over a mile. The Federals were visibly losing ground—and the battle.

What was visible on the monitor was by no means visible to the generals who were trying to direct the fighting; they were unsure of what was happening outside of their own limited field of vision. Until Grant arrived there was no one on the Union side to coordinate the efforts of the five divisions engaged (General Smith was bedridden at Savannah, fighting a losing battle with infection). Albert Sidney Johnston had overall command of the attacking forces, but the battle was not evolving according to his plan. He had wanted the leading edge of the Confederate advance to sweep the Federals away from Pittsburg Landing, while in fact the other end of his line was advancing more rapidly and driving them back toward the river. Along the line units became mixed and scattered in their sporadic movements, so that brigade and even regimental commanders were reduced to trying to direct those in their vicinity, whether they were in their command or not.

But in fact the fighting seemed to be guided more by accident of terrain or happenstance than anything else. The monitor would reveal that early in the afternoon the upper portion of the line began to bend back upon itself. In fact where the bend occurred the Confederates had run into a seemingly immovable obstacle, a mass of Union soldiers, including survivors of Prentiss's command and Prentiss himself, who had taken refuge in a spot where a worn road and thickets of brambles offered some protection—a spot that would become known as the "Hornets' Nest." In the course of the afternoon the Confederates launched as many as a dozen assaults on this salient, one of them led by General Johnston himself–it was nearly his last act in life, for he would soon receive a mortal wound. The Hornets' Nest became increasingly isolated as other parts of the Union line were driven back. Ultimately it was completely surrounded. This sealed the fate of its defenders, but at the same time reduced the pressure on the rest of Grant's army.

For over an hour the reduction of the Hornets' Nest absorbed almost the entire attention of the Confederates. Grant profited from the brief respite by strengthening his lines and concentrating artillery. He now had some unusual reinforcements: two Federal gunboats had taken up position at Pittsburg Landing and could add their firepower to the contest. Finally, General Buell had arrived and more infantry was on the way. It was after five o'clock when General Prentiss surrendered his command. General Bragg launched one ineffectual thrust at Pittsburg Landing, and then General Beauregard, who had succeeded Albert Sidney Johnston, decided not to renew the assault on Grant's lines. The firing died down and that evening both armies bivouacked amid the wreckage of battle; Beauregard slept in Sherman's tent.

The guns commenced again the next morning at five o'clock. On the monitor the line reappeared where it had been the evening before; it began the same undulations and had its fits and starts, but now the direction of its movement was reversed. With 25,000 fresh infantry, Generals Grant and Buell held a decisive edge; by four o'clock that afternoon the line on the monitor was back approximately where it had been when battle started. About then General Beauregard started a retreat behind the protection of a holding force; the Confederate Army took the road back to Corinth. In two days of battle it had lost 11,000 men, killed, wounded, and missing. The Union forces had suffered a loss of nearly 13,000. So ended the greatest battle that had ever been fought on the North American continent–but while Shiloh would go down as a memorable battle, its record would soon be surpassed.

Sherman's conduct during both days of battle was exemplary.

Such was the opinion of those who fought at his side and such is the judgment of history, based largely on their recollections. Historian James McDonough wrote of Sherman: "Observers said he seemed to be where he was needed, to have a grasp of what was taking place all along the line, and sometimes to anticipate what would occur before it happened." These are clear signs of superior generalship. His presence—and that of Prentiss— were decisive.

Such was the judgment of his superiors. Halleck wrote General Ethan Allen Hitchcock that Sherman's "skill and good judgment" had counted for much in the Union victory. More important, he wired Lincoln: "It is the unanimous opinion here that Brig. Genl. W.T. Sherman saved the fortunes of the day on the 6th and contributed largely to the glorious victory of the 7th." Halleck asked that he be elevated to the rank of major general of volunteers, to date from April 6.

Though Halleck did not mention it, Sherman had also been something of a model to the officers and men under him, almost all of them new to combat, in that he seemed indifferent to danger. He directed the battle mounted, accompanied by a staff also on horseback, so he and his party were inviting targets. His orderly was killed at his side by a shot Sherman always believed was intended for him. He himself was shot in the right hand and suffered a bruised shoulder from the glancing blow of another projectile; he summoned no surgeon, but simply wrapped a handkerchief around his bleeding hand and continued his direction of the fighting. He had two, three, or four horses killed under him, depending on what source one reads. (After the battle he made application to be reimbursed for only two mounts, which the Treasury Department graciously honored—in 1868.)

News of the great battle did not become widespread among the American public for several days. Though there were reporters with the armies, Northern journalists did not have access to the telegraph any closer than Cairo; a reporter from the *Cincinnati Commercial* probably filed the first account from there on the evening of April 8. First reports, vague as to details, stressed the intensity of the fighting and the ultimate Union victory. Thereafter further journalists' dispatches arrived to fill in details, among them a number that were pure fantasy—for example, reports that Grant had personally led a bayonet charge and that a cannonball had passed through Sherman's hat.

The more disturbing aspects of the battle soon dominated the news, notably the far from satisfactory conduct of some Northern units under fire. Whole regiments had apparently "shown the white feather," scattering in panic under the enemy's fire or even at his appearance. The public learned that by the end of the battle several thousand Union soldiers had fled the field and taken refuge at Pittsburg Landing, where they remained deaf to the threats and entreaties of their officers; some, in their panic, even tried to swim to the other bank of the Tennessee. The *Cincinnati Commercial* reported that the Sixteenth Iowa had "scattered in the wildest confusion," but took issue with the Chicago *Tribune* for doing "great injustice" to the men of the Seventy-seventh Ohio by reporting that the same thing had happened to them. But soon stories circulated that whole regiments were being visited with official censure and even disgrace.

There had to be some other explanation, and newspaper editors soon found it. The *Commercial* pioneered in this effort. On April 23 its editor intoned: "Ohio does not breed poltroons or cowards." He explained to the uninitiated reader of war news that even the best troops could be taken by surprise. The *Cincinnati Gazette* had already described the complete surprise achieved by the Confederates, who bayoneted Northern soldiers in their beds. Now the *Commercial* called for an investigation into the negligence of certain Union commanders that would "strip epaulets from some unworthy shoulders."

Some of the journalists would have taken away Sherman's epaulets. The *St. Louis News* published a highly improbable story about an Illinois chaplain who had learned of the impending Confederate attack at Paducah, of all places, and who had rushed to Pittsburg Landing to warn Sherman, only to be reprimanded for leaving his unit without authorization. According to the *Commercial* of April 18, a major had tried to warn Sherman of the impending attack: "General Sherman, with a smile of incredulity, said 'My dear sir, it is impossible that they should think of attacking us here, at the base of our operations–mere skirmish, sir.'"

A Confederate cavalry probe that carried off a half-dozen pickets on April 4 had led Sherman to issue Order No. 19 that same day, giving his brigade and regimental commanders what proved to be most timely guidelines about what to do in case of a sudden attack: they were to form in order of battle on their parades (he had already positioned his division so that when the brigades and regiments assembled on their respective parades they would be facing south, the direction from which an attack would most probably come).

He took no further steps to be forewarned of the enemy's approach, though there is testimony that he planned to lead a major reconnaissance on April 6 to see just what was in front of him by way of enemy strength. But when the Confederate forces swept forward early on the morning of that same day, Sherman's line of pickets was probably not more than a couple hundred yards in advance of the division. There is no indication in his order book that he advanced his pickets or urged on them any increased vigilance after the firing on the fourth. He acknowledged in his after-action report that until the battle began he still believed he was facing only a force posted to keep him under observation.

Yet he vehemently denied that he had been surprised. To Charles Ewing he wrote: "As to the cock and bull story of surprise it is absurd—we had been skirmishing for two days, and on the morning of the battle every regiment was armed and equipped & in line of battle, every battery harnessed in position, and cavalry saddled up—not a knife or bayonet wound among my 1200 wounded and yet the newspapers make out we were surprised & men slaughtered in their beds." His assertion that his division was in order of battle when the enemy appeared is true enough, for it is corroborated by other witnesses.

There are degrees of surprise and there are degrees of preparation, and the fact is Sherman could have arranged for a more timely warning of the enemy's approach and he could have made better preparation to receive an attack—statements equally applicable to all the other divisional commanders at Shiloh. To be sure there were extenuating circumstances: Grant, who had overall command of the divisions encamped at Shiloh, remained at Savannah and had designated no one to act in his place; the ranking general among the five division commanders was John A. McClernand, with little experience in war and no professional training for it. Neither before nor during the battle did he bring his seniority into play; in fact during the battle he sought Sherman's advice.

The most telling evidence of the measures Sherman failed to take on his own divisional "front" before the battle are those he introduced immediately after it was over, and which were duly indicated in his order book and correspondence. On the subsequent approach to Corinth Sherman's troops were covered by an extensive screen of pickets and guards. In some instances pickets were posted as far as six miles out, and beyond them ranged mounted patrols. These troops were stationed at such a distance that Sherman himself complained about the time consumed in going out to check on them, which he did in person. And if he found something amiss on the post-battle picket lines his wrath could be formidable; on one of his trips forward he encountered a body of his horsemen taking their ease far from where they should have been and the officer who commanded them absent. Sherman charged the officer with a "criminal offense," ordered his arrest, and sent word to the thunderstruck colonel of the regiment involved to have his entire command in the saddle by four o'clock the next morning for a trip to division headquarters, where a dressing-down no doubt awaited them all.

The other important measure Sherman might have taken was to have his troops "dig in," to throw up field fortifications. Though these techniques were known, at least in theory, they were only rarely employed. Initially there was considerable prejudice against them since it was thought that giving soldiers something to hide behind tended to make them overly timid. It would be hard to get them to leave their trenches to advance on the enemy; if they were driven from their ramparts it would be even harder to stop their retreat without similar barriers to shelter them. Judging from his remarks early in the war, Sherman seems to have subscribed to these views. Months after the battle, while serving as witness in a trial by court-martial he ordered for one of his subordinates, he argued with curious logic that "to have erected fortifications would have been evidence of weakness and therefore would have invited an attack." Nevertheless in the weeks following Shiloh his division acquired a considerable amount of experience in field fortification, entrenching seven times on the approach to Corinth; Sherman's circular of May 5, 1862, stipulated: "In front of the whole line underbrush will be cut to a distance of 300 yards, and heavy logs felled as a breastwork along the front of the artillery and camps."

Why Sherman did not show more care and concern about the defense of the position around Pittsburg Landing is not easy to say. The almost obsessively defensive mindset of his Kentucky command a few months before is no longer in evidence. Yet he was not experiencing one of those periods of soaring self-confidence, his "whip the creation" mode, as John Chipman Gray described it. That spring he was still talking of staying in the background, of leaving honor and glory to Halleck, Grant, and others more worthy. James McDonough, whose work on Shiloh has been cited earlier, suggested that "he was perhaps overcompensating in an effort to avoid the kind of criticism he encountered in Kentucky." This may explain Sherman's negligence—and it must be called that. It does not explain that of Grant, who was perhaps even more culpable; in his case it was probably a string of successes that led him to take so few precautions.

But the newspapers had still another question for the generals they held "responsible" for Shiloh: Why had they committed the supreme folly of placing most of their forces at Pittsburg Landing in the first place, a location where they had a very large enemy force to their front and a river at their backs? Here Grant came in for much criticism, but the inquiring journalists showed even greater interest in Sherman. His was the first Union force to take up position there, so he should have seen its weaknesses before anyone else.

Sherman had a simple explanation that he repeated in letters to family and friends. "Whether we should have been at Pittsburg Landing or on the other side of the Tennessee is none of my business. I was ordered to disembark here." The landing, he maintained, was selected by General C. F. Smith. Subsequently Sherman would expand on this theme: not only the choice of Pittsburg Landing but likewise the placement of the various divisions there—also the subject of much adverse comment by analysts of the battle—all was done "by order of General Smith."

This assertion is somewhat short of the full truth. General Smith was immobilized by his painfully swollen leg. Whether he ever saw the landing is questionable; if he did it was from the *Hiawatha*, which had brought him from Fort Henry to Savannah, where he would succumb to raging infection on April 25. That he could have walked over the considerable expanse of ground was a sheer impossibility. What he knew of the area bordering the landing could only have come from what Sherman told him and showed him on a map he had made while there.

If Smith approved the spot as a point of concentration for the arriving divisions it could only have been on the recommendation of Sherman— who was in fact telling Grant about the same time that he was "strongly impressed with the position's advantages." Though the landing itself was much reduced in periods of high water, the place could hold 100,000 men and was "of easy defense by a small command." This thesis makes all the more understandable Sherman's vehement insistence in later years that the position had been an extremely strong one, even as he asserted that its selection had been none of his doing.

It was no doubt Sherman as well who fixed the locations of the divisions as they arrived at the landing, despite his later disavowal. On March 20 he wrote the commander of the Second Division, then arriving, that Smith had "requested" that he "give the directives for encamping the troops as they arrive," which he then proceeded to do.

As revelations in the press appeared and public clamor continued, Grant and Sherman reacted very differently to the brickbats that came their way. Grant affected indifference. He made no response to his critics. Perhaps wisely, he never submitted a full report on the battle; even when he wrote his memoirs two decades later he found words with which to glide over the question of his mistakes or negligence at Shiloh. Sherman, on the other hand, felt every barb, smarted at every intimation of negligence or incompetence. At first he vented his anger in letters to those close to him, but from time to time he was provoked into outbursts that brought even more abuse down on his head. Before the summer was out he was trading insults in letters exchanged with his critics, some of these being placarded in the press.

In the hue and cry over Shiloh he saw the same phenomenon—"the same game," as he put it—that had followed Bull Run: "Men run away, won't obey officers, won't listen to threats, remonstrances, and prayers of their superiors, but after the danger is passed they raise false issues to cover their infamy." And the press, which always catered to such democratic-anarchical masses, broadcast about the country their version of what had happened; politicians, anxious to curry favor with those same masses, followed suit.

His anger was particularly acute against the reporters with the armies, "the most contemptible race of men that exist, cowardly, cringing, hanging round and gathering their material of the most polluted sources." He was convinced that they were out to injure him "in every way in their power." If they were, there were any number of possibilities in the climate of recrimination following Shiloh.

Colonel Rodney Mason of the Seventy-first Ohio, under something of a cloud himself after the poor conduct of his regiment in the fighting, published a letter in one of the Cincinnati papers claiming Sherman and his entourage were prejudiced against Ohio soldiers and showed marked favor to the Fifty-fifth Illinois. To meet this attack Sherman prevailed on his brother-in-law, Philemon Ewing, to write a retort, sending him an eight-page letter explaining his side of the case.

When the *Cincinnati Gazette* took the line that Sherman was chiefly responsible for the surprise at Shiloh, it was Thomas Ewing, signing himself "E," who published a refutation in the *Louisville Journal.* John Sherman launched a counterattack in a letter he talked the *Commercial* into publishing on April 30.

John also made a speech in the Senate that he had printed up and distributed; in it he wove together two themes: the admirable bravery of Ohio's soldiers and the equally admirable generalship of his brother.

Eventually Sherman himself could not resist being drawn into an epistolary duel with Ohio's Lieutenant Governor Benjamin Stanton, whom he charged with "deliberate and malicious falsehood & calumny against officers and gentlemen capable of vindicating themselves." The exchange was over the "surprise" at Shiloh and the comportment of troops in battle, and did credit to neither man. It ended only when Halleck wrote Sherman to stop. Though Sherman accordingly withheld his latest letter from publication, he made several copies of it and had friends pass them about; one went to Lieutenant Governor Stanton.

Hardly had the repercussions of this exchange subsided when Sherman was drawn into another controversy over the mass of Union soldiers who had left the battlefield and taken refuge at the landing. He regarded many of them as simply cowards and was furious that great numbers of them were carried off in the makeshift medical evacuation effort set up after the battle. At some point he purportedly said to a physician named Cook involved in bringing back the evacuees, "We don't want the cowards from the hospitals." Cook took exception to his remark and word of their exchange spread.

Sherman received letters that must have stung him. One was from an indignant young lady, a Miss Elbit of Columbus, Ohio. Her letter has not survived, but Sherman's reply has: "I did not say the absent soldiers were all cowards, but I did say what is true, that the bulk of the cowards did escape from Pittsburg Landing under pretense of being sick. They are not confined to Ohio . . . Don't distress yourself about 'the men of Ohio,' but get you a good husband and mind your own business."

If his official acts and orders are any indication, the post-Shiloh Sherman was a stern and demanding taskmaster, determined to improve the combat performance of his men. His order of April 12, 1862, announced that during battle "any officer or soldier who is out of his place in the ranks is worse than an open enemy and must be shot."

Regiments that had given way before the Confederate assaults of April 6 got special attention. Sherman had the officers and men of the Fifty-seventh Ohio assembled so that he could "pitch into them," as he put it, denouncing their shortcomings in scathing detail. He paid a similar visit to the Fifty-third Ohio, where he unburdened himself for nearly an hour. An Indiana soldier who happened to overhear the general's harangue was profoundly impressed. Sherman called both officers and men "cowards, dastards and every low name he could think of, telling them they were a disgrace to the nation and finally wound up by promising them that at the next battle they should be put in the foremost rank with a battery of Artillery immediately behind them and then if they attempted to run they would open on them with grape and canister."

The officers of these regiments were given the closest scrutiny as part of a larger effort to cull out incompetents. If some of the units had performed poorly, then there was a leadership problem at regimental and company levels. Grant's General Order, No. 40 of April 18, 1862, set up machinery for a general housecleaning, with inquiries into the "capacity, qualifications, propriety of conduct and efficiency" of any commissioned officer; performance during the recent battle was to be especially scrutinized.

Sherman prepared extensive lists, sending in the names of no fewer than four of his twelve infantry colonels. One of these courts-martial did no credit to Sherman. It convicted Colonel Thomas Worthington of the Forty-sixth Ohio Infantry, son of an Ohio governor and West Point graduate, class of 1827. When war came Worthington had raised a regiment that became part of Sherman's division. He had performed creditably at Shiloh and Sherman had even commended him in his official report. He had also asked his brother John to support Worthington's promotion to brigadier-general.

But Worthington was of a thorny and contentious nature and a "know-it-all," prompt to criticize; he also drank too much. After the battle he showed around a diary in which he claimed he had recorded Sherman's mistakes before and during the fighting. When he undertook to publish it Sherman lost patience and charged him with conduct unbecoming an officer and drunkenness on duty.

In preferring charges, naming the members of the court, and appearing as a witness against Worthington, Sherman had exceeded his authority, violating a law of 1830 that had brought changes to court-martial procedure–a surprising error for an officer with a penchant for the legal details. The guilty verdict was set aside on "technical grounds," but Worthington felt he was not fully vindicated. He wrote his own history of the battle in which he said, "Sherman ruined everything he handled or meddled with." Ragged and penniless, Worthington haunted the halls of Congress for years, seeking vindication and–to Sherman's fury–attracting sympathetic reporters.

The two days of fighting at Pittsburg Landing were a major turning point in Sherman's life. His biographers have generally presented it as a sudden change of fortune, a "spectacular resurrection." He had salvaged his career by dint of hard fighting against the Confederates on April 6 and 7, 1862–and in the months following against the envious colleagues and the carping critics who were his enemies in the Union camp. The lesson was not lost on him; in future meetings with both adversaries he would lay on with a will.

# 12

## MEMPHIS

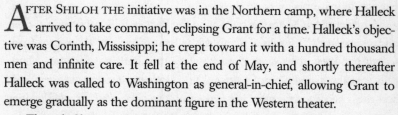

AFTER SHILOH THE initiative was in the Northern camp, where Halleck arrived to take command, eclipsing Grant for a time. Halleck's objective was Corinth, Mississippi; he crept toward it with a hundred thousand men and infinite care. It fell at the end of May, and shortly thereafter Halleck was called to Washington as general-in-chief, allowing Grant to emerge gradually as the dominant figure in the Western theater.

Though Sherman looked forward to serving under Grant, he wrote the departing Halleck a letter of rare effusiveness: "I cannot express my heartfelt pain at hearing of your orders and intended departure." He went on: "You took command in the valley of the Mississippi at a period of deep gloom, when I felt that our poor country was doomed to a Mexican anarchy, but at once arose order, system, firmness, and success in which there has not been a pause." But now, with Halleck's departure, Sherman professed to see trouble ahead: "I now fear alarms, hesitations, and doubt. You cannot be replaced out here, and it is too great a risk to trust a new man from the East." There is something excessive about these lines, and a recipient less susceptible to flattery than Henry Halleck would have discounted their sincerity. But Sherman knew his man; for the next three years he would have no greater champion in Washington.

As for Sherman, he had easily weathered all the controversy that followed Shiloh; by the summer he was on something of an emotional high. "That single battle had given me new life," he recalled a dozen years later, "and now I was in high feather." Ellen wrote her brother Charles as much: "Cump feels fully vindicated from the miserable slanders of last winter and is again in fine spirits." As time passed Sherman even began to see the painful episodes in Kentucky in a new, more favorable light. "Time has proved the truth of my representations," he wrote an Army colleague that August. And reminiscing in a letter to Ellen, he could appreciate in retrospect his own resolve and daring there: "Now one year ago today we occupied Muldro's Hill with a corporal's guard and . . . by bluster we deterred Buckner from approaching Louisville nearer than Green River . . ."

After the fall of Corinth Sherman's command—his division and that of General Hurlbut—spent its time repairing rail lines, then in July he was ordered to west Tennessee, fixing himself at Memphis. The city had fallen to the Union Army at the beginning of June and had had a quick succession of military rulers, but Sherman would stay for four months, long enough to leave his mark on the city.

The people of Memphis had found the Yankee military regime a sober and severe one. The sale of alcohol was banned, as was the use of Confederate money; the town's economy was thus hard put for cash. Memphis was also something of a prison, since to leave it a pass was required, and these were given only "in cases of urgent necessity." The authorities pressed the population to take the loyalty oath. Able-bodied white males had to take it or leave town.

On his arrival, July 20, 1862, Sherman inspected the town thoroughly, spending most of the day there. He had his troops encamp outside, since the town had a serious water shortage; moreover he believed that garrison life enervated troops, and he wanted to keep the two divisions under his command in fighting trim. He would keep his troops active over the next four months with endless working parties, patrols, and sorties.

Sometime that first afternoon the new commandant of Memphis encountered a group of local citizens who were curious to know what his policies would be; in answer to their questions he delivered an impromptu and somewhat rambling address, and it was taken down by a *Commercial* reporter. The crowd "hoped he was going to pursue a friendly policy and not do anything to exasperate the people." What they got was vintage

Sherman: "He said he thought Memphis was a conquered city . . . He had not heard that there had been any terms at the capitulation. They [he and his army] didn't come here to visit their friends. The people of the city were prisoners of war. They may be Union, and they may not. He knew nothing about that. One thing was certain: they had not fought for the Union, so far as he had heard . . . Memphis was a military post, a point of military importance he was bound to hold. He had nothing to do with social, moral or political questions; he was a soldier and obeyed orders, and expected to have his orders obeyed."

Jarring as this introduction was, the citizens of Memphis were pleasantly surprised by Sherman's first actions. He sent a gracious letter to Mayor John Park and the Board of Aldermen, assuring them that he had "unbounded respect for civil law, courts and authorities" and desired them to continue their efforts to preserve life, liberty and property. Regrettably the war made it necessary for the military to be "superior" to the civil authority, but the town's government had important functions and these should be extended. He favored the restoration of municipal courts and an effective police force to handle matters "which the military authority has neither time nor inclination to interfere with."

Sherman's Order No. 61 of July 24 abolished the system of passes: travel in and out of Memphis would be "as free and unobstructed as is consistent with the state of war." There were some limitations: travel could take place in daylight hours only; travelers would have to use one of five designated roads and pass through checkpoints where their possessions might be subject to search.

The general also contributed to the revival of a busy and animated city by removing the ban on alcohol. "To deny it absolutely," he informed the Memphis Board of Trade, "would lead to a hypocritical violation of the rules or orders proscribing its use." Bars and brothels might expect good business, for Sherman's soldiers could come into town on passes. The town would soon be protected by a garrison and fortified; Fort Pickering would be erected within the city limits.

The new commandant of Memphis was also sympathetic on the matter of municipal revenues, which were insufficient. The mayor and Board of Aldermen proposed an annual $2 poll tax on each white male, a similar levy on each dog in town, and "privilege taxes" on bars, pool halls, market stalls, livery stables and "each house suspected of being a bawdy house or house of ill fame." After an exchange of views between the general and a delegation of the Board of Aldermen the taxes on dogs and bawdy houses were dropped and the rates for the other taxes fixed.

Toward the end of his stay Sherman moved his headquarters to the Gayoso House, the town's best-known hotel, which he described to Ellen as "the loafing center of Memphis." One day a visitor came to call on him there—Rear Admiral David Dixon Porter, who had heard much about the general but had not met him; luckily Porter was a diary keeper, and he affords us an excellent look at the ruler of Memphis at work.

Porter had himself announced and as he waited he noted the brisk, businesslike atmosphere of the place, and was at first reminded of the ambience at McClellan's headquarters; then he began to notice differences. Here the rooms were plainly furnished and there were no carpets on the floors. The officers had a "bronzed and weatherbeaten" look and wore plain uniforms. And each officer "wore a serious expression on his countenance as if he were engaged in something that would require all his time and attention." Porter would later say that the difference between McClellan and Sherman was the difference between "fuss and feathers" and "rough and ready."

Still, Porter had waited a full hour when Sherman finally appeared. "He seemed surprised to see me, when I introduced myself, and informed me that he did not know I was there." Then Sherman turned to deal with one of his quartermasters. "I was not, I confess, much impressed with Gen'l Sherman's courtesy." But Sherman soon finished his business with the quartermaster: "He turned to me in the most pleasant way, poked up the fire, and talked as if he had known me all his life . . . He told me all he had done, what he was doing, and what he intended to do, jumping up every three minutes to send a message to someone." Thus began a friendship that would last as long as both men lived.

In his first days in the city Sherman received a parade of white Memphians, almost all of them wanting his help with their "servant" problem. As he explained it to Ellen, "Miss Nancy, raised in her mistress' bosom and nursed like one of her own children, has run off and the whole family must rush to Gen. Sherman." In truth blacks fleeing Memphis were relatively few; the town was filled with those seeking a haven from their bondage.

Sherman had no quarrel with that bondage: "Personally I have no hostility to slavery or any of your local laws," he assured a local magistrate. "I would they had never been disturbed."

He needed laborers at Fort Pickering; working there for the army, able-bodied black males could be fed, clothed, and given a pound of tobacco a month. Their modest wages were held back until their legal status—and that of their putative masters—could be determined. In late September Sherman estimated that there were about 1,500 blacks working on the fort and about an equal number who were not; these latter, he said, "either find employment or sponge on the others." But by then increasing numbers of black women and children were coming into his lines: "I cannot give them employment for we have no work for them, nor can I restore them to their masters. So the poor devils suffer."

The most important and also the most troublesome aspect of the city's commercial life that summer was the trade in cotton, for which there was enormous demand as the Union war effort moved into high gear. Northern mills clamored for the fiber, prices rose dramatically, and the Federal government made strenuous efforts to obtain it in the South.

The military in general were appalled at the excesses this trade in cotton produced. Grant denounced the buyers as "Jewish speculators" and gave his provost marshal orders to ban Jews from trains heading south, an action that brought a swift rebuke from Washington. Sherman's overriding concern was with the flow of money into the South, where he was sure the Confederacy would use it to purchase arms.

He protested to the adjutant general: "If the policy of the government demands cotton, order us to seize it and procure it by the usual operations of war . . . This cotton order is worse to us than a defeat. The country will swarm with dishonest Jews who will smuggle powder, pistols, percussion caps, etc. in spite of all the guards and precautions we can give." In Memphis Sherman decreed that no cotton could be sold in direct exchange for gold or even treasury notes; but these could be deposited with the quartermaster, who would hold them, "subject to the order of the vendor after the war is over." He was soon ordered to drop this hasty and ill-conceived measure.

The *Commercial*'s Memphis correspondent pointed out that "the Treasury Department has seen fit to open the trade and it is none of Sherman's business." Halleck wrote to remind him that "money is of no more value to the rebels than cotton, for they purchase military munitions with both." He added that an adequate supply of cotton really was critical to the Union cause, for without it the War Department would be unable to equip the new levies with tents.

Sherman had in the meantime trimmed his sails. He wrote Halleck that he would "religiously carry out any line of policy as to trade the proper authority dictates, and with absolute confidence in its right as soon as I feel you are at the helm . . ." To a colleague he confided, "I think Halleck will soon stop all this nonsense."

The cotton trade became one of his crotchets; he could be all too easily provoked when the subject came up. A government official discovered this when he went to Sherman for permission to land a cotton gin, press, and 30,000 pounds of cotton in the seed: "He flatly refused & treated me in a rude manner, said very unpleasant things and entirely uncalled for. I wished all cotton in hell and the general too." A second official who was present tried to intervene, but "he [Sherman] pitched into him like thunder too. Is it necessary for a general in charge of a department or division to be a brute in order that he may be considered a soldier?"

His efforts to prevent any movement of goods south to the Confederacy probably absorbed more of his time than anything else, and there is little evidence that he was successful in this endeavor. He sometimes wondered if the Confederates had not given up the city on purpose to have it as a nearby source for smuggled supplies: "I have no hesitation in saying that the possession of the Mississippi River by us is an enormous advantage to our enemy, for by it and the commercial spirit of our people they, the enemy, get directly or indirectly all the means necessary to carry on the war."

What angered him most was this "mercenary spirit" that led Northern merchants to inundate Memphis with vast quantities of goods needed by the Confederates. Cincinnati businessmen he regarded as the prime culprits: "Cincinnati furnishes more contraband goods than Charleston, and has done more to prolong the war than the State of South Carolina. Not a merchant there but would sell salt, bacon, powder and lead if they could make money by it . . ." That summer his old friend Braxton Bragg was in Kentucky with a sizable Confederate army, and Sherman confided to

Ellen that he would find a grim satisfaction in seeing Bragg cross the Ohio, march on Cincinnati, and "bring the war home to them."

Smuggling was rampant. In addition to the five authorized roads there were half a hundred other roads and paths leading in and out of town that could not be kept under constant surveillance. The smugglers were tireless and inventive. Military supplies were known to have left the city in coffins and in hog carcasses. Searches at the checkpoints frequently turned up prohibited items on their way to the Confederacy; Sherman wrote Halleck, "we find clothing, percussion caps and salt concealed in every conceivable shape."

That summer the most critical item in this illicit movement of goods was probably salt, for the commodity was growing scarce in many parts of the Confederacy and salt meat was a staple in the Southern soldier's diet. The quantities being ordered by Memphis merchants were quite heavy—so heavy that in August the U.S. Treasury Department banned further shipments to the town. The Memphis correspondent of the *Commercial* reported that local farmers had gone into a lucrative salt-running business; they could make two trips to town each day, since checkpoint personnel were changed at noon, and could carry out each time a barrel of salt for themselves and one for each man who accompanied them as passenger in their wagons: "The consequence is that the whole Confederacy is in a fair way to be supplied with salt through what may be termed the Memphis sieve."

Sherman acknowledged that he had a problem. The solution was to increase the penalties. He classified salt, medicines, and chloroform as contraband of war and allowed the death penalty for those convicted of trafficking in them. Most of those charged went before a military commission whose three members Sherman described as "kind and humane officers." All death sentences were reviewed in Washington, so when the commission condemned to death a man named Miller for smuggling arms that August, Sherman wrote to assure the adjutant general that it was an open-and-shut case, the culprit having confessed in open court. "I hope the President's known humanity will not interpose in this breach of the well known laws of war."

But on occasion the commission could be lenient with smugglers and so could Sherman. When Martha Allen and Frances Jones tried to leave

Memphis with a trunk containing some unspecified contraband, the commission decided an appropriate punishment would be confiscation of the trunk and its contents and banishment of the women to Arkansas—"but in consideration that Mrs. Allen is rather an aged lady and Mrs. Jones is *enciente* [sic], the Commission recommends them to the mercy of the Commanding General." And Sherman showed them mercy: "The Commanding General directs that the goods except the trunk and two dresses be confiscated. The ladies will go home and not attempt this again."

The most systematic press critique of Sherman's administration in Memphis could be found in the published letters of the *Commercial*'s local correspondent, but other papers from time to time dedicated articles to the subject, and when they came to Sherman's attention he was usually not pleased. Such was the case when his brother John sent him an article from the *New York World* announcing that "abuses of a flagrant character have grown out of the loose administration of public affairs under Gen. Sherman." The basic charge leveled by the *World*, and taken up by the *Commercial* and other papers, was that Sherman had simply confirmed in power the same people who had run Memphis when it was a Confederate town, thus ignoring the interests of the Union element.

Sherman had no choice but to endure these criticisms from the "outside press," but the local newspapers he followed closely. To take the place of the *Appeal*, one of Sherman's predecessors had set up the *Union Appeal*, run by the Reverend Samuel Sawyer, temporarily detached from his function as regimental chaplain. It joined one preexisting paper, the *Argus*, which had adjusted its editorial policy to conform to the new order of things. Sherman read both papers carefully, as well as a third, the *Bulletin*, which appeared later; he did not hesitate to address the editor when he found something amiss.

Running Memphis took a great deal of his time; though detachments of his troops went outside the town frequently on various missions, he seems never to have accompanied them. He worried about a sudden attack on Memphis, particularly before Fort Pickering was completed, and this consideration alone would have kept him in the city. He managed to write Ellen about a letter a week; in one of them he reveals that he had assigned two members of his staff to write her at regular intervals, sparing him the task of chronicling everyday happenings in Memphis. He wrote the children from time to time, his letters often containing curious confidences along with expressions of endearment. Ellen was anxious to visit him in Memphis, but had to cancel her trip more than once at his insistence that it was not safe; she finally came down in October, bringing Tommy with her.

Sherman's life in Memphis had its social dimensions. He encouraged the return of the theater to the town and regularly attended its performances. He served as protector and guiding spirit of the city's Union Club, which he addressed from time to time and nurtured as though it were some delicate plant. In October he told Ellen that it had 600 members, "but very timid and fearful"; at year's end, with his encouragement, they paraded through the streets of Memphis. He occasionally attended church, notably at Calvary Episcopal, where there was a woman whom he told Ellen sang remarkably well. That fall he made acquaintance with the prominent Episcopal cleric James H. Otey, Bishop of Tennessee; the two struck up a friendship and visited back and forth.

Then there was Felicia Shover, widow of an officer Sherman had known in the Third Artillery. The Shermans and the Shovers had met at West Point in 1850, so Sherman mentioned her in passing in a letter to Ellen, saying she had "presumed on my favor which she did not get" and "insisted" that he call on her, a chore which he dutifully accomplished that same day. But in fact he did a number of favors for Felicia Shover, including passing through the lines a letter she had written to a Confederate general. And he apparently saw her a number of times that fall; at some point during his stay in Memphis she offered a dinner in his honor. The two exchanged letters for some years thereafter, and wrote about meeting again, though it is not clear whether they did so. While the surviving correspondence is hardly scandalous, the banter and teasing in Sherman's letters indicate that the two were very good friends; for Ellen the letters would have made painful reading.

If Sherman's personal life in this period must remain somewhat in the shadows, in his public role as the ruler of Memphis he was constantly in the spotlight. His tenure is well-documented and also highly revelatory. This was the first time that Sherman the soldier found himself responsible for a large civilian community; the attitudes and policies he adopted in Memphis would reappear countless times during the war.

In his impromptu speech that first day in Memphis he had posed as the rough soldier who had neither the time nor the interest to take care of civilian matters. He indicated as much in his first letter to the mayor and council of Memphis; he was happy to confirm them in functions that were properly theirs. But he warned them that in wartime the military power was "superior" to the civil one; his exercise of that power gradually gave him control over virtually every aspect of the city's life.

Though a profound admirer of the written law and above all the Constitution, Sherman claimed that he derived his powers over the city from that uncodified mass of usages known as "the laws of war." Thus he held essentially the powers acquired by the right of conquest. He was accountable to his superiors and bound to observe the Articles of War and army regulations, "but with regard to all others his power is unlimited"; as for those he ruled, for them "the law of war is the will of the commander." Sherman could, in his own view, even put an end to the city, driving out its entire population. He mentioned the possibility several times in his letters; if he desisted, it was not for any lack of power but simply because it would "make the place a desert."

The structure of justice in Memphis had the same duality as that of government, with the military branch dominating, and for the same reason. As Sherman explained it, "only such parts of our civil system of jurisprudence can be allowed as are consistent with the laws of war." A military commission composed of three officers became the most important court in Memphis. Its purview was described variously by its creator as punishing violators of "civil or state offenses" and hearing cases under "the laws of war." Judging from the variety of crimes it dealt with, its area of competence was both vast and vague. It could impose the gamut of punishments: fines, imprisonment, death, and banishment, defined as being "sent away," either north or south. Sherman added the stipulation that "no prisoner may be released unconditionally who does not take the oath of allegiance."

The military commission heard cases involving the misbehavior of city policemen and local journalists. It tried and convicted a British subject named John Fogg for "using disrespectful language against Maj. Genl. W.T. Sherman, an officer of U.S. service"; Fogg had called the general "the most ornery man he ever saw" and said "it was of no use for anyone to apply to him for any purpose, as he would be treated rudely." Fogg got off with a warning, perhaps because of his nationality; but in the same session the commission sent W. E. Copewood to a Northern prison for the rest of the war for publicly stating that "he would aid and abet the Confederacy at all times when possible for him to do so."

The military commission was in no sense of the word an independent judiciary. Sherman's letter book correspondence makes it clear that its president consulted him on delicate and doubtful cases. Sherman's advice was very often translated into the commission's verdict. The proceedings of the city's civilian courts were followed by its military commandant. Judge John T. Swayne and Sherman were friends who shared the same conservative political views; they stayed in contact long after Sherman had left Memphis. But in one complicated case Sherman was quite unhappy with Swayne's charge to the jury and told him so plainly: Swayne had a "too delicate sense" of his official obligations; after all, there was a war going on.

Perhaps the most bizarre instance of Sherman's intervening in the judicial matters is to be found in a report of his provost marshal, Colonel D. C. Anthony: "September 30. General Sherman in person tried Samuel Mansfield, David Jennings, wholesale druggists, charged with selling quinine and knowing it to be for the Southern Army, and ordered them to be sent to Alton, Illinois under arrest. Same day the General tried R.D. Ward and W.R. McClelland, druggists, upon same charge, and ordered them to be sent to Saint Louis, to report to the Provost Marshal."

Predictably, Sherman's relations with the Memphis press were not very good, leading to considerable problems for the newsmen. Reverend Samuel Sawyer was the first to incur the general's wrath when, with the best of intentions, he offered the *Union Appeal*'s readers a capsule biography of their new commandant.

"Personalities in a newspaper are wrong and criminal," Sherman informed him, "you make more than a dozen mistakes of facts, which I need not correct as I don't desire my biography until I am dead" (nevertheless the general listed the errors later in the same letter). He assured Reverend Sawyer that he would be a true friend of the press if he found the journalists of Memphis "actuated by high principle and a sole devotion to their country," but if he found them "personal, abusive, dealing in innuendoes and hints at blind venture, and looking to their selfish aggrandizement and fame, then they had better look out . . . "

The newspapers were expected to publish official notices and announcements, but on Sherman's orders they ceased publication of anonymous letters; any published battle report had also to carry the name of its author. In August the general pointed out how the press might be a force for good by stressing certain themes in its columns. In a long letter to the editors of the *Bulletin* and the *Appeal* he suggested "these views will give you points for some paragraphs." With guerrillas and partisans increasingly active outside Memphis, he wanted to get the message to the local population that it could only lose by harboring or tolerating such bands, for in doing so they invited destruction of their own property by the Union Army: "They lay themselves clearly liable to all the risks of war without any of its excitement."

A month later Sherman had had enough of the *Union Appeal*. He wrote Colonel Anthony that the journal had "for some days published material calculated to discredit the government of the United States and injure the cause for which we are fighting." Anthony was to close up the building after allowing the occupants to remove their possessions. Reverend Sawyer was to be packed off to Arkansas to rejoin his regiment. His chief assistant, whom Sherman styled "an impudent and useless fellow" and a "conceited ass," was to be sent under arrest to Cairo.

The turn of the *Argus* came in November. Its editor's transgressions were similar to those of the *Union Appeal*. He had been lingering over "every little disturbance of the peace, & by design, I think, wants to establish the idea that all of Memphis is dangerously disorderly. Arrest the editor, make him give good bonds to answer for publishing false & mischievous nonsense & let the trial be before the Military Commission."

It would be wrong to paint Sherman's tenure in Memphis in somber colors only. The evenhanded and effective rule of a hostile, sullen population would have been a challenge to any Civil War general. Without being especially laudable, Sherman's record is hardly the worst; it compares well with General Ben Butler's controversial rule over New Orleans that same year. Though affected by his vagaries, his administration was generally sound, honest, and efficient. Complaints addressed to his superiors were uncommon; only one is known to have reached the secretary of war.

When he discovered that passes to Arkansas were being sold by one of his men in the provost marshal's office, he made a public acknowledgment of "shame and mortification" and invited those with knowledge of other derelictions in his administration to come forward–but no anonymous letters would be accepted.

Memphis had more than its share of destitute people in the summer and fall of 1862; Sherman mobilized ward committees and a central board to work with the poor and homeless, and set aside buildings to be used as shelters. He issued a personal call to all the units in his command, which were bountifully supplied with foodstuffs, to set aside meat, flour, bread, rice, and so on for those in need. The general himself supplied twenty-five cords of wood a month for cooking and heating and a thousand dollars from his Confiscated Property Fund.

For Memphis prosperity lay in trade and commerce, and Sherman felt he was largely responsible for its revival. Shortly after going to his next assignment he wrote Secretary Stanton: "During my administration of affairs in Memphis I know it raised from a condition of death & gloom & darkness to one of life and comparative prosperity. Its streets & stores & buildings were sad and deserted as I entered it. When I left it life & business prevailed . . ." As in much of the general's prose, the contrasts are a bit forced, yet there is no doubt that the trade which was the city's lifeblood revived during his tenure as commandant.

His letter books and other evidence of his rule in Memphis do not reveal an evenhanded, open-minded and judicious administrator laboring steadily for the welfare of the population he has been given to rule, nor do they reveal that pattern of relentless malevolence that would distinguish the Sherman of postbellum Southern folklore. There is, in fact, no pattern at all, and this is perhaps the most troubling aspect of his rule. One has the impression that each pronouncement, each act, was inspired by the mood, the vagary of the moment, that the ruler of Memphis was making up the script as he went along. A kind gesture would be followed by a harsh one; there were sudden, wrenching shifts in policy; when problems appeared he often met them with draconian pronouncements in the peremptory style of the "official" Sherman.

He boasted that whatever else Memphians might think of him, they all would admit, "I stick to the law." Given his reverence for the written law, one would expect him to be punctilious in this regard. In fact he was not. Here also he improvised or invented, with *pronunciamentos* inspired by little more than his feelings of the moment. These, for example, are his grounds for ordering the arrest of the editor of the *Argus* and his translation before the military commission: "The rule is that Editors have no more right to falsify & exaggerate than anybody else, and the publication of false and mischievous news paragraphs should be punished."

Then when several local men appeared before the military commission charged with the murder of a private of the Fourth Illinois Cavalry, the crime having been witnessed by several people who offered testimony, he had the commission display an astonishing leniency: he advised its president to simply put the accused under bond and let them go home. All the witnesses were black and he didn't want to "use negro testimony when white men are concerned."

If there is a conclusion to be drawn from Sherman's tenure in Memphis, it is that he was a man badly out of place, a good soldier ill-equipped for the direction of civilian affairs, trusting to his instincts and confusing the impulsions of his own nature with the dictates of law and justice. The correspondent of the *Commercial*, most probably an officer in Sherman's own command, admired the general as a soldier. But of the six military rulers Memphis had known, he rated Sherman "the best man in the field and the worst in the chair that can easily be found."

It is manifest that Sherman lacked that ability to grasp or understand public sentiment in that distracted city or to influence it for his purposes, despite several heavy-handed attempts in that direction; with "the better sort of people," Mayor Park, Judge Swayne, Bishop Otey, and others, his relations were marked by unfailing cordiality and respect. But with people in the mass Sherman could rarely strike the right chord.

In justice to the man it must be said that he did not relish the assignment to Memphis. He was never comfortable administering civilians and he said as much; he admitted from time to time that he did not understand civil administration. But that did not mean he regarded his rule in West Tennessee as a failure—quite the contrary. He believed his direction of affairs in Memphis was a distinct success. While he was coming to recognize in the Southern people generally a bitter and implacable foe, he saw in Memphis the beginning of a reconciliation for which he could rightly take credit. He told John that he had been successful in "allaying all opposition without lowering the dignity and pretensions of our national government."

Such was his belief in the success of his policies in Memphis that he began to think of himself as having a special gift for the reconciliation of disaffected populations. He considered himself something of an expert on the South, understanding its ways and customs; then, as he explained to John, he benefited from being a Westerner rather than a Yankee, since the South had a "violent antipathy" to New Englanders. In Louisiana General Ben Butler of Massachusetts was obviously not having the same success that Sherman enjoyed. If Butler left, John could tell Secretary Stanton that his brother would be willing to go there and lead Louisiana back into the fold. He shared the same bright vision with Ellen, but added that first he had another assignment. Louisiana would be waiting, he told her, "if I can take Vicksburg."

# 13

## VICKSBURG

IN THE SPRING OF 1863 IT WAS common to refer to Vicksburg, Mississippi, as the "Gibraltar of Dixie." At the time the description was an apt one—both sites were strongholds dominating vital waterways and both seemed virtually impregnable. Sitting on a range of bluffs, Vicksburg commanded the Mississippi at one of its major bends, giving its guns an unrivaled field of fire and making it difficult to assail.

While Vicksburg limited the enemy's use of the river, it also linked the two halves of the Confederacy: a line coming from the east via Meridian and Jackson terminated at Vicksburg; just opposite, on the Louisiana side of the river, was the terminus of the Vicksburg, Shreveport, and Texas line running west. Under the protection of Vicksburg's guns steamboats came and went endlessly, shuttling cargoes and passengers from one rail line to the other.

Vicksburg would be the major goal of the next Federal campaign in the Mississippi Valley. In October 1862 Grant assumed command of the Army of the Tennessee, acquiring with it responsibility for territory on the east bank of the great river—and Vicksburg. Close study of maps of the region indicated difficult terrain to the north of the town, swamps and bayous associated with the Yazoo-Sunflower river system. The best avenue of approach was one running south through Holly Springs and

Oxford, Mississippi, to Jackson, then west to Vicksburg, following rail lines and converting them into supply lines. Approaching from the east, the assailants could begin the investment of Vicksburg on the solid terrain lying between the Yazoo to the north and the Big Black River, which emptied into the Mississippi below the town. This was the plan advocated by Halleck; in mid-November Grant and Sherman met briefly in Columbus, Kentucky, to talk about it. And they probably talked about John A. McClernand; Grant had learned that Lincoln had authorized this political ally and novice at war to mount an expedition to take Vicksburg, so Grant was eager to get his own campaign under way.

On November 24 Sherman and his force left Memphis, paralleling the movement of Grant's columns and connecting with them as they approached the Tallahatchie River. There Confederate General John C. Pemberton was said to be waiting in an elaborate fortified line. But Grant found the Tallahatchie abandoned; General Earl Van Dorn was defending the Yalobusha, some forty miles further south.

On the morning of December 8 Grant was at the university town of Oxford, Mississippi, when he got a telegram from Halleck; it prompted him to dispatch a message to Sherman, a few miles to the west: "I wish you would come over this evening," the message read, "I want to talk with you about this matter." The move to take Vicksburg could be accelerated: Grant would continue his approach, holding the enemy's attention and drawing most of his troops. Sherman would hurry back to Memphis, taking one of his divisions with him. There he would collect and embark a large force, sail down the Mississippi with a gunboat escort, then up the Yazoo where it entered the Mississippi just above Vicksburg; there, at Chickasaw Bayou, he would debark his troops and attack the town from the north. With some 32,000 troops, he might well surprise and overwhelm the Vicksburg garrison; failing that, his advance, added to that of Grant, would put Vicksburg between the hammer and the anvil. With help from the south, where N. P. Banks had replaced Butler, Vicksburg could be put under close siege.

Sherman hurried out of College Hill, carrying orders to sail his force down to Vicksburg and "proceed to the destruction of that place in such manner as circumstances and your own judgment may dictate." By December 12 he was back in Memphis and immersed in preparations,

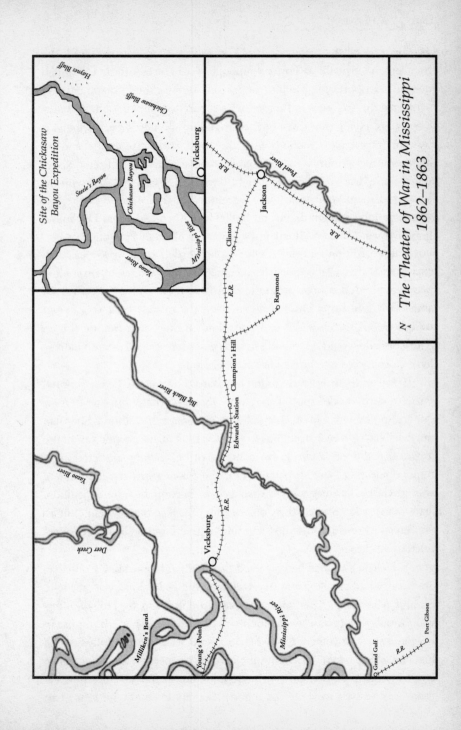

The Theater of War in Mississippi
1862–1863

Site of the Chickasaw
Bayou Expedition

Haynes Bluff

Chickasaw Bluff

Steele's Bayou

Chickasaw Bayou

Yazoo River

Mississippi River

Vicksburg

Yazoo River

Deer Creek

Milliken's Bend

Young's Point

Vicksburg

Big Black River

Champion's Hill

Edwards' Station

Clinton

Jackson

Pearl River

Raymond

R.R.

R.R.

R.R.

R.R.

Mississippi River

Grand Gulf

Port Gibson

N

sweeping up all the troops he could, including many earmarked for the McClernand expedition. Porter's gunboats would be escorts, but Sherman needed transports for 31,000 troops, plus supplies and equipment. He explained this to Lewis B. Parsons, who supervised shipping on the Western waters, giving him three days to round up the necessary shipping. It was an impossible deadline; with the most strenuous efforts and an extraordinary authority to commandeer boats, Parsons collected sixty-seven transports in the space of eight days—and Sherman sailed with his leading elements that same day, December 20. He stopped at Helena, Arkansas, long enough to pick up 10,000 men, then Christmas Day found his convoy at Milliken's Bend, just a few miles north of Vicksburg.

The expedition had sailed with explicit orders on how to deal with two threats that preoccupied its commander: newspaper reporters and people who fired at boats from the riverbank. If the convoy encountered small arms fire in its passage down river, the nearest boat was to put ashore troops to "clear out all opposition"; if there was cannon fire an entire brigade would disembark. In both cases houses and other buildings in the locality were to be set afire in retaliation.

If civilian reporters were found on board they were to be conscripted into the army on the spot. If they were found with anything written for publication they would be charged with espionage. In addition, Sherman may well have given instructions to A. H. Markland, his postmaster on the Yazoo expedition, to check the contents of bulky outgoing letters and remove any that were dispatches destined for newspapers; that charge was made by journalists who managed to accompany the expedition. "Sherman could not whip the Confederates," recalled one of them, "but he was equal to opening the mail bags and extracting therefrom the letters relating his defeat."

At Helena Sherman had received a secondhand report that Confederate cavalry had attacked and overwhelmed the Federal garrison at Holly Springs, Mississippi. The report no doubt troubled him, for Holly Springs was Grant's supply base, with mountains of rations and materiel for use in his southward movement toward Jackson and Vicksburg. If Holly Springs had really been taken the loss of supplies could jeopardize Grant's advance and consequently Sherman's own operation up the Yazoo. Unfortunately there was no way to verify the report; telegraphic communications had been interrupted across the region—an ominous sign in itself.

The report was in fact true. The Confederates had reduced Grant's depot to ashes, while swarms of their cavalry moved over northern Mississippi and Tennessee, sowing confusion and destruction, tearing down and carrying off miles of telegraph wire, burning bridges, and wrecking sections of the rail lines that were to sustain Grant's army. Grant had no choice but to halt his army and start a withdrawal back to the line of the Tallahatchie—and he could not get word to Sherman.

On the twenty-third, when Sherman gave his division commanders the details of the upcoming operation, he mentioned that raids on Grant's rear might "disconcert him somewhat," but made their own mission all the more important. He also passed them copies of a map of the area around the Yazoo "compiled from the best sources"; he and Admiral Porter had copies, so with everyone using the same map there would no chance for confusion.

A complication of a different sort was developing in Illinois. On December 17 General McClernand was in Springfield when he got wind of Sherman's expedition. He rushed to prepare his own departure and sent off a flurry of telegrams and letters, including a wire to Lincoln that said, "I believe I am superseded. Please advise me." Lincoln saw Stanton and conferred with Halleck. Words of reassurance went to McClernand, while Halleck wired Grant that it was "the wish of the President" that McClernand have immediate command of the Yazoo expedition, under Grant's general direction.

McClernand thus had a clear mandate to take charge of the force sailing toward Vicksburg—if he could catch up with it. He wound up his business hastily and took a chartered steamer south on December 26. Here and there he put in and inquired for news of the Yazoo expedition. At Memphis he found that Sherman had passed through, stripping the town of nearly all its troops. McClernand wrote Lincoln in considerable anger: "Either accident or intention has so conspired as to thwart the authority of yourself and the Secretary of War and to betray me."

On the day after Christmas Sherman's convoy was at the mouth of the Yazoo when someone came aboard Sherman's ship, the *Forest Queen*, with news that McClernand was on his way south to take over. The word spread over the *Forest Queen*; Quartermaster Parsons, who had developed an admiration for Sherman in those frantic days of preparation, was struck by the general's muted reaction to the news: "Of course Gen. Sherman must have felt unpleasantly, but he does not show it in the least and bears it like the true soldier he is."

This show of equanimity must have taken some effort on Sherman's part, for in the intimacy of his circle he heaped abuse on McClernand. He had known him at Shiloh and had not been impressed; Sherman despised the man as a person and as democracy's answer to the need for military leadership: a glad-handing demagogue with no more claim to the major general's stars he wore than a six-year-old.

But for the moment it was the Yazoo that dominated Sherman's thoughts. Porter had probed the river earlier in the month to determine its navigability and test its defenses, and had lost the ironclad *Cairo* to one of the enemy's "torpedo floats," primitive but effective mines that the Confederates had planted here and there in the watercourse. As a consequence Sherman's flotilla could not sail up the Yazoo until it was cleared of mines; this was done by men in rowboats who used grapnels to fish for the "infernal machines," as Sherman called them; the van of the flotilla was composed of ironclads capable of dealing with enemy batteries, though in fact they encountered none that would pose a problem; then came the fleet of transports, interspersed with gunboats, moving single file since the Yazoo was too narrow for more than one column. Thirteen miles up the Yazoo the transports reached the designated point of disembarkation and the bulk of the troops went ashore, disposed of enemy pickets, and set up camp—all of these actions being accomplished by the evening of the twenty-sixth. The expedition was beginning reasonably well.

It would not end that way. In the covering letter for his official report Sherman later told its story with singular brevity and candor: "I reached Vicksburg at the time appointed, landed, assaulted, and failed." After relating succinctly what had happened, he added with a gracious touch: "I assume all responsibility and attach fault to no one, am generally satisfied with the high spirit manifested by all."

The expedition had lost considerable time while the Yazoo was being cleared of torpedoes. Once they were fished up they had to be disposed of by detonating them; those detonations attracted the attention of the enemy, who was soon banging away at the unfortunate men working the grapnels; but in fact the delay and the noise were probably not critical. The Confederate authorities suspected something was afoot well before Sherman arrived. By the time he was ready to make a serious assault on the defenses of Vicksburg, General John C. Pemberton, who commanded Confederate forces in the region, had on hand 14,000 men; holding the high ground and standing on the defensive in prepared works, as they would at Chickasaw Bluffs, they would be more than sufficient to balk Sherman.

The chief handicap for the attackers was the terrain; the map that had been so carefully prepared "according to the best sources" failed to show one critical detail: the ground chosen for the landing and the advance toward Vicksburg and its defenses was not solid. Sherman described it as "Mississippi alluvion," which presented itself in the form of sand bars, swamps, and bayous, of which the most prominent was Chickasaw Bayou; in one spot the same body of water might be up to a man's ankles, and in another over his head. Often the men could advance only by means of pontoon bridges or by sinuous paths that had to be discovered by trial and error; any forward movement tended to funnel troops through obligatory passageways, compacting the advancing columns and making them more vulnerable to the enemy's fire. And only men could advance in that fashion. Wagons mired down, as did artillery pieces.

Admiral Porter had tried to talk the general into making a landing at another spot on the river, Haynes' Bluff, where there was high, firm ground. There were also Confederate batteries there, but the admiral was confident he could silence them with his gunboats. Sherman would not consider it; Porter got the impression he wanted to make his assault without the assistance of the navy.

The immediate objective was Chickasaw Bluffs; added to their natural defenses, formidable in themselves, were others the enemy had created: abbatis—tangles formed by felled trees that could slow the infantry's advance to a crawl, a levee in the form of a parapet, rifle pits, artillery emplacements, sharpshooters, and the bluffs themselves, lined with defenders. Between the levee and the bluffs was a road along which the Confederates could shift their forces rapidly and with impunity.

Only after overcoming these various obstacles could Sherman's men hope to meet the Confederates on even terms; until then their fighting would be uphill in every sense of the word. And finally, Sherman lacked something a modern-day commander would have considered indispensable—an accurate weather forecast. Here the generals of his era had to trust to luck, and Sherman's luck was bad. On the second day after the landing, when the Federals were still exploring the terrain between them and the enemy positions, that terrain became heavily shrouded in fog. It rained intermittently, then the downpour began in earnest toward the end of their stay. Sherman had noticed with some concern high water marks a dozen feet up on the trunks of trees; now the Yazoo began to rise rapidly, and Admiral Porter took to tethering his vessels to tree trunks. At night the men slept in their sodden clothing, denied even the benefit of a campfire.

After two days of struggling forward, on the evening of December 28 Sherman's units were within striking distance of the Confederate defenses. All along he had been hoping that he would get word that General Banks had made an appearance, or that he would hear distant cannon fire announcing that Grant too was rapping on the gates of Vicksburg, but all he could hear now was the whistles of trains arriving in Vicksburg with supplies and reinforcements for its defenders. Nevertheless, on the night of December 28 he wrote out orders for the attack: "the whole line will move as nearly east as the ground will permit, simultaneously attacking the crest of hills in their front, Morgan's division securing a lodgement on the top, M.L. Smith the face of the hill, and A.J. Smith the country road . . . "

The attack was made the following day and failed. One brigade became confused and took the wrong direction; here a column was caught in a crossfire of artillery and driven back with heavy losses; in another place the men found shelter from the storm of fire and could not be coaxed or threatened into resuming the advance. Morgan's division, on which Sherman was counting above all, failed to attain its objective, and for this Sherman would blame its commander, General George W. Morgan. That evening the rain intensified; the Yazoo continued to rise and threatened to put the army's miserable encampments under water. There was no choice now but to withdraw.

It was a profoundly dejected Sherman who went aboard the *Black Hawk* that evening. He would have been even more dejected if he had known that in the regiments that had made the fruitless assaults there was considerable muttering about the insanity of their commander. That evening Porter saw in Sherman all the signs of physical and mental exhaustion: "He had shared the privations of his men, bivouacking out in the cold with them. No general could long conduct an army that way . . ." Porter claimed that night was the only time in the war that he saw Sherman despondent; "all unhinged" was the term he used. "He sat down in an armchair, and leaning his head upon his hands, seemed lost in thought. He sat thus, speaking to no one, until at length I said: 'Why General, where are all your spirits tonight? What's the matter with you?' 'My spirits,' he said, 'are on Chickasaw—Admiral, I never was so cut in my life. This has been a dreadful disaster.'"

Porter told him he was surprised to hear the fighting described thus and asked him how many men he had lost. He answered 1,700 killed, wounded, and missing; Porter said that didn't sound like very many. Soon Sherman took on a more detached tone; God for some reason had not wanted them to win, he said, but later they would. Porter told him that before the war was over he would think nothing of losing 1,700 men. This remark seemed to console him. He cheered up and asked the steward to bring him a toddy. The subject of attacking Snyder's Bluff was brought up again, no doubt by Porter, and this time Sherman showed interest in it; he had some of his subordinates come aboard, maps were passed about, and Sherman decided to give it a try the next morning if the weather was favorable. But the dawn brought a day that was completely opaque; the oldest hands on the river said they had never seen such a fog. There would be no further attempt to break through the enemy's defenses along the Yazoo.

But Sherman's cup was not yet full. Word came that General McClernand had arrived and was at Young's Landing; Sherman dutifully called on his successor, explained the need for withdrawal, and gained McClernand's agreement; McClernand told him what had happened to Grant. The question now was what to do next and for that discussion McClernand joined Sherman and Porter on the *Black Hawk* on January 3.

There was the possibility of going up the Arkansas River and taking a Confederate installation known, confusingly, by three separate names: Fort Hindman, Post of Arkansas, and Arkansas Post. It was regarded by the Union commanders in the area as a tempting target; an earlier expedition against it had been foiled by low water. Now Porter and his two guests examined the maps and agreed to undertake the operation. McClernand remarked that it would be good for the troops, who had been "demoralized by the late defeat," while Sherman, looking daggers at him, remained silent.

The next day McClernand renamed his force the "Army of the Mississippi"; corps designations had just been authorized by Washington, so Sherman found himself commanding the Fifteenth Corps. Thus reorganized, the expedition sailed for the Post of Arkansas. After considerable reconnoitering and skirmishing, on the afternoon of January 11 McClernand, Sherman, and Porter had their forces properly positioned for the assault. The fighting had lasted only three and a half hours when white flags appeared on Fort Hindman. Union losses were fairly heavy—a hundred or so killed and nearly a thousand wounded—but this was now the going price of infantry assaults on entrenched defenders. The outcome was never in doubt. The assailants outnumbered the defenders six to one and could count on the heavy cannon carried on Porter's gunboats to batter down the Confederate works.

The taking of the Post of Arkansas is not considered a major military event; it is often listed as an "engagement," rather than a battle. Yet in his *Memoirs* Sherman dedicated six pages to recounting the fighting there, considerably more than he wrote on Bull Run. In his account of the fighting McClernand appears as a peripheral figure who made no attempt to guide events but stayed in the rear, with "a man up a tree" to watch the fighting. Sherman wrote John: "I led the columns, gave all the orders, and entered the place, when he came along and managed the prisoners and captured property." According to Sherman's *Memoirs* when McClernand learned of the victory his subordinates had won for him, he exclaimed, "Glorious! glorious! my star is ever in the ascendant!" The testimony of Porter tends to corroborate that of Sherman, describing McClernand as an "ornament" to the expedition.

McClernand was not surprised that the "professionals" should try to deny him his laurels. He wrote Lincoln that his success was "gall and wormwood to the clique of West Pointers who have been persecuting me for months." Unflattering McClernand stories circulated on the army grapevine: about his cowardice at Shiloh, about the bride he had the bad taste to bring with him on campaign, about the half a hundred staff officers in his suite. The stories would continue as long as McClernand was with the army.

McClernand pleaded with Lincoln for an independent command, perhaps there in Arkansas, where he could follow up on his victory unhindered—an idea Lincoln wisely rejected. So McClernand would stay under Grant, in whose circle he would find a wall of hostility. On January 30 Grant assumed direct command of all the forces dedicated to the effort against Vicksburg, effectively clipping McClernand's wings and reducing him to the function of corps commander, on a par with Sherman.

Infighting of this sort has always been an all too common feature in the circles of high command. What is of interest here is the role played by Sherman. In his *Memoirs* he cast himself as the author of the Post of Arkansas project, proposing it to McClernand, who made "various objections." Sherman added that he didn't think McClernand had "any definite views or plans of action" of his own, but only spoke grandiloquently of "cutting his way to the sea." But McClernand was well-aware of the Post of Arkansas as a potential objective—he had discussed it with Federal officers at a stop in Helena on the way down. What is more, in Porter's version of the meeting aboard the *Black Hawk* it was McClernand, not Sherman, who proposed the operation.

Nor is it possible to credit the account of either Sherman or Porter that during the attack McClernand was simply a timid spectator. Recent research indicates that he carried out a preliminary reconnaissance of the Confederate positions personally, that he drafted the plan for the assault and supervised its execution. He was entitled to a share of whatever laurels there were. However modest McClernand's talents—and he probably had no business commanding an army in the field—he did not simply bring up the rear at the Post of Arkansas.

McClernand's comportment in this whole affair is of no particular relevance here, but that of Sherman is highly revelatory. His defeat at Chickasaw Bayou—and it was incontestably a defeat—then his replacement by, and subordination to, a rank, obnoxious amateur such as McClernand, were severe blows to an ego that had already taken considerable battering in Kentucky and in the hue and cry that followed Shiloh. Ellen, who had seen him through that crisis just a year ago, now feared the worst.

This time calamity did not plunge him so deep into despair; only once or twice did he mention to Ellen that little corner of obscurity the two of them might seek out. This time his recovery was rapid, and the emotion driving him in those first months of 1863 was not so much shame as anger. To anyone who would listen he portrayed McClernand as a cowardly, posturing incompetent, in the process doing some violence to the truth. To Ellen, his most faithful and credulous correspondent, he related highly colored accounts of McClernand's all-around worthlessness. At Post of Arkansas the man had made a blunder and then had come running to beseech his subordinate's help: "Sherman, what shall we do now?" At Shiloh "he showed the white feather . . . his troops reported to me for orders in his presence . . . he hung around me like a whipped cur."

Sherman also gave vent to his anger at Lincoln, writing John that he had a message for the president: "If you see Mr. Lincoln tell him I appreciate the reproof in his ordering McClernand to command me, and that I will consider it an honor when he says he can spare my services." Now he might indeed quit—"slide out," he called it—or so he wrote several times to Ellen and John, but it would not be in shame and humiliation—and he would pick his own time if he did. He thought he would at least wait to see the Mississippi opened—unless McClernand replaced Grant, in which case he would be on the next train to Lancaster. By the time spring came he was no longer writing seriously about leaving the army.

At the same time his letters reveal considerable effort at self-justification, at reaffirming the correctness of his actions at Chickasaw Bayou. He decided that he could have taken Vicksburg if only Grant had carried out his part of the plan, "but it was never dreamed by me that I could take the place alone." As it was, he had performed an important service: "I was on time, made a sufficient attack to draw their troops from Bragg, Pemberton and even Virginia, relieving those points to that extent." When in late May he had the occasion to revisit his old battleground at Chickasaw Bayou, he hastened to write John the good news: "the correctness of every step I then took is now verified by actual possession." As for the Post of Arkansas, "not a soldier in this army, an officer of the Navy who handles a gunboat but knows who planned all."

In a long letter to a friend from California days he explained that, appearances to the contrary, he occupied a significant place in the army:

"I am still unambitious in this war, bandied about as all sorts of fool, ass, and crazy by our 'special correspondents on the spot.' Still I find myself always in advance; where danger threatens and others despair, they lean on Sherman." And he lectured Ellen in the same vein: "I know one fact well: that when danger is present or important steps are necessary, Sherman is invariably called on."

His relations with his chief critic, the press, underwent something of a change in this same period, though hardly for the better. He was following his usual habit of reading every newspaper that came his way. He was clipping out articles concerning him in some way—"slips," he called them—and sending them in batches to Ellen for her "archives." He found much there that was unpleasant reading: "Complete Repulse of Sherman" for example, the *Indianapolis Daily Journal*'s comment on Chickasaw Bayou, or its word portrait of General McClernand, lauding his "chivalric bearing and courteous manners."

Here too Sherman went on the offensive. "Somebody must make the issue," he wrote Ellen, "and I cannot avoid it." He drew up a proposed press policy, a "rule" according to which all articles proposed for publication by journalists with the army would have to be examined and approved by "an officer specially appointed for that purpose," and "none containing praise or blame should be allowed." He tried without success to get the *Cincinnati Commercial* to publish this proposal. To give the scheme added weight he sent it to Thomas Ewing to deliver in person to Stanton, along with Ewing's endorsement. Ewing did so, and that was the last ever heard of it. Sherman wrote his old friend Henry Turner, then in St. Louis, to protest about a letter critical of him published in the *Missouri Republican*. He contacted St. Louis attorney Thomas Gantt saying he wanted to sue the paper for libel. Gantt wrote back advising against it. The case would not be tried for at least a year; besides, Missouri juries made light of such suits for libel. Sherman should regard a press attack as a compliment.

Among the correspondents who sailed on the Chickasaw Bayou expedition had been Thomas Knox of the *New York Herald*, masquerading as an assistant steward. After his lengthy reportage on the fiasco somehow vanished in the mails, he wrote a second draft; after the expedition returned he personally took it to Cairo to mail, and it appeared in the newspaper's January 18 issue. Knox had come ashore on Chickasaw

Bayou and conducted hasty, scattered interviews with anybody he could catch, mostly underlings with little knowledge, to whom the actions of those above them were baffling; the result was a picture of widespread confusion in which an "exceedingly erratic" Sherman figured prominently, issuing contradictory orders or no orders at all, producing a communications breakdown that left his subordinates in the dark and revived talk among them about his sanity.

Knox reported that after the failed assault of December 29 the battlefield was littered with Union casualties, but Sherman at first refused to arrange a truce to succor the wounded and inter the dead. This was true, for General Morgan made the same charge some years later, claiming Sherman was afraid to let the enemy see his heavy losses; by the time he relented most of the surviving wounded were in Confederate hands.

Knox's article was soon in Sherman's hands; so was Knox himself, for he had remained with Sherman's forces. Sherman had him charged with espionage and other offenses, and then had Grant send him before a court-martial with the intention of having him executed if possible. Sherman claimed, with no little exaggeration, that the reporter's execution would "save ten thousand lives." But the court would not go that far, sentencing the reporter to be banished from the army, on pain of arrest and imprisonment if he returned, and Lincoln reduced even that punishment.

But Sherman learned something valuable from this affair. In the course of his questioning of Knox the reporter had told him in candor: "I have no feeling against you personally, but you are regarded as the enemy of our set, and we must in self-defense write you down." Here was an explanation—and a flattering one—for why he drew the thunderbolts of the press: they assailed him not because they had any real quarrel with his conduct in the field but because he stood up against them—a theme he would expound on in coming battles with the fourth estate, quoting Knox to make his point.

And to this idea there was an intriguing corollary. Grilling Knox, Sherman learned he had had many conversations with Frank Blair on board the *Continental*. Blair, the politician turned soldier, was now the commander of one of Sherman's brigades. Sherman drafted an inquisitorial letter with twenty-two questions for Blair to answer. There was nothing in the responses that might derail Blair's military career, but to Sherman a kind of collusion was obvious. One hand washed the other: Blair catered to the rabble-rousing press, which in return sang his praises, writing him "up" while they wrote Sherman "down," each of the partners benefiting from this sordid arrangement.

So he explained it to Ellen: "Frank Blair and his ilk fought not for the real glory & success of the nation but for their own individual aggrandizement." McClernand too was of that ilk: "his master is Illinois and personal notoriety." As for patriotic men who served selflessly in the interests of the nation, their achievements were either ignored or "written down." Sherman was of this honorable if unsung group: "I prefer to serve the whole United States and to check the gnawing and craving appetite for personal fame and notoriety which has brought our people into a just contempt with Foreign nations." Thomas Gantt was right: he should take press accusations as a sign of his own merit. He took to making ironic comments to that effect in letters to his friends: he must be doing well, since the reporters were after him again. A good general who did his duty would need to develop a thick hide. But Sherman never did; he felt every pinprick.

The struggle to take Vicksburg went on without any perceptible progress; Sherman was now once again subordinate to Grant, whose divisions had been grouped into corps; three of these corps, the Thirteenth under McClernand, the Fifteenth under Sherman, and the Seventeenth under James Birdseye McPherson, would do the bulk of the fighting at Vicksburg. When Grant resumed direct command the army was concentrated in and around Young's Point and Milliken's Bend, Louisiana, just upriver from Vicksburg. Grant's immediate problem was that his army was on the wrong side of the river, but beyond that lay another problem that would take a long time to solve: just how to get at Vicksburg.

The subject was much discussed at Grant's headquarters. Even the most straightforward and heroic of approaches was briefly considered: a cross-river assault, with troops coming ashore on the town's waterfront, escalading the bluffs, and storming the enemy positions that crowned them. Here was general agreement that the chances of failure would be astronomically high, and so would the casualties. At the other extreme was the slow, roundabout route that Grant had started on back in December, before the Holly Springs fiasco. It was in accord with the principles laid down by

Jomini; it had the endorsement of Halleck, and it had the support of McPherson and Sherman, who quoted Jomini in his arguments.

But others at Grant's headquarters were of a different mind: Lieutenant Colonel James H. Wilson, who had joined Grant's staff as a specialist in engineering; John A. Rawlins, a prewar friend of Grant who served as his adjutant general despite his lack of military background; and Charles A. Dana, a prominent journalist serving as assistant secretary of war, Stanton's personal representative and, as Sherman put it, "accredited spy." This group took the larger view. If the war was being fought in a kind of vacuum, where only strictly military considerations counted—numbers, terrain, logistics, and so on—then Sherman's solution was probably the most sensible. But there was another aspect, and an important one. Sherman's "Oxford Plan," as he called it, would mean packing up, marching away from Vicksburg, and returning to square one. In military terms it might be described as a change of base, but in layman's language it was retreat; given the manifest importance of Vicksburg, the impact on the Congress, the press, and the people could be great. It might wreck Grant's career, though no one mentioned that in his presence.

Grant tended to be a simple listener when these two strategies were being discussed. His own preference may have been impelled as much by natural inclination as by any arguments he heard. He wrote afterward: "One of my superstitions had always been when I started to go anywhere or to do anything, not to turn back, or stop until the thing intended was accomplished."

Between the end of January and the beginning of April Grant made several unsuccessful attempts to move his army around to high ground east of Vicksburg; Porter had his gunboats probe the complex waterways around Vicksburg extensively, putting them at considerable hazard in the process; on one occasion Sherman and his men had to come to their rescue at night, moving through dense forest by candlelight. Each of the routes explored was found to be effectively blocked either by Confederate forces or natural obstacles.

As early as January 1863 an intriguing idea had surfaced: Grant should simply pack up and move his land and naval forces downriver, well south of Vicksburg. Wilson, Rawlins, and Dana were partisans of the idea, which called for some derring-do: the gunboats and transports would have to run past the Vicksburg batteries. The troops would move below by a land route and their materiel by a canal roughly paralleling it. Once well below Vicksburg, Porter's vessels could easily ferry the army across the river. Grant listened to the arguments for and against—and began allocating manpower to the road and canal projects.

Early in April Sherman sent Grant a long letter urging him to ask his subordinates for their ideas, and outlining in detail a version of the "Oxford Plan"; Grant made no response, at least none that is recorded. From Sherman's letter to Grant and from letters he wrote John and Ellen, one might compile a considerable list of reasons why reverting to the original or "Oxford" plan was preferable to the shift of forces southward, but they could be distilled into a single argument: the one was sure and natural, the other was difficult and "hazardous in the extreme." Whether Sherman really thought he could sway Grant is debatable, but in any event he put himself on record as opposing the shift of forces down the Mississippi in the event it ended badly.

Eventually Union engineers opened a fairly direct water route for shallow-draft vessels by connecting natural bodies of water with canals, and they improvised a roadway for men, artillery, and trains. Porter's gunboats and transports ran the gauntlet past Vicksburg's batteries—an operation Sherman called "one of the most dangerous in war." To reduce risks the vessels made the attempt in two groups, both under cover of darkness. The first contingent sailed on the night of April 16. They were detected and the Confederate batteries opened a spectacular show of pyrotechnics; to light up the river the Confederates set fire to houses along the bluffs. According to Assistant Secretary Dana's count the rebel artillery fired 525 rounds. All the vessels but one made it through; a few nights later the second contingent ran the gauntlet with moderate loss.

By then Grant's army was marching south, McClernand's corps was in the van, followed by McPherson and his troops; Sherman would to bring up the rear—hardly the post of honor, but he confided to Ellen "I don't object to this, for I have no faith in the whole plan." Before departing, Sherman's troops could render service by making a "demonstration" to distract Pemberton, and for that purpose he went back up the Yazoo and landed troops briefly, then disengaged after some skirmishing.

Grant's complicated movement did not go smoothly, but he persisted, and by May 7 he had brought his army well south of Vicksburg and had established it on the eastern bank of the river below Grand Gulf. Once there, he found himself in a quandary. General Banks was mired in operations of his own and unable to join forces with him; on the other hand Confederate General Joseph E. Johnston was busy concentrating troops at Jackson, forty miles east of Vicksburg, so that Grant risked being caught between two fires. He also faced a serious logistical problem: he was at the end of a long and tenuous supply line stretching down from his base at Milliken's Bend; if he advanced into the interior of Mississippi he would stretch the line further and expose it to attack by the redoubtable Confederate cavalry, whose handiwork he had seen at Holly Springs. Sherman warned one of his brigadiers: "I apprehend great difficulty in the matter of food." To Ellen he confided: "I see on the horizon the first clouds that threaten Grant's fair fame . . . "

But here Grant made what has been acclaimed as one of the boldest decisions of the war. He resolved to campaign without a supply line, subsisting his men almost exclusively on the food they could find in the countryside. Other generals on campaign had been forced to this expedient in the short term for sheer survival, but to convert what was a temporary expedient into the basis for an extensive campaign by a 40,000-man army deep in the enemy's territory—that was an intrepid act.

Grant had Sherman bring up 120 wagons loaded with coffee, sugar, and salt and fallback rations of bacon and hardtack; each regiment took with it two wagons loaded with provisions and one with munitions. Fortunately Grant was campaigning in a favorable time of the year, through what was after all a rich agricultural area. The army thrived. Sherman noted that men who went ten days with two days of army rations now gained weight in the process. Soon he found them spurning army rations altogether for "the fruits of the country."

Now mobile and with no supply line to worry about, Grant accompanied Sherman with part of his army to take Mississippi's capital and drive off Joseph E. Johnston's army; then with the other he defeated Pemberton at Champion Hill and moved on Vicksburg. There, with his army united on good ground, he could conduct a proper siege. He made two attempts to storm Vicksburg; both failed, but at least gave him a reason to pack off the incompetent McClernand. The siege would be memorable chiefly for the hardships that the town's inhabitants endured; the besiegers heard stories of cats and rats being offered in Vicksburg's butcher shops.

Then on July 3 there was an interruption in the heavy pounding by Federal artillery; officers of high rank appeared at the picket lines and there were unaccustomed contacts with the enemy. Rumors raced through the army and were soon confirmed: Pemberton was going to surrender an army of nearly 30,000 men and the city of Vicksburg. The surrender took place at ten o'clock on the morning of July 4. At that same hour, 900 miles to the northeast, Robert E. Lee's army was packing up for the march back to Virginia, leaving behind the battlefield of Gettysburg, including a gentle slope–still strewn with the debris of battle–that tour guides would point out to later generations as "the high water mark of the Confederacy."

Sherman was not present at the surrender ceremony, for Grant had ordered him to deal with a renewed threat from Joseph E. Johnston's army at Jackson. He told Ellen that he did not relish the assignment, but to Grant he gave the response of a good soldier if not an enthusiastic one: "I did want rest, but ask nothing until the Mississippi River is ours." Grant, for his part, tried to lighten the burden imposed on his friend by giving him a free hand: "I have no suggestions or orders to give. I want you to drive Johnston out in your own way, and inflict on the enemy all the punishment you can."

Sherman advanced toward Jackson in a period of heat so intense that his force switched to night marches. Water was scarce; many of the streams were polluted with the carcasses of dead animals. Sherman reached Jackson on July 9 to find that Johnston had once again fortified it. He made no assault but shelled the town with three cannon, firing every five minutes, day and night.

On the night of July 16–17 Johnston's forces withdrew east. Thereupon Sherman occupied Jackson and dedicated his efforts chiefly to wrecking the rail lines radiating out of it. Beyond sending a force to monitor Johnston's retreat for a dozen miles or so he took no steps to pursue the enemy host. Concerned over the security of his line of communications to Vicksburg, he asked Grant for another division to cover it.

Halleck was following the situation as well as he could; the view in Washington was that Johnston and his army should be "disposed of."

Grant had given him assurance on the day Vicksburg fell that Sherman would "drive Johnston from the state." Yet Grant could not or would not send Sherman orders to that effect. Instead he sent suggestions. A flurry of messages passed between Vicksburg and Jackson. On July 17 Grant wired: "If Johnston is pursued, would it not have the effect to make him abandon much of his trains and many of his men to desert?" But that same day he sent a second message: Sherman could stay where he was and hold Jackson—and then a third telling him to return when he wished. Sherman clearly had no desire to pursue Johnston or linger in Jackson, so he headed back toward Vicksburg, probably stirring some criticism among the Young Turks in Grant's circle.

Was the criticism justified? Impossible to say with surety, but Sherman must have expected it, for he took measures to answer it. On the evening of July 17 he wired Grant three times within the space of an hour and a half, giving six different reasons that he would not pursue Johnston or remain in Jackson: extreme heat, dust, fatigue, casualties, officers "determined on furloughs," and the condition of General W. S. Smith, who was "really quite ill." He had taken to writing Porter about actions he took that might be questioned—in the Knox affair, for example—for Porter could always be counted on for a sympathetic ear and a vigorous word of support. Now he wrote the admiral again, explaining that he would have "ruined" his command if he had pursued Johnston further: "On a review I think I have fulfilled all that could have then been reasonably expected." And Sherman made an unusual request of Ord, whose corps had come with him to Jackson: "As explanatory of the reasons for not pursuing and pushing Johnston's retreating forces, I am desirous that you should reduce to writing the substance of what you said yesterday about the physical condition of your corps."

But there would be no rancorous charges or recriminations after the fall of Vicksburg. The great prize had been won; now the mood in the army was one of both elation and exhaustion.

# 14

## CHATTANOOGA

MOVING BACK TOWARD VICKSBURG, Sherman and his exhausted, footsore troops went into camp along the Big Black. The next few weeks marked a striking change, an interlude of perfect peace and repose after a long, hard campaign. During this respite a cascade of good news arrived at Sherman's headquarters: Halleck wrote to tell him that at Grant's urging, he and McPherson had both been made brigadier generals in the regular army. Halleck was warm in his praise: "If you continue, my dear General, as you have begun, no one at the end of the war will have a more brilliant record." Sherman's old West Point teacher, Dennis Hart Mahan, wrote to compliment him on Vicksburg: "European warfare can produce nothing equal to it since the 1st Napoleon."

In his camp Sherman now played the genial host. He wanted visitors and issued invitations left and right. He invited John, who could not get away, but Ellen came with the four oldest children. Sherman installed them in tents pitched in a grove of enormous oaks; the encampment on the Big Black took on the air of an endless picnic. There was a Ewing family reunion of sorts, for Charles had become Sherman's inspector general and Hugh led one of his divisions. A battalion of the Thirteenth U.S. Infantry was nearby; Willie had been made an honorary sergeant in the regiment and Tom an honorary corporal, and the children often visited there. Minnie and Willie frequently rode with their father to drills and reviews.

This idyllic interlude would shortly end in crisis and tragedy. The seeds of disaster sprouted to the east, on the Georgia-Tennessee border. There, in the mountainous terrain around Chattanooga, Sherman's old friend Braxton Bragg, commanding the Confederate Army of Tennessee, opposed General William Rosecrans's Army of the Cumberland. The two collided in the Battle of Chickamauga, a complex battle fought September 19 and 20, 1863. When it was over Rosecrans and his badly battered army took refuge in Chattanooga, where they faced surrender or slow starvation.

Sherman got first news of the disaster on September 22, when Grant asked him to send one of his divisions to Chattanooga, via Vicksburg. The next day he summoned Sherman himself, ordering him to Tennessee with two more divisions. Since Rosecrans could hardly maintain his own force on his slender line of supply, Sherman would rebuild the railway from Memphis east to Athens, Alabama, creating his own supply line as he advanced. In the camp on the Big Black all was movement; by the twenty-seventh all the troops were gone.

Ellen and the children were busy preparing to return to Lancaster. Just as Sherman was taking the family aboard ship for the trip upriver they noticed that Willie was sick. A physician with the Fifty-fifth Illinois saw signs of typhoid fever and was anxious to get the boy to Memphis where he could consult some colleagues. They reached Memphis on October 2 and carried Willie to the Gayoso. There his condition worsened and the attending physicians pronounced his case hopeless; Ellen bent over her son and told him that he was going to heaven. At five o'clock in the afternoon of October 3 Willie breathed his last. The scene was still vivid in his father's mind when he described it a dozen years later: "Mrs. Sherman, Minnie, Lizzie, and Tom were all with him at the time, and we all, helpless and overwhelmed, saw him die."

In death Willie Sherman received full military honors. Soldiers of the Thirteenth U.S. Infantry, rifles reversed in sign of mourning, escorted the casket from the Gayoso to the steamer *Grey Eagle*, bound for Cairo. Sherman saw Ellen and the children aboard, then threw himself into the complexities of moving an army across the country. He found another outlet for his grief in writing. He wrote Porter, Halleck, and any number of others. To Porter he confessed: "He was my pride and hope in life and his loss takes from me the great incentive to excel." In a letter to the battalion commander of the Thirteenth Infantry he laid out before a young captain his feelings of guilt: "Consistent with a sense of duty to my profession and office, I could not leave my post, and sent for the family to come to me in that fatal climate and in that sickly period of the year, and behold the result!"

Sherman spent a week in Memphis in ceaseless effort to find locomotives and cars to transport his force, then started east. By mid-October the work on the railroad was fairly begun and Sherman had progressed into northwestern Alabama. In Washington, meanwhile, decisions had been made that would produce important changes in the Federal high command. On October 16 Grant took over the newly created Military Division of the Mississippi, the theater-wide command Sherman had long favored; consequently Sherman would accede to command of the Army and Department of the Tennessee, while George H. Thomas would replace Rosecrans at the head of the Army of the Cumberland.

Grant was under great pressure to redress the military situation in Tennessee; not only was Rosecrans under semi-siege in Chattanooga, but Burnside was soon to be threatened in Knoxville by a Confederate force brought from Virginia by General James Longstreet; and for once the Confederate forces in a zone of operations were as numerous as the Federals—so Sherman and his troops would be needed. The message he received from Grant on October 27 was terse and imperative: "Drop all work on Memphis & Charleston Railroad, cross the Tennessee, and hurry eastwards with all possible dispatch toward Bridgeport, till you meet further orders from me." By November 1 Sherman was across the Tennessee; he took his army into Tennessee, then pushed on ahead of his men; when he reached Bridgeport, Alabama, some thirty miles from Chattanooga, he found another message from Grant, urging him to press on to Chattanooga for deliberations on the coming fight, for Grant was determined to give battle. Grant, Thomas, and others had already studied the terrain around Chattanooga and agreed upon a plan of action, but much would depend on Sherman, so Grant wanted his opinion.

He reached Chattanooga on the evening of the fourteenth; the next morning he joined Grant's party on a trip to Fort Wood. "From its parapet," Sherman recalled, "we had a magnificent view of the panorama. Lookout Mountain, with its rebel flags and batteries, stood out boldly, and an occasional shot fired toward Wauhatchee or Moccasin Point gave life to the scene . . . All along Missionary Ridge were the tents of the beleaguering force; the lines of trench from Lookout up toward the Chickamauga were plainly visible; and rebel sentinels, in a continuing chain, were walking their posts in plain view, not a thousand yards off."

Grant outlined the plan of attack, then Sherman and William F. "Baldy" Smith, Grant's chief engineer and author of the battle plan, went to another vantage point. As they faced east, Chattanooga was behind them, tucked into a pocket of land formed by a bend of the Tennessee. Stretching out before them, at a distance of perhaps two miles, was a long prominence called Missionary Ridge, running roughly from north to south. The most important part of this ridge was its northern end; there, near where Chickamauga Creek flowed into the Tennessee, was the vital rail line that brought Bragg his supplies and reinforcements; there also were the roads that afforded him connection with Longstreet to the east. This northern extremity of the Ridge was to be the primary objective of Sherman's forces.

He would be given a sizable force, over 20,000 men, and other support would be within reach if needed. He was to move his forces discreetly to the Tennessee north of town, then begin crossing it under cover of night. After he had got his entire force across and deployed, he would storm the northern end of the ridge, then sweep south along it, turning and driving the rebels before him. General Thomas, posted in front of Chattanooga, and General Hooker, south of the town near the Confederate stronghold of Lookout Mountain—clearly visible to the south—would help by tying down enemy forces in their sectors; at the proper time General Thomas's forces would join those of Sherman in driving the rebels. Smith asked Sherman if, under the conditions just outlined, he could carry out the plan. He recalled that Sherman answered with an emphatic yes.

The date was fixed: Sherman would have his troops at the crossing of the Tennessee by November 19, cross that night, and assault Missionary Ridge the following day. Sherman left Chattanooga that same day in order to "hurry up" his divisions; he traveled all night, reaching Bridgeport by dawn of the sixteenth.

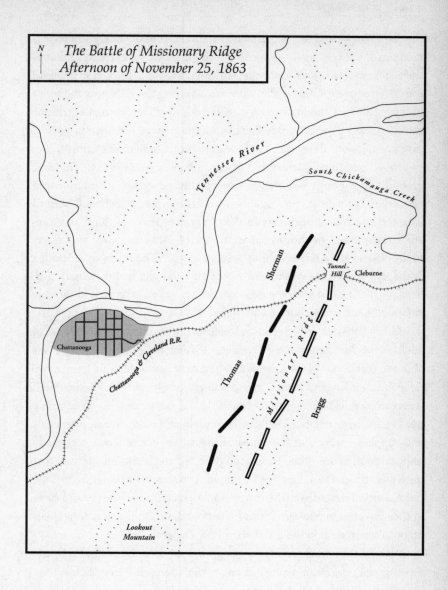

N

The Battle of Missionary Ridge
Afternoon of November 25, 1863

Tennessee River

South Chickamauga Creek

Sherman

Tunnel Hill

Cleburne

Chattanooga

Thomas

Missionary Ridge

Bragg

Chattanooga & Cleveland R.R.

Lookout Mountain

What happened to Sherman and his divisions in the next eight days is recounted in the general's *Memoirs* in a single sentence: "The condition of the roads was such, and the bridge at Brown's so frail, that it was not until the 23d that we got three of my divisions behind the hills near the point indicated above Chattanooga for crossing the river." If Sherman's divisions were to observe the timetable fixed by Grant, they would have to cover a distance of about thirty miles in four days, not a demanding effort under ordinary conditions. Here the conditions were extraordinary. The area around Chattanooga was rugged and even in peacetime its unimproved country roads were a challenge. Now the region had felt the full blast of war; the inhabitants driven away, their houses repeatedly plundered, livestock slaughtered, fields long since stripped of crops and fence rails. Sherman spoke of their route as "that desolate gorge." The roads were sodden and deeply rutted, kept that way by late fall rains and heavy military traffic. They were lined with the debris of war, wrecked wagons and the carcasses of innumerable horses and mules.

Given the state of the roads it is difficult to understand how Sherman could write McPherson on November 18 that his Fifteenth Corps was about to make "one of the longest and best marches of the war." Even more difficult to understand is the order of march for Sherman's divisions: each was accompanied by its wagon trains. This arrangement was commonly used and convenient, ensuring that each division would have its provisions and supplies handy. But the speed of the whole column was necessarily reduced to the pace of the trains, and over bad roads and hilly terrain that pace was reduced to a crawl. Where rapidity of movement was essential the trains were all relegated to the rear and left behind; the soldiers would carry in their haversacks cooked rations for several days. On the march to Chattanooga Sherman's divisions did not follow this practice.

Nor did they reach the rendezvous point for their river crossing by the time agreed upon; not a single division was there on November 19; the hindmost division, that commanded by General Charles R. Woods, was all the way back at Bridgeport. By then Grant had moved the date back, and he would be obliged to do so again—not until the evening of November 23 did Sherman have enough troops present to begin the crossing of the Tennessee. Grant's patience was sorely tried in the intervening days; James H. Wilson, now a brigadier general on Grant's staff, noted in his diary that his chief was having difficulty concealing his anger at Sherman. On the twenty-first Grant had Rawlins send him a terse, unequivocal order: "You will have your troops pass your transportation and move up at once." The next day, when he learned that General Woods was struggling along far behind schedule, he took the unusual step of bypassing Sherman and sending Woods a direct order to pass around the trains in front of him.

Sherman's failure to keep to the timetable could be partly attributed to what Von Clausewitz called "friction"–the myriad little delays, contretemps, and "glitches" that bedevil any complex human undertaking: General Osterhaus was absent from his division, and General Woods, his temporary replacement, could not act with the same surety and celerity; a courier went astray, so for want of a movement order Hugh Ewing's division marked time. But by far the most egregious blunder was that which chained Sherman's divisions to their lumbering trains. Charles A. Dana was at Grant's headquarters when the full implications became known: "Grant says the error is his; that he should have given Sherman explicit orders to leave his wagons behind, but I know that no one was so much astonished as Grant on learning they had not been left, even without such orders."

On November 23 Sherman sent Grant a letter of apology: "I need not express how I felt, that my troops should cause delay." According to Dana, Sherman did furnish some explanation, though there is no extant text from Sherman's hand to support it: Frank Blair had reported all of his command at Stevenson, Alabama, before they had arrived, and this led Sherman to make erroneous calculations. "But the fault of marching with trains Sherman attributes to himself," Dana noted, "Grant's orders that he should get all his troops here by Friday night having been positive, and it was his own duty to see that nothing hindered his arrival." Sherman's message of the twenty-third concluded with a solemn promise: "No cause on earth will induce me to ask for longer delay, and tonight at midnight we move."

He was almost as good as his word. At 2 A.M., November 24, his first troops were being rowed across the Tennessee not far from the northern extremity of Missionary Ridge. Their crossing was unopposed and apparently undetected. All morning the ferrying of troops continued, essentially unchallenged by the enemy, while "Baldy" Smith's engineers worked on a pontoon bridge. By early afternoon Sherman was solidly established on the east bank; there was still little reaction from the enemy, scattered picket firing and an occasional cannon shot; without much difficulty his men had pushed forward and taken up what seemed to be good positions on the ridge.

The day was overcast and the November afternoon drawing to a close. Around 3:30 or 4 P.M., in the fading light, Sherman called a halt. He believed he now held a good position on the northern extremity of Missionary Ridge, a springboard for the next day's task of driving the enemy southward. He still had troops crossing the Tennessee, and like any prudent general he saw to the consolidation of what he had taken; his men got orders to halt and entrench and spent much of the night at it.

Here he made a serious error. The terrain was more complex than his maps had indicated, a confusing jumble of knolls and depressions, slopes and gorges. His men did not yet hold Tunnel Hill, at the northern end of Missionary Ridge—it was separated from his forward positions by a deep valley or ravine. The Confederate force defending it was composed of a single under-strength Confederate brigade; had he persisted in his advance, with his advantage in numbers on hand—perhaps eight to one— he might have swept them away like chaff.

In Baldy Smith's plan the battle would open with Sherman's advance at dawn or shortly thereafter, but it was eight o'clock before they started. General Wilson, who had taken a place where he could watch Sherman's attack, was highly critical. As Wilson watched time pass he grew indignant at the seemingly endless delays. From Sherman's dilatory ways, from his "slow and ineffective" movements, Wilson drew this conclusion: "The simple fact is that Sherman, for all his brilliancy, was not the man for such bold and conclusive movements."

In his diary and in his memoirs Wilson was so often critical of Sherman in this period that one wonders if there was some sort of animus at work. But in the case of Sherman's conduct at Chattanooga there is other evidence. Historian Peter Cozzens has sifted it all for a meticulous dissection of the affair, and his criticisms echo faithfully the complaints of Wilson. Cozzens describes Sherman's comportment during the day of the twenty-fifth as marked by hesitation, by "uncommon trepidation," and by movements conducted "at a snail's pace."

When Sherman's troops renewed their advance they met rock-hard resistance; in the interim the defenders had been reinforced; moreover the dynamic, enterprising General Patrick Cleburne had taken command and had strengthened the Confederate defenses. The assaults of Sherman's troops were steadily repelled with considerable cost to the assailants. The struggle went on for most of the afternoon. General Carl Schurz, whose division had been sent to reinforce Sherman, reported to him about two-thirty and found the general watching his troops struggling unsuccessfully to mount a slope while the rebels above them poured on a galling fire. Schurz, who had never met Sherman before, received a cordial welcome, but Sherman's "unhappy frame of mind" over his predicament was very evident. Schurz recalled that "he gave vent to his feelings in language of astonishing vivacity."

Sherman's problems were visible to Grant, who was with General Thomas, confronting a section of Missionary Ridge to the south. With Sherman stalled and the battle plan not working, Grant improvised. He ordered Thomas to send his men forward. The order was given verbally, and no one recorded Grant's exact words. It has been widely believed that Grant wanted the men to drive the Confederates from their rifle pits at the base of the riddge, but not advance further—and that the men stormed and took it on their own initiative. Recent research indicates that orders to take the ridge had indeed been given. Further to the south General Hooker had succeeded in taking Lookout Mountain, leaving Braxton Bragg with only the option of saving his defeated army.

Grant was happy to have won a victory, even in unexpected form. That evening he wrote Sherman a remarkable letter: "No doubt you witnessed the handsome manner in which Thomas' troops carried Missionary Ridge this afternoon, and can feel a just pride, too, in the part taken by the forces under your command in taking first so much of the same range of hills, and then in attracting the attention of so many of the enemy as to make Thomas's part certain of success." Sherman gladly embraced this version of events. In his report on the battle he described waiting most of the twenty-fifth for the assault Thomas was to make on Missionary Ridge, having him-self done his part by drawing to his flank "vast masses of the enemy."

This view of the battle gained in authority in later years when Grant and Sherman took up the pen to write personal accounts of the war years destined for the American public. There both men clearly tinkered with the truth in relating various episodes in the campaign for Chattanooga, but Grant's efforts were the more deft: he generously wrote that Sherman's march to Chattanooga "was conducted with as much expedition as the roads and season would admit of"; as for his own order to have Thomas advance his troops to the base of Seminary Ridge, a discreet footnote told the reader that the move was "preparatory to carrying the ridge," which indicates it was his intention all the time.

In his *Memoirs* Sherman offered a bold rewrite of the battle plan for November 25: "The object of General Hooker's and my attacks on the extreme flanks of Bragg's position was to disturb him to such an extent that he naturally would detach from his centre as against us, so that Thomas's army could break through his centre. The whole plan succeeded admirably . . . "

If this was the plan, then it made no difference that Sherman's men had not been able to drive the rebels from the northern extremity of Missionary Ridge. But it made a difference to Sherman, for in an artfully worded passage in his official account of the battle he undertook to refute "the report . . . that we were repulsed on the left." "Not so," he explained, "the real attacking columns of Gen. Corse, Col. Loomis, and Gen. Smith were not repulsed. They engaged in a close struggle all day, persistently, stubbornly, and well." They just didn't advance.

The days following the battle Sherman was not there to enjoy the afterglow of their success. Here, as after Vicksburg, there were loose ends to be taken care of, in this case the relief of Knoxville, where Burnside was besieged and about to run out of rations. The job of relieving him fell to Sherman, who turned his columns northeastward. Grant sent James H. Wilson to accompany the movement–and perhaps to prod Sherman along. On November 30 Wilson noted in his diary "Gen'l Sherman not very fast on the move"; on December 4 he wrote: "Saw Sherman. Conversation about necessity of more rapid movement." But the next day the need for urgency ended: contact was established with Burnside, who reported that Longstreet's Confederates had abandoned the siege and departed.

In this instance Sherman probably made the movement as quickly as he could, given the condition of his command. The general and his men had been on the road for a month and a half, covering several hundred miles in that time, much of it on execrable roads in bad weather; on arrival at Chattanooga they had immediately been committed to the fighting there, then sent on to the relief of Burnside. They were now living off what they could find in the countryside, with uniforms in rags and shoes worn out. The men were also worn out; as for their leader, his energy seems for once to have been depleted. His correspondence was apparently affected; there are very few of his extant letters in this period, with a notable hiatus in those addressed to the two correspondents he wrote to most assiduously, Ellen and John. Ellen must have been much relieved to receive a brief telegram from her husband on December 8, even if in it he confessed, "Am pretty well worn out by hard marching and exposure."

If in the end things came out well, Sherman was not at his best in the fall campaign in Tennessee. One may suppose that Willie's death still haunted him, perhaps preoccupied him, robbing him of celerity and decisiveness in his actions. Such a distracted state might explain the gaffe over his trains on the march to Chattanooga, which otherwise is inexplicable in a general of his caliber and experience. But Sherman's performance during the whole year of 1863 is somewhat spotty.

The explanation may likely lie in his uncomfortable role as subordinate to Grant, despite the two men's frequent and public expressions of friendship and esteem for each other. While it is true that Sherman's personality meshed easily with Grant's, the same could not be said of the two men's ideas about how to make war. While Sherman always said positive things about Grant in this period, he was careful not to attribute to him superior generalship. The day after Vicksburg surrendered, when Grant might have merited such an accolade, Sherman wrote Ellen: "Well, I thank God we are far from Washington and that we have in Grant not a great man or a hero, but a good, plain, sensible, kindhearted man."

Sherman's letter to Grant on the occasion of Vicksburg's fall is more remarkable still; He managed to extol the victory without complimenting its author: "Did I not know the honesty, modesty, and purity of your nature, I would be tempted to follow the examples of my standard enemies of the press, in indulging in wanton flattery." Instead, he had a word of advice for his friend: "As a man and soldier and ardent friend of yours, I warn you against the incense of flattery that will fill our land from one extreme to the other." Yet worked into this strange letter was an accolade of sorts for Sherman himself: "Though in the background, as ever I wish to be in civil war, I feel that I have labored some to secure this glorious result."

Even at West Point Grant had not been much of a student of warfare in the formal sense—Dennis Hart Mahan remembered him as "never a reader." In 1863 Sherman was still quoting from Jomini and talking about convergent lines of operations and such, and one suspects that like the bookish Halleck, he was troubled by Grant's sketchy acquaintance with and lack of concern for the "rules" supposedly followed by the great masters of warfare—and even more troubled by Grant's improvisations and casual, slapdash ways.

Then, while one might admire Grant's unconventional notions in the abstract, in practice they could get his subordinates in trouble. The ill-fated Chickasaw Bayou expedition was Grant's scheme, but it blew up in Sherman's face. In subsequent operations, the massive shift of forces south of Vicksburg, the advance into Mississippi, the march on Jackson in July, Sherman is a dutiful subordinate, but scarcely a hard-driving, enthusiastic, and enterprising one. The impression is inescapable: for a good part of the Vicksburg campaign the man's heart was not completely in his work and he was not at his best.

Nor is his performance better in the fall campaign in Tennessee. The snaillike progress of his needlessly encumbered columns toward Chattanooga was not at all typical of the man; nor was his lack of dispatch and perseverance in committing his troops against Missionary Ridge on November 24 and 25. There is some evidence that these accumulated shortcomings—for such they can be labeled—may for a brief time have once again shaken Sherman's confidence in himself. At the end of the day's fighting on November 25 he did something completely uncharacteristic, something he later claimed he had never done in his entire career: he called a council of war. There he polled his subordinates about the course to follow now that Bragg was beaten. Hugh Ewing and one other general were for immediate and vigorous pursuit; all the others present, Sherman included, were opposed. Sherman's inaction may have made it easier for Bragg to escape into Georgia with his battered army intact.

If Grant felt disappointment over Sherman's performance he kept it to himself. But on one occasion at the end of 1863 he seems to have indicated that his subordinate and friend had his limitations. In late December Lincoln, Stanton, and Halleck were considering a possible replacement for General George Meade, head of the Army of the Potomac (in the end, Meade kept his job). His was widely considered the premier army command on the Union side, and also the most challenging, involving operations in the critical hundred miles separating Washington and Richmond and contending with the Confederacy's best army and its most renowned leader. If there was a general who could determine the fate of the Union in an afternoon, it was the commander of the Army of the Potomac.

Through Dana, who was going back and forth between Washington and the Western Army, the president and his associates sought Grant's suggestions. Grant proposed two men, Sherman and Baldy Smith. Pressed to give his preference, he confided to Dana that he would choose Smith. That was also the view in Washington, as Dana later wrote him: "The President, the Secretary of War, and General Halleck agree with you in thinking that it would be, on the whole, much better to select him than Sherman."

The much-touted comradeship and collaboration between Grant and Sherman did have a basis in fact, but it would function best after Grant went East early in 1864 and left Sherman largely his own master in the West.

# 15

## GOOD WAR, BAD WAR

THE UNION VICTORY AT Chattanooga was the last major event of 1863, bringing to a close the pivotal year of the war. It had opened with reverberations from the Union defeat at Fredericksburg; the spring brought another major setback in the east, at Chancellorsville, while Grant seemed mired before Vicksburg. But with mid-year fortune seemed to change camps, bringing the simultaneous successes at Vicksburg and Gettysburg; now the solid victory at Missionary Ridge provided a springboard for the invasion of Georgia. To officers in Grant's army who were of reflective bent, the war seemed to be finally going somewhere, taking the right direction and acquiring a certain momentum.

But the war had been evolving in another way too, departing from the image presented in the antebellum classrooms of West Point. There Henry Halleck, Dennis Hart Mahan, and other instructors had demonstrated to future officers that war was an art that had undergone great refinement over the centuries. Waged in an enlightened manner, a trial by arms could resolve disputes that had defied peaceful solution, settling them quickly and without the wholesale destruction or indiscriminate violence and slaughter of past ages. And one could cite several examples of this "good war" in other parts of the world in the 1850s and 1860s, three of them in Europe alone.

In some ways the great contest of 1861–65 conformed to this image; officially the two belligerents made distinctions between combatants and noncombatants, though admittedly there were some gray areas—not uncommon in civil wars. The war was overlong and the slaughter great, excessively so; but it was essentially confined to soldiers, and they died in what Sherman liked to call "open, manly warfare." And with all the butchery there was a concern for propriety, a persisting civility. Not without some reason did Winston Churchill call the conflict the last gentlemen's war.

In keeping with his legal interest, Sherman was particularly well-versed in the laws and usages of war. As early as 1846 he had his own copy of Emmerich de Vattel's *The Law of Nations*, then the standard work on the laws of war. When Grant proposed taking the parole of some Southern civilians it was Sherman who reminded him that by long custom the practice was limited only to warriors, whose word was guaranteed by their soldierly honor; and it was Sherman who pointed out that Confederate General Stephen D. Lee, who had given his parole at Vicksburg and then assumed another command, was in violation of "the laws of capitulation."

Sherman's concern for doing the honorable, soldierly thing went beyond matters of punctilio. He corresponded with Confederate General John Pemberton in an effort to end the tit-for-tat killings that were part of a feud between Union cavalry stationed in Memphis and Confederate horsemen around Hernando, Mississippi. In the late summer of 1863 he exchanged letters frequently with General W. H. Jackson, CSA; the two established a demilitarized zone in the area between the Big Black River and the Mississippi Central Railroad. In Mississippi he took a number of humanitarian actions: after occupying Jackson he provided several thousand rations to the state insane asylum, and provided food for the inhabitants of Clinton, Mississippi, as well. "As representatives of the United States," he explained to his subordinates, "we cannot permit people to perish for want of food if we have any, and the fewer conditions imposed the purer the charity."

This man whom several generations of Southerners would regard as the prince of plunderers found the practice personally abhorrent. Asked by an official of the Western Sanitary Commission to provide some "war trophies" for a benefit auction, he refused, saying the request "embarrassed" him; by law all items taken from the enemy were government property. When some of his men came to present to him a rebel flag they had taken, he at first refused, then decided he could accept it: since the Confederacy had no legal existence, the flag had no proper owner (he let the flag be auctioned off by one of Ellen's charities).

If Sherman shunned "trophies" and other fruits of war, from the outset his men were of a different mind. In the first weeks of fighting "bad war" made its appearance. The government had hoped that a Federal military presence might stimulate the loyalty of many Southerners, but the misbehavior of its soldiers produced just the opposite effect. After Bull Run Sherman concluded, "henceforth we ought never to hope for any friends in Virginia."

To some degree the army itself was at fault; its field rations, designed for ease of transportation and conservation, were reasonably palatable but monotonous; soldiers inevitably looked for variety in the gardens, smokehouses, and henhouses they encountered. Then the army did not use field kitchens; the men built their own cooking fires, taking whatever wood was at hand—usually fence rails and occasionally planks pulled off buildings. From habitations they might take food, other articles of convenience—cooking pots, plates, bedding, and the like—and things that simply struck their fancy.

Some argued that these and even worse transgressions, vandalism and arson, were the work of a few "rotten apples," but Sherman put the blame squarely on the mass of volunteer soldiers. When the secretary of war wrote him about the Eighth Missouri Infantry, which was suspected of several cases of arson in Memphis, he answered that the men of the Eighth, "like all our volunteer soldiers, have been taught that the people, the masses, the Majority are 'King' & can do no wrong. They are no worse than other volunteers, all of whom come to us filled with the popular idea that they must enact war, that they must clear out the Secesh, that they must waste and not protect their property, must burn, waste and destroy." By the end of 1863, on the eve of those campaigns that would bring Sherman and his army both fame and infamy, his troops—and all troops—were thoroughly acquainted with "bad war" in its many manifestations.

It was the great mass of these volunteers who gave the armies of both sides their personality and their mindset. And as it became apparent that their behavior could not be changed without provoking wholesale mutiny

or worse, those who commanded them–Sherman included–had ultimately to accept them for what they were. By 1865 the only one of Sherman's army and corps commanders who was still fighting theft, pillage,
and vandalism with undiminished zeal was the saintly General Oliver Otis
Howard.

As for Sherman, his order books and correspondence indicate that for
the first two and a half years of the war, into the fall of 1863, he personally
carried on a serious and sustained effort to rein in the predatory and
destructive tendencies of his soldiers. He harangued his troops verbally
and issued written appeals to their honor and patriotism as well as threats
of hard labor and summary execution. He had the stern penalties and pro-
hibitions of the Articles of War read to them once a month. He tried having roll call three times a day, with severe punishment for unauthorized
absence. He set up mounted patrols around his camps and authorized
them to shoot down pillagers. He ordered surprise inspections of his entire
command, carried out with the rigor and thoroughness of a prison shake-
down. He requested and obtained authorization to deduct from the pay of
an entire regiment or brigade sufficient sums to compensate for property
damage done by one or more of its members. In October 1862 he made an
assessment of this sort on a cavalry regiment, noting as he did so, "I know
of no other way to stop the spoliation of private property wantonly and
wrongfully." Still, he confessed, "it is a herculean task to impress on our
men that whilst the U.S. may rightfully take anything, it is wrong for them
to clean out the Secesh." For him the sanction of the law made the difference between day and night; why others could not see that difference
always mystified him.

He continued his efforts long after he felt there was any hope of bringing
rebellious Southerners back into the fold by kind and generous treatment,
hopes he had largely abandoned by the end of 1861. Why did he persist as
long as he did? Because he believed–as did every military professional–that
the habits of marauding, theft and barn-burning would engender other forms
of indiscipline and convert an army into an ungovernable mob.

On July 24, 1863, on the march back from Jackson, his anger at the
destructiveness of his troops boiled over in a spectacular way. Someone in
the general's party called attention to a cotton gin that had just burst into
flames, then to a soldier who could be seen running away from it. Sherman

spurred ahead and stopped the soldier. He learned from him that he belonged to the Thirty-fifth Iowa and had been ordered to burn the gin. Pressing on, bypassing the column, Sherman caught up with the Thirty-fifth Iowa—and with Captain William B. Keeler, who admitted he had ordered the gin burned.

Sherman had him arrested on the spot. Normally officers had the privilege of placing themselves under arrest, but not in this case. Sherman had the captain surrender his sword and remove the emblems of rank from his uniform. He charged Keeler with arson, violation of a general order, and neglect of duty. He put him under guard and named a court to try him.

Sherman appeared before the court and gave his testimony, while General James Madison Tuttle, Keeler's division commander and a man of some mettle, appeared on the captain's behalf. Tuttle praised Keeler as a valuable officer; he said that cotton, gins and such were routinely burned along the line of march, and that he had heard Sherman say he preferred to see such things put to the torch rather than see cotton used in sordid speculation.

The court deliberated and found Keeler not guilty. In a written commentary on the case Sherman disapproved the verdict, which he could not overturn. He wrote Grant that Keeler had escaped punishment only because of a "volunteer court-martial, tainted with the technicalities of our old civil courts." He asked his chief to dismiss Keeler dishonorably from the service, that being the only way left to punish him. Grant made no reply, or at least none that is recorded. On Keeler's court-martial file in the National Archives there is the notation: "Findings and acquittal confirmed. By order of Maj. Gen'l. U.S. Grant."

While his regiments bivouacked behind the Big Black, Sherman worked on another project to combat indiscipline. He collected all the various laws and regulations he felt related to the soldier's comportment, among them the Third and Fourth Amendments to the Constitution, selected items from the Articles of War, and various acts of Congress; these were followed by a four-part exegesis from Sherman's hand. He had this compilation, styled General Orders, No. 65, printed up—in the original it runs to six pages of fine, close print—and read to each company at evening parade; like the Articles of War, it appears to have been read to the troops periodically.

The army that settled into its camps along the Big Black that August was exhausted and in need of recuperation; Sherman too came back tired—and more than that, disheartened. He wrote John that Grant had offered to assign him to Natchez, a charming river town that had come through the recent campaign intact, but he had declined. He did so because Natchez had not yet been looted by Union troops. He explained to John: "This country [the west bank of the Big Black] is cleaned out, and at Natchez we would have the same to go through . . . I am sick and tired of the plundering and pillaging that marks our progress."

It may well be that he was also tired of the fruitless efforts to curb his soldiers' predatory instincts, for General Orders, No. 65 proved to be his last major attempt to deal personally with the crimes and depredations being committed by his own troops. Two and a half months later, on October 24, 1863, he succeeded Grant as commander of the Army and Department of the Tennessee. His first directive to his new command instructed corps and post commanders to "maintain the best possible discipline, and repress all disorders, alarms and dangers in their reach"; in effect he passed those thankless chores to his subordinates. Five months later, when he took over command of the Military Division of the Mississippi, with its three armies and three departments, he distanced himself even more from involvement with everyday matters of discipline.

There was probably a conscious decision on his part to pass the burden to others; it would not be the only time he had passed responsibility in difficult or contentious matters over to subordinates. In any event he reserved the right to intervene; he did so in June 1864, with a special field order that was doubtless drafted on the spur of the moment, after several days of hard fighting during which he discovered alarming numbers of officers and men wandering about the army's rear areas with the apparent goal of staying out of the fighting. His Special Field Orders, No. 17 was aimed at those found "shirking, skulking and straggling" while fighting was going on. It is written in unusually violent language, even for Sherman: "The only proper fate of such miscreants is that they be shot as common enemies to their profession and country; and all officers and privates sent to arrest them will shoot them without mercy on the slightest impudence or resistance."

Five months later, on November 8 and 9, 1864, he issued Special Field

Orders, Nos. 119 and 120, containing guidelines on discipline and other matters in preparation for the March to the Sea—this was his last major pronouncement of the war on the subject of discipline; its implementation he would confide entirely to his subordinates.

By 1863 heavy-handed and relentless foraging had become a major manifestation of "bad war." In Kentucky Sherman had made strenuous efforts to limit the damage done. Well into 1862 he warned his subordinates that "looseness" in foraging operations could have bad effects on the soldiers, leading to pillage and general indiscipline that had "proved fatal to whole armies." Those charged with foraging operations were "to use all possible forbearance, explaining to the party the necessity, and giving a receipt for quantity and price, with a promise to pay at the pleasure of the U.S., on proof of loyalty." Later Sherman would look back in amusement at his own naiveté in those first campaigns: "I, poor innocent, would not let a soldier take a green apple or a fence rail to make a cup of coffee."

From the legal standpoint Sherman and Grant were on solid ground during the Vicksburg campaign in having their men live off the enemy population. Henry Halleck had written in his *Elements of Military Art and Science* that the practice was sanctioned by the laws of war, was widely used in Europe, and could give far greater "velocity" to an army than if it drew stocks from its own country, maintaining a chain of magazines behind it and extending that chain as it advanced. And if "regulated to repress pillage and . . . levied with uniformity and moderation," this system could be "relied upon with safety in more populated countries."

In Europe the system was well-known and used by all belligerents. The commander of the invading army made his wants known to local functionaries—mayors, provincial governors, and so on—and they arranged the collection of the supplies. This tax in kind was a feature of "good war," enabling the husbandman and the merchant to pursue their livelihoods in conditions of relative peace, while the occupying army's collection and distribution system functioned with little waste or disorder.

But in America this kind of imposition was virtually unknown. Then Southern officials, fearing charges of treason, fled or went into hiding at the arrival of the Yankees, so there was no authority with whom the invaders could treat; they had to go themselves directly to the source—the region's farms, mills, granaries, and so on—and take what they wanted. If the Northern armies were in movement, as was often the case, it was not practical for commissaries or quartermasters to try to amass supplies centrally and make proper distribution; instead, the army's brigades would send out foraging parties each day to find and take what they needed. This practice was wasteful and gave rise to any number of abuses and even crimes, as a predatory soldiery was repeatedly turned loose on the countryside. As a result by the fall of 1863 famine was appearing in what had been some of the richest agricultural regions of Mississippi.

By then Sherman was less condemnatory, for he discovered that heavy foraging in an area could have other, unanticipated advantages. He noticed that the passage of an army of any size had the effect of so reducing food supplies that no similar force could pass that way again anytime soon, for simple lack of sustenance; in the parlance of the day the region had been "eaten out." He saw a potentially useful weapon here. By late 1863 his army was in fact eating out sections of eastern Tennessee and it occurred to him that he could do this systematically and create a sort of buffer zone to shield his command. He talked about marching his army back and forth to create a "belt of devastation." It might be hard on the locals, but it had military utility, and to his mind was thus sanctioned by the laws of war. This became his general rule.

If the "eating out" were pushed far enough, the natives themselves would have no sustenance and would be obliged to flee. Here was a weapon the general did not hesitate to brandish, sending word to the population of one county or another that he would come and eat them out if, for example, they played host to partisans, or later to Confederate cavalry.

These partisans were a Southern contribution to "bad war," and one guaranteed to make the conflict more ruthless and indiscriminate. So active were partisans around Memphis in the late summer of 1862 that Sherman did not even attempt to reopen the rail lines: "The country is so full of guerrillas that it would be labor lost." When he reopened the telegraph line, the wires were cut almost every night.

Among the first units in the region organized under a Confederate statute of April 21, 1862, was the First Mississippi Partisan Rangers, who figure from time to time in the correspondence of Sherman and his lieutenants. This band was commanded by Colonel W. C. Falkner, whose exploits would provide inspiration for his great-grandson, William Faulkner.

Once launched, the partisan movement grew apace. In October 1862 the Confederate Congress created a local defense system using the stay-at-homes. Groups of twenty or more armed men over the age of forty-five could constitute themselves a "military company," electing their own officers and fixing their internal rules. They would then be considered—by the Confederate authorities at least—as belonging to the Provisional Army of the Confederate States, serving without pay or allowances. And according to the statute they would be "entitled, when captured by the enemy, to all the privileges of prisoners of war." As for subsistence, the partisans would simply "find it in the neighborhood," which meant they would take it from the same farmers and planters as the Yankees.

Sherman had serious problems with partisans when he took command in Memphis. They created various disorders in the countryside and occasionally clashed with his cavalry. The cavalrymen reported that they labored under considerable handicap because the inhabitants were either unable or unwilling to supply any information about the enemy horsemen. Not content with these incursions, the partisan bands began to attack river traffic. On September 23, 1862, the steamboat *Eugene* was fired on and nearly captured by band of thirty-odd partisans as it was making a stop at the village of Raleigh, not far from Memphis.

Ever sensitive about the Mississippi and its commerce, Sherman made a rapid and severe response. He ordered Raleigh burned, with one house left standing to mark the spot. Initially he had some hesitation about using this method, for he sent a letter of explanation to Grant: "I would not do random mischief or destruction, but so exposed are our frail boats that we must protect them by all the terrors by which we can surround such acts of vandalism."

Grant approved; confirmed in this "random mischief" policy, Sherman extended its use. To those who protested their innocence when the punishment fell on them the general had a response: "In war it is impossible to hunt up the actual perpetrators. Those who are banded together in a cause are held responsible for all the acts of their associates."

He assigned Colonel William Bissell to undertake many of these missions, supplying him with a well-equipped force, the steamboat *Crescent City*, and elaborate instructions: "Now the way to treat such fellows [partisans] is for every such outrage to land a force anywhere, proceed back a

few miles and burn the house, barn and fences of some known secession-
ist and leave word or notice for every outrage on the river retaliation will
be made ... "

While Bissell was in the neighborhood he was encouraged to correct
other injustices as well: "When any Union man is dispossessed of his property, force the neighbors to buy him out or be made prisoners 'til indem-
nity is made. When life is taken and the rightful party escapes punishment,
take five neighbors as hostages for the delivery of the murderers."

The general didn't seem to care who was picked out for punishment
in these expeditions; he once sent this directive: "Make somebody suffer
for the break in the railroad at Acworth; somebody over in the direction
of Roswell or McAfee's Bridge." This attitude reflected a basic conclusion
he reached while in Memphis: the entire white South, its "six and a half
millions," was now the enemy. In August 1862 he advised Treasury Secretary Salmon P. Chase that "the government of the United States may now
safely proceed on the proper rule that all in the South are enemies of all in
the North: and not only are they unfriendly, but all who can procure arms
now bear them as organized regiments or guerillas." The month following
he assured another correspondent, "The people of the whole South are
now on duty as soldiers."

The increased activity of Confederate partisans in Mississippi in 1863
seemed further proof of a mass mobilization on the part of the rebels. In
actuality, following the surrender of Vicksburg a goodly number of the
30,000 paroled Confederate soldiers opted to take service in partisan units,
though their legal status would be unsure in case they were captured.

Their legal status was clear enough to Sherman, judging from the
terms he was using to describe them: "bandits," "irregular cavalry out-
laws," "pests," "wandering Arabs," and "murderers under the assumed title
of Confederate soldiers." To the general they had no place in "open, manly
warfare," a point on which he was sure his Confederate counterparts
would agree with him: "Joe Johnston would never sanction such dogs as
call themselves guerillas in Kentucky," he wrote a friend from that state,
"nor would Lee or Bragg."

For a time he seems to have been unsure just how to proceed when
suspected partisans were apprehended; they were doubtful cases legally,
which for the time being he referred to higher authority. In a sense Sher-

man's hands were tied. The obvious penalty in these cases was death, but he was constrained by the requirement of getting presidential approval for executions. Early in 1864 he wrote Judge Advocate General Joseph Holt that he had a backlog of condemned spies and guerrillas who needed execution badly and that he would like to proceed with the business "without bothering the President." Such people should be "hung quick, of course after a trial." Holt should know that among Sherman's men, frustrated by these delays, there was a growing practice of "losing prisoners in the swamp." By the time Sherman wrote those lines (April 1864), he had begun writing orders suggesting he didn't want to be bothered by guerrilla prisoners.

Though the general was much occupied with the guerrilla problem, he refused to divert his army from more conventional military operations. He kept his forces concentrated, in the event they had to contend with an enemy army, which was after all their proper foe. He regarded fortified outposts scattered about to protect the countryside as so many signs of weakness; he called them "trading posts" and said they afforded the soldiers in them too many opportunities for plunder. If he learned a band of partisans was in a locality, he sent a strong cavalry force, a practice he recommended to General Stephen A. Hurlbut, who later succeeded him in command at Memphis. Hurlbut tried it but was unable to catch the partisans: "If I send cavalry they break up and scatter and my own cavalry commit depredations in following them."

But Sherman's goal now was not so much to catch the guerrillas as to teach their hosts a lesson, for it had become an article of faith with him that there was "always an understanding between these guerrillas and their friends who stayed at home"–the guerrillas stayed among willing hosts. This was proved to his satisfaction in November 1863 when a youthful clerk of his was seized by partisans near Florence, Alabama; the youth was released after Sherman threatened to punish three prominent men of the town in retaliation.

If partisans pursued by Federal cavalry should disperse and blend in with the local population, the officer leading the pursuers would be "perfectly justifiable in retaliating on the farmers among whom they mingle. It is not our wish or policy to destroy the farmers or their farms, but of course there is and must be a remedy for all evils. If the farmers in a neighborhood encourage or even permit in their midst a set of guerrillas they cannot escape the necessary consequences." They were culpable and therefore liable; they were, Sherman the legalist ruled, "accessories by their presence and inactivity to prevent murders and destruction of property."

Since in the general's view a locality harbored partisans by choice, they could be made to change their minds. If enough pressure could be brought to bear on the population—for example, by seizing prominent figures in the community as hostages and threatening to act against their persons and property—the community would make the partisans in its midst go somewhere else, or at the very least report them to Northern authorities. This seems highly unlikely; moreover General Hurlbut, who had considerable experience in these matters, was convinced such a policy would not work. It is not likely that he shared his conclusion with his chief, for it made all of Sherman's reasonings idle and vain: "the people of the country are more afraid of the guerrillas than of our troops, and therefor will not report them." Yet Sherman persisted in his notion: "They can suppress guerrillas, I know it." And he looked for further ways to pressure the population into doing just that.

In that summer and fall of 1863 a strange phenomenon appeared among the desperate population of Mississippi: meetings and petitions calling for the end of hostilities in the region by one means or another. A patrol sent out from Natchez in mid-July came back with the report that a large public meeting had been held in the community of Hamburg "to consider the question of abandoning the Confederacy." Grant wrote Washington that he was receiving petitions and delegations, and such movements were occurring throughout Mississippi. He felt they were genuine: "The people are ready to accept anything." Grant did what he could to encourage the movement, giving orders that troops moving through the state do so "in an orderly manner, abstaining from taking anything not absolutely necessary for their subsistence . . ." It was now his policy "to make as favorable an impression upon the people of the state as possible." Sherman's reaction was more skeptical. He wrote his friend Ord at the beginning of August, "I have in my hand papers with a long string of names, asking us both to stay away—the people are tired of war—they want a ration, a wagon, etc. The old Seminole story."

Halleck was intrigued by this popular movement, but far from the scene; he asked field commanders for their views and opinions. Would it be wise to set up a civil government in Mississippi and possibly elsewhere, subordinated for the time being to the military authority? Grant was inclined to favor the idea, but Sherman was emphatically opposed: the introduction of civil government anytime in the near future would be "simply ridiculous." He then entered into an analysis of Southern society for Halleck's benefit. There were the large planters, the ruling class, to whom, with certain restrictions, the control of affairs could eventually be confided. "I know we can manage this class," he assured Halleck. The small farmers, merchants, tradesmen, and so on he considered "hardly worth a thought"; they would follow wherever the planters led. The third class, the Unionists, he disdained for their lack of backbone in the secession crisis. "They allowed a clamorous set of demagogues to muzzle and drive them as a pack of curs . . . I account them as nothing in this great game of war."

The fourth class was essentially one of Sherman's invention. This element he called the "young bloods": hot-tempered, hard-riding young men possessed of good pedigree but with no interest in gainful employment. They had found life's ultimate pleasure in fighting on horseback: "War suits them, and the rascals are fine, brave riders, bold to rashness and dangerous subjects in every sense." He did not hesitate to call them "the best cavalry in the world." If the Federal government hoped for any kind of peace it would have to take these redoubtable cavaliers into its own service or kill them. The black element in the South's population, a third of the whole, Sherman did not take into consideration.

Halleck was apparently impressed by Sherman's arguments; for the time being the political status of Mississippians living between the Pearl and Big Black Rivers would remain that of a nominally hostile people in political limbo. The Union army continued to furnish food to the destitute east of the Big Black, but Sherman was quick to describe it as "pure charity to prevent suffering just as we would to the Indians on the frontier or to shipwrecked people."

In the period between two campaigns, from July to September 1863, Sherman was dealing extensively with civil populations. His communications and pronouncements are in style very similar to those he produced while in Memphis; there is quite a bit of scolding and moralizing, and as always with Sherman, reference to the law, or at least what he held was the law. In a long communication to the citizens of Warren County, Mississippi, he pointed out that the slave owners had lost title to their slaves "by rebelling against the only earthly power that insured them the possession of such property," and "*ex necessitate* the United States succeeds by act of war to the former lost title of the master."

While steadfastly refusing to involve himself in local affairs, he urged the people of Warren County to simply "make a government" themselves. "You have only to begin and form one precinct, then another: soon your county would have such organization that the military authorities would respect it." The example of one county would inspire others "by compound ratio," so that soon the whole state would be reorganized and strong enough to "defy any insurgent force."

Here lay the chief reason for Sherman's interest in letting local civil authorities go back into business—to combat the partisans. He considered such work beneath him and unsuited to his army. As he explained his policy to a confidant, "I will encourage the people to organize for self defense in their own way and let order come out of chaos in a natural manner. I think if the military authorities will confine their attention to military matters, that civilians will in due time attend to the rest."

At first sight Sherman's idea has a distinctly modern ring, for recent thought on counterinsurgency warfare has stressed wooing the population away from the guerrillas and encouraging them to resist their incursions. But Sherman did precious little wooing. The locals would have to do it all by themselves: set up a government, maintain order, and drive out the partisans in their midst without his assistance. He made no offers of arms, equipment, technical assistance, or other aid. And left to themselves, the desperate, distracted Mississippians could do nothing.

Conquest, by its very nature, implies the assumption of rule; the Union Army was the only force then capable of maintaining law and order and mounting anything like a credible, functioning civil administration in the areas where its armies operated. In the commands of some other Union generals the provost marshals were allowed to fill the vacuum, developing into rudimentary occupation governments. But Sherman refused to involve himself or his subordinates; when McPherson suggested they help the two races resurrect the region's agriculture, Sherman said it was none of their business. It was well-known in Washington that Sherman "simply ignored occupation responsibilities as much as he could." In so doing he needlessly condemned the people in his zone of operations to life in a Hobbesian state of nature.

When Confederate cavalry under Colonel Wirt Adams moved into the area around Clinton, Mississippi, in August 1863 and began collecting provisions, Sherman denounced their action as a "violation of all the rules of fair, open, manly warfare." As a measure of retaliation he stopped supplying rations to populations east of the Big Black, telling them they would have no more until they drove the rebel horsemen out. The inhabitants would have to "earn good treatment."

In September he was turning a deaf ear to the pleas of people in the destitute areas. "As long as their country is traversed by these bands we have no interest in them. I tell all I see that we don't care what they do." He boasted he could easily chase the enemy horsemen out, but "the rebels are serving our cause in making the people of Mississippi hate their rule." Then he changed course again; he replaced indifference with hostility, decreeing that in districts where the people tolerated guerrillas–or Confederate cavalry–"horses, mules, wagons, corn, forage etc. are all means of war and can be taken." "Punish the country well," ran another of his orders.

As the year 1863 closed Sherman was back on campaign in East Tennessee, showing increasing impatience with guerrilla bands he encountered and with the inhabitants of localities they frequented. When General Grenville Dodge passed through Pulaski, Tennessee, Sherman instructed him: "Let your mounted men hunt out the pests that infest that country. Show them no mercy, and if the people don't suppress guerrillas, tell them your orders are to treat the community as enemies."

He began to speak of treating civilians as combatants, to talk increasingly of "eating out" troublesome areas. He suggested that treatment for the region around the Elk River: "You will find abundance of meat and corn up and down the valley of Elk River, which use of freely, leaving barely enough for the inhabitants, and let them feel and know that by breaking our communications they force us to eat them out."

At the end of 1863 Sherman himself was virtually eating out the regions in Tennessee through which his army was passing: he wrote General George Crook: "I find plenty of corn, cattle, hogs, etc. on this route, but I don't think there will be much left after my army passes. I never saw such greedy rascals after chickens and fresh meat. I don't think I'll furnish them anything but salt." By late 1863, then, the general seems to have accepted in some fashion what he had formerly found intolerable: his men would never learn the obedience and the good conduct of the regulars, and on occasion their depredations could have a certain utility, as his policy toward civilians became more frankly punitive.

In that summer of 1863 Sherman again demonstrated his persistent obtuseness, his inability to understand the mindset of the Southern population. When he learned that former Governor Moore of Louisiana had been taken prisoner, he dashed off a letter to Thomas Ewing: Moore should be tried in a U.S. district court for theft of government property–those rifles he made Sherman store at the seminary. Moore would be found guilty and appeal to the Supreme Court; that august body, in confirming his conviction, would declare secession unconstitutional. Though the war had been going on for over two years, Sherman was convinced the court's decision "would still be greatly respected by thousands at the extreme South and by millions in the Middle or Border states." Nothing further was heard of this project.

January 1864 saw another attempt at psychological warfare, Sherman style. As early as 1862 the general had spoken of transplanting or deporting "young bloods" and other irreconcilables, and now it seemed to him that letting Southerners know what might be in the offing would have a sobering and salutary effect on them. So on January 31 he addressed his adjutant, R. M. Sawyer, on this subject. He wrote of how the British had expelled Irish troublemakers from Ireland and replaced them with Scots, and said a similar measure could be taken against those Southerners "who persist too long in hostility." Sherman invited Sawyer to show the letter around and even "assemble the inhabitants" to hear its contents. He sent John a copy, saying he wouldn't mind seeing it published, since "it is something new and it is true." John didn't think it was a good idea. But the letter got the wide circulation Sherman had hoped for. The *Nashville Times* published it and the Confederate authorities had it printed up in a broadside format addressed "To the Soldiers of the Army of Tennessee." It carried a commentary pointing out what they and their families might expect from the "the tender mercies of the infamous Sherman."

# 16

## ATLANTA

Though the year 1864 will always be associated with Sherman's campaign in Georgia, it was preceded by an overture of sorts—the Meridian "raid," which was launched at the beginning of February, bringing more destruction to the much-battered state of Mississippi as Sherman continued his campaign against the Southern rail system. He started from Vicksburg with 20,000 men, passed through Jackson—his third visit to the town—and after several skirmishes reached Meridian on February 14.

Though his campaign was marred by the failure of a cavalry-infantry rendezvous, Sherman's men wrecked the lines radiating out of Meridian and the facilities of the rail center itself; the general came away predicting that it would take the enemy a year to restore the damage done to their rail network. The raid was unusual in that it lasted a month and the troops involved were mostly infantry; it was the precursor of Sherman's great expedition to come.

The wrecking of Mississippi's rail lines meant it would be difficult if not impossible for the enemy to concentrate large numbers of troops there in the near future, thus reducing the danger to Federal strongholds such as Memphis and Vicksburg, and to the Mississippi itself. Now it would be easier for Grant's Western army to shift part of its strength to the east. He and Sherman had already talked over the possibilities in December.

Sherman considered the Valley of the Mississippi the dominant geographical and military feature in the country; as for that portion of it to the west of the great river, he regarded the war there as practically over; Union forces had only to tighten their hold on Arkansas and Louisiana, while Texas could simply be jettisoned: "Assert the absolute right to the Valley of the Mississippi and leave the malcontents to go freely to Texas . . ." The Southwest would become the battlefield in an unending war between Texas and Mexico—a suitable fate for a land he did not consider worth the bones of a single Union soldier.

When Sherman and Grant had talked over possibilities for the coming campaign back in December it had seemed likely that Washington would want the movement of Western troops into the Eastern theater to take place through East Tennessee into Virginia. Sherman had not been happy about the prospect, preferring a plan of operations that would take him into Georgia. So matters stood at the beginning of March, when he had just wound up the Meridian operation; then he received a letter from Grant with momentous news. Congress had revived the rank of lieutenant general, previously held only by George Washington, Grant had just received word to report to Washington, and it looked as though he would be confirmed in the new rank; he wanted Sherman and McPherson to know of his coming advancement, and to thank them for all they had done to support and assist him.

Soon Sherman was summoned to Nashville, where he found his friend wearing three stars on his shoulder straps. There was good news for Sherman too: he would inherit the Military Division of the Mississippi. Grant would be moving east, leaving Sherman as his proconsul for the theater between the Appalachians and the Mississippi, comprising four military departments, those of the Ohio, the Cumberland, the Arkansas, and the Tennessee, the last of these being commanded by his friend McPherson (the Department of the Arkansas was shortly transferred to another command). Grant would oversee generally the operations of the armies in the several theaters, but would stay east.

Sherman accompanied Grant on the train as far as Cincinnati, so the two could talk about the upcoming campaign. There both generals checked in at the town's best-known hostelry, the Burnet House; Mrs. Grant was along and Ellen came down from Lancaster. While the ladies visited, the two men closeted themselves in one of the rooms with maps strewn about the floor; when they came out they had a plan of campaign for 1864.

Their plan contained no timetable and it specified no route of advance; but the two armies would begin operations simultaneously, and they would keep pressure on the enemy so as to prevent his shifting forces from one theater to another; May 5 would mark the beginning of the campaign. Each general would have an adversary: Grant would campaign against the redoubtable Robert E. Lee; Sherman would face Joseph E. Johnston, with whom he had already sparred in Mississippi. And each general had a destination: Grant's was Richmond and Sherman's was Atlanta. In point of fact Grant's description of their plan of campaign made the two enemy armies their objectives; they were to be assailed and ultimately "broken up."

Sherman would have a force far greater than he had ever led before, just under 100,000 men, comprising the Armies of the Cumberland, the Ohio, and the Tennessee. He would be facing a Confederate host second only to Lee's Army of Northern Virginia. Its commander, Joseph Eggleston Johnston, was a career soldier who had graduated from West Point seven years before Sherman entered. Curiously, though the two had spent years in the same tiny officers' corps, they had never met, but Sherman knew Johnston to be an able and resourceful commander. Johnston's Army of Tennessee was no longer in Tennessee, but was within sight of it, camped in and around Dalton, Georgia. It was inferior in numbers, with a peak strength that spring of a little over 70,000 men; it was nonetheless a formidable opponent, as it had shown a few months earlier at Chickamauga.

That April logistics was the main subject on Sherman's mind. He liked to work with figures, and the future needs of his armies provided plenty of them. He spent much time with calculations: he estimated that a locomotive could haul a payload of 160,000 pounds; he figured that that sixty-five carloads a day would sustain an army of 100,000 men and 30,000 animals, but he strove to reach 120 cars per day, "for accidents and accumulation."

Those accidents could easily happen. He said he was "never easy with a rail road," finding them fragile; he would have preferred supply by water. Thus he gave the railways his first attention; ten days after he succeeded

Grant he had already completed a personal inspection of the rail links in Tennessee that would be feeding his army. He was not happy with what he saw; he had noticed at least a dozen locomotives and "an abundance" of empty box cars sitting idle at Stevenson, Alabama when they could have been helping with the logistical buildup for his campaign. He complained to Grant that rail superintendent Daniel C. McCallum was too preoccupied with observing timetables.

Ahead in Georgia would be a rail problem of still unknown proportions. The Western and Atlantic, which connected Chattanooga and Atlanta, was a single track affair known to be in bad state of repair, and there was no doubt it would be in even worse condition when the Confederates gave it up. It was particularly fragile, with a score of wooden bridges and trestles between the Tennessee-Georgia line and Atlanta, in country that would be swarming with Confederate partisans. Then much of North Georgia was thinly populated and its cultivated land limited; if the rail line failed, then even with the most ruthless foraging, a hundred-thousand-man force would eat the country out in short order.

Throughout the campaign Sherman would continue to interest himself in all matters of transport and supply. General O. O. Howard, who had close and frequent contact with him during 1864 and 1865, recorded that "Easton [Langdon Easton, Sherman's quartermaster] went to him for the solution of transportation problems ... He would demonstrate to his Chief Commissary the number of rations that would support his armies for a whole week or a month." When he wrote to Philemon Ewing in October 1863, Sherman had already identified in himself this unusual range of expertise: "Without being aware of it, I seem to possess a knowledge into men & things, of rivers, roads, capacity of trains, wagons, etc., that no one near me professes to have. All naturally & by habit come to me for orders and instructions."

Through April and into May he watched carefully the rate at which supplies accumulated in his depots and he took a number of steps to augment them. Determined to carry on a vigorous but spartan campaign, he ruthlessly trimmed the army's needs. Perhaps at the suggestion of Quartermaster General Montgomery Meigs, he cut his army's "burdensome tentage," allowing only one headquarters tent per regiment; the men could make do with simple tent flies that they could carry themselves.

He began issuing orders restricting rail transport of persons, a habit he continued through the summer. There were to be no passenger cars south of Nashville and no civilians either, if he could help it. He refused to allocate transport for sending food supplies to East Tennessee Unionists, despite President Lincoln's plea. He told Assistant Secretary of War Charles A. Dana that when he got a travel request from a civilian his usual answer was "show me that your presence at the front is more valuable than 200 pounds of powder, bread, or oats." Few were able to do so to the general's satisfaction. That was the case with commissioners from the various state governments sent to check on the welfare of the men from their state; Sherman claimed agents were even sent "to count the boys' teeth." The restrictions on railway passengers covered travel in both directions and applied to both the quick and the dead; the bodies of men killed at the front could not be disinterred or sent north before November 1, 1864 (a notable exception was made when General McPherson was killed).

It was May before the Federal Army began its offensive. Northern soldiers were astonished to find North Georgia's corn and wheat crops so far advanced (there would be no traditional harvesting of these crops, for the 35,000 horses and mules in Sherman's armies would be turned into the fields of ripening grain as the army advanced). Sherman opened the campaign with a flourish, launching a major flanking movement that might have decided the contest for North Georgia right there, had it succeeded. While he and most of his force moved up against Johnston's extensively fortified positions before Dalton, he dispatched General McPherson with 20,000 men by another route to the rear of the Confederate Army, with the intention of having "Mac" seize and sever the Western and Atlantic at the town of Resaca. On May 9 Sherman got a message McPherson had written earlier in the afternoon; he had passed unchallenged through an obscure cleft in the mountains known as Snake Creek Gap and was within a mile and a half of the rail line. "We were all jubilant," Sherman recalled in his *Memoirs.*

But McPherson never made it to the rail line. When he neared Resaca he found it defended–by a minuscule and nondescript force, as it turned out later; but he decided to retire and entrench. Sherman was more sorry than angry at his subordinate's lack of initiative. "Such an opportunity does not occur twice in a single life," he later wrote of this episode; Mac had been too timid.

Ten days after McPherson's failure, at the little town of Cassville, General Johnston prepared an unpleasant surprise of his own. In its forward progress Sherman's vast army was obliged to follow the vagaries of the North Georgia roads, which could separate considerably his advancing columns; placing troops in proximity to one of these routes, Johnston hoped to surprise and destroy a part of the Federal Army; but he himself was surprised by the arrival of Union troops from an unexpected direction. There was some indecisive fighting, then in the night the Southern Army withdrew south. The war in North Georgia would not be decided by any decisive stroke or brilliant coup.

The campaign was to be one of endless maneuvering. In its progress from Dalton to Atlanta—a hundred miles as the crow flies—Sherman's army covered almost twice that distance. But in Georgia maneuvering was not a substitute for fighting. Though many of the collisions were minor in terms of the number of troops involved, there was a great deal of combat and sizable casualty lists to show for it. The compilers of the *Official Records of the Union and Confederate Armies* could find in the first month and a half of the Georgia campaign only one contest sufficiently important to be called a battle, that of Resaca; but they identified by name some sixty other clashes of lesser importance, and in addition an unknown number of armed collisions "not identified as to number or place." So frequent were they that to some participants they all seemed to merge into one, the "Battle of Georgia," and Sherman himself described the spring campaign as "one grand skirmish."

The number and nature of these collisions were determined largely by the pace and the aggressivity of the invading army, the relentlessness with which it pressed the Confederates and compelled them to give up ground—and this relentlessness was imparted by its leader. The whole campaign thus bears the personal stamp of William Tecumseh Sherman.

A pattern of sorts developed: the Union forces would approach the Confederates' entrenched positions and by one means or another render them untenable, by a flanking movement or the threat of one, by intense bombardment, or—less frequently—by frontal assault. By this period in the war the two armies entrenched in one, two, or even three lines, screened by a picket line and sometimes with cleared zones of fire before them. The Confederates might fall back to a second line already prepared or they might withdraw south during the night. In the latter case the Federals would usually follow up the next morning, though if Sherman got wind of the enemy's withdrawal during the night he might well rout his soldiers from their beds and follow immediately. In either event his troops would not go far before they encountered the enemy again, entrenched once more on good ground, and Sherman would immediately set about levering the Johnnies out of their new positions.

That spring and summer Sherman the commander could be seen at his best in the conventional warfare of his day. He was now free to direct as he wished the efforts of a hundred thousand men. While he was tied by telegraph to Washington, that line was in no sense a leash. Grant had promised him he could run his campaign as he pleased; he kept Henry Halleck advised by almost daily telegrams, but since Halleck had been eclipsed by Grant, he was now simply an administrator, moving paper but not armies; he sent no orders in return. The commander of the Military Division of the Mississippi was essentially master in his own house. That spring he had just turned forty-four. His keen mind had been further honed by three years of war and he was still possessed of his remarkable stamina. In Georgia he would reach the zenith of his career.

And it must be said that in the 1864 campaign his competence seemed to expand effortlessly to fill the broader responsibilities before him. There was no more mention in his family letters of self-doubt, of holding back and letting others do the leading; now the dominant tone is confident, even expansive. And along with the confidence is more than a hint of omnicompetence; his attempts to assert greater control over his transport system and the provisioning of his army are indicative; his flat rejection of Lincoln's plea on behalf of the Tennessee Unionists was the act of a man sure of his position.

In past campaigns he had manifested a tendency to keep close track of his subordinates, to stay informed of what was happening everywhere in his command. Admiral Porter, who also knew Grant very well, noted that Sherman differed from his superior in that respect: "Sherman attended to all details himself, while Grant left them to others." While in Memphis Sherman had planned cavalry sorties in the most minute detail; with an army of 100,000 men he could not command with such exactitude. Still, the need to verify, to see for himself, combined with the great breadth of his interest in things military and his restless, impatient nature, made of Sherman one of the most mobile and peripatetic of commanders. "I must see everything," he wrote Ellen. He might startle a signalman by climbing to the top of the signal tower and studying the surrounding terrain through his glass; he might visit an artillery emplacement or suddenly materialize before a picket—a somewhat dangerous practice, given the high state of alert that sometimes prevailed on the picket line.

The terrain his army encountered was for him a subject of endless fascination. There were usually two topographical engineers on Sherman's staff, and they worked under his close supervision. Porter recalled that "the General's great delight was to pore over maps, and he seemed to take in all the roads, fields, and rivers, as if they were good to eat and drink." Porter also claimed Sherman had a flawless memory when it came to topography: "He never forgot a house, a road, or a bayou." The Army of the Cumberland had specially constructed photographic equipment that he used to change the scale of maps and copy them by photo reproduction. Periodically during the campaign his cartographic experts "published" new editions. He even had them printed on linen for his cavalrymen.

Sherman had no intelligence officer, this military specialty then being largely unknown, but assumed that function himself; he kept himself informed with the reports of scouts and reconnaissance parties he sent out and to a lesser extent from accounts of prisoner interrogations that reached him. He read the Southern press diligently. When he was before Atlanta, local Unionists volunteered their services; he had the Atlanta papers delivered to him through the lines every day. He had special funds for the hiring of spies, and by 1864 he had a number of operatives whose activities he directed; judging from the names that survive in the records, he seems to have had a preference for women in this line of work.

Those who worked most closely with Sherman on a daily basis during the Georgia campaign were the members of his staff, which he scrupulously kept within the limits prescribed for his rank. At the beginning of the Georgia campaign they numbered a dozen. In the Army of the Potomac generals often served in staff capacity, their high rank reflecting the importance of their functions. Not so with Sherman's staff, where only two men wore stars, his chief of artillery, William F. Barry, and Inspector General John M. Corse. Barry, Quartermaster Langdon Easton, and Commissary Amos Beckwith were rough contemporaries of Sherman and acquaintances from the old army; other members of the staff were considerably younger. Chief Engineer Orlando Poe served in his rank as captain in the regular army.

Life in the general's official family was a rigorous, spartan one, for his close associates of necessity adopted his habits, including irregular hours. He slept generally for short periods, both night and day. A story ran through the army that several soldiers had come across their commander one day sound asleep in a fence corner and had taken him for a drunk sleeping it off. One of his staff asked him about the story and he said it was essentially true; he had explained to the "boys" that he had been up most of the night.

It was apparently his custom to handle paperwork in the hours preceding reveille, and for that purpose he often used a room in a house if one were handy, preferably one whose owner was absent; he might sleep on the porch in wet weather. He had no headquarters tent and in the field often slept under a simple fly—a sheet of canvas without enclosed ends. He often slept in his clothes—during the siege of Vicksburg he claimed he did so for two months running. He tended generally to neglect his dress. Fortunately he had a black manservant named John Hill—"the invariable Hill"—who had been recruited by Ellen sometime before Bull Run. Hill washed his socks, coaxed him into clean clothes, and tried to ration his cigars and his whiskey. In letters to Ellen he made mock complaints about Hill's tyranny; in a May 1863 letter he boasted "I have slept on the ground the past two nights to Hill's disgust." His only other servant was an equerry whom he referred to as "my nigger Carter."

At Sherman's headquarters liquor sometimes flowed in the evening—bourbon was his preference—and while there is no testimony that he was ever drunk, he did become mellow. His brothers-in-law Hugh and Charles Ewing would sometimes join him, and one suspects that the three may have been in their cups one evening when Sherman wrote Ellen an uncharacteristic letter. They had been talking about her and had reached this conclusion: "We agree that you have bottom and in the race of life will hold out against many a nag of more speed."

Beyond his staff Sherman's most frequent contacts were with the commanders of his three armies. He visited one, then another. His correspondence is full of little notes for rendezvous: "I will be over before three (3) o'clock," "I am just starting for your flank," etc. Then there are the terse queries that call for immediate answers: "How far out did Cox go today? Did he observe anything different from usual? Have you heard anything from Kilpatrick?"

As his armies approached Atlanta and the country became less rugged the army's tactical communications underwent a revolution: it became possible to replace the dashing couriers bearing hastily scribbled notes with a special "field" telegraph line linking Sherman's headquarters with those of his three army commanders, who could often be separated from one another by ten miles or more. Sherman, always in something of a hurry, used the telegraph extensively. On June 27, the day of the Battle of Kennesaw Mountain, he wired Thomas no fewer than sixteen times. And almost every day Sherman telegraphed a brief report to Washington, and as previously agreed, Halleck relayed these reports to Grant.

As Sherman's legions pushed on southward the terrain became what he called "sub-mountainous," with isolated prominences such as Lost Mountain, Pine Mountain, and Kennesaw Mountain, the last of which posed a special problem, since by late June it stood squarely in Sherman's way. He had two choices: he could attack it and try to take it, or pass around it, but in the latter case he would have to abandon temporarily his lifeline, the Western and Atlantic. And as he explained to Halleck, "we are already so far from our supplies that it is as much as the road can do to feed and supply the army."

As the mountain loomed nearer Sherman was initially of the opinion he should avoid an assault. In his letter to Halleck of June 13 he stated flatly, "we cannot risk the heavy losses of an assault at this distance from our base," but three days later he wrote that he was "studying" the matter and was "now inclined to feign on both flanks and assault the center. It may cost us dear but the results would surpass an attempt to pass around."

By June 24 he had made his decision: his Special Field Order, No. 28 stipulated that General Thomas would attack the enemy positions before Kennesaw Mountain at precisely eight o'clock on June 27, while Generals Schofield and McPherson "feigned" on their portions of the Union line.

Thomas let Sherman know that his heart was not in the assault: he had studied the enemy's positions closely and they were extremely strong. "The troops are also much fatigued in consequence of the continuous operations of the last three or four days."

The Union attack came off as scheduled. Sherman had installed himself on a hilltop with telegraphic connections to each army headquarters. Thomas's assault was a costly failure; despite renewed efforts his men made almost no headway. At 2:25 that afternoon Sherman inquired how things were going and Thomas answered that his men still held what little ground they had gained; but the enemy's works were so formidable that they could not be taken "except by immense sacrifice"; he added, "one or two more such assaults would use up this army." This dialogue by wire continued through the afternoon. When Sherman asked if Thomas was willing to abandon the railway and bypass Kennesaw he answered that it was "decidedly better than butting against breastworks twelve feet thick and strongly abatised." When the day was over Union casualties totaled 2,500, of which 2,000 were in the Army of the Cumberland.

Just why Sherman chose to fight is not easy to say. If he really believed the chances of success were good, he was virtually the only one to hold that view. Certainly, if achieved, the breakthrough he sought could have had important consequences. Johnston was fighting with his back to the Chattahoochee River; if beaten and pursued, his force might have dissolved in panicky flight, throwing open the door to Atlanta. The parlous state of Sherman's supply system just then was probably a factor. It was a bad time to swing around Kennesaw Mountain and his rail line; he would need supplies to tide him over, and these were only trickling in. June had been a bad month on the Western and Atlantic with a number of attacks on the trains and considerable interruption of traffic; moreover the month had seen extraordinarily heavy rains in North Georgia, which bogged down the horse-and-wagon link used to connect the camps with the railhead.

There was another consideration, a psychological one, that Sherman mentioned before the battle and tended to stress after it was over, perhaps because it was a way of winning even when he lost. He argued that win or lose, his army would have demonstrated that it was formidable on the battlefield. Did he really believe in this psychological dividend? Apparently so, for he had justified the heavy losses at Shiloh with essentially the same argument: he insisted that not a single Union soldier had died there in vain, for the heavy losses manfully sustained had shown to friend and foe alike that the Western army was made of the sternest stuff. Under the right circumstances, as at Shiloh and Kennesaw, heavy losses could thus be regarded as a kind of investment in future success. This way of looking at the 3,000 casualties at Kennesaw Mountain may help explain why Sherman was neither depressed nor conscience-stricken by them. He wrote Ellen three days after the battle: "I begin to regard the death and mangling of a couple thousand men as a small affair, a kind of morning dash, and it may be well that we become so hardened."

On the morning of July 3 Sherman had given orders for the move around Kennesaw and was scanning the mountain with his telescope when he spotted figures ascending its slope. They proved to be some of his own men, scaling the heights that had been abandoned by the Confederates. Johnston had detected the Federal Army's movement and withdrawn under cover of night. The next natural obstacle to be overcome was the Chattahoochee, which the first of Sherman's troops crossed on July 8. Within ten days his forces were across and within sight of their goal–Atlanta.

In Richmond there was increasing dissatisfaction with the Fabian tactics of General Johnston, who had retreated a hundred miles without a major effort to check or turn back the invader; even in his own army, where he was generally popular, there were jokes about his collecting pontoons for an eventual withdrawal to Cuba. Now his giving up the line of the Chattahoochee was for President Jefferson Davis the final disappointment; on July 17 Davis had him relieved of command. Johnston had been seriously wounded in Virginia two years before and had undergone a long and painful recuperation; it could be argued that this ordeal in some fashion made him hesitant at the prospect of combat, hence he failed to come to grips with Sherman.

That was certainly not the case with his successor, John Bell Hood. Having had an arm shattered at Gettysburg, then losing a leg at Chickamauga, by 1864 he was reduced to the condition of an invalid (it took four men to strap him onto his horse), yet immediately after he replaced Johnston his army brought on three fierce battles in the space of ten days: Peachtree Creek (July 20), Atlanta (July 22), and Ezra Church (July 28).

Hood's army came off second best in each of these encounters.

All three battles came as something of a surprise to Sherman, though he had been sufficiently prudent in the disposition of his armies that the attacking Confederates found little by way of weakness they could exploit, and Sherman's own plans for the investment of the city suffered only a brief check. The battle of the twentieth was particularly costly in lives, for the rebels struck portions of General Thomas's troops after they had crossed Peachtree Creek but before they had sufficient time to throw up much by way of breastworks–thus opposing swarms of infantry collided in the open field, a phenomenon increasingly rare in this third year of the war.

In the Battle of Atlanta, actually fought in the neighborhood of Decatur, the Southerners did manage to get on the left flank of Sherman's aligned forces, where McPherson's army was posted. At one point in the battle when McPherson's army was heavily engaged there was a possibility of striking the enemy in the flank. Sherman, who was on the spot, rejected the idea. As he put it in his *Memoirs,* "I purposely allowed the Army of the Tennessee to fight this battle almost unaided." His reasoning: "If any assistance were rendered by either of the other armies, the Army of the Tennessee would be jealous," by which he evidently meant resentful of outside help. It has been suggested, quite plausibly, that Sherman wanted his favorite army, once led by himself and then by his best friend McPherson, to win its laurels unaided. That this army held a special place in his heart would be amply demonstrated after the war, when he communed with its veterans in the emotionally charged reunions of the Society of the Army of the Tennessee, his favorite of all the gatherings of old soldiers.

Sherman's own role in the three July battles was minimal. At the Battle of Ezra Church his chief intervention was to send General Jefferson C. Davis's Division off to strike the rebels' flank by a roundabout route. What might have been a decisive stroke came to naught–according to Army gossip because the phrase "to East Point" in Sherman's order was copied as "toward East Point," and Davis's division floundered about far into the night.

Sherman's lack of involvement in these battles is all the more striking considering the keenness with which he watched over the course of events in the Battle of Kennesaw Mountain, but there he initiated the engagement and had crafted the battle plan, and its success or failure would reflect on him most directly. Not that he would accept any blame for failure–in his view the assault on Kennesaw had good chance of success but the troops failed to achieve it. In postwar conclaves he would tell his veterans that he did not hold their failure against them; if General Charles Harker had not been killed in the attack, with his leadership they might well have succeeded.

For Sherman the Battle of Atlanta would be forever memorable because McPherson fell on that July afternoon. His grief at the loss of his friend approached in intensity what he had felt at Willie's death. He rushed to the house where McPherson's body had been taken, and there he paced the porch where the corpse lay; each time he stopped to look at his fallen friend tears brimmed in his eyes. He had a physician examine McPherson's wound and give him the reassuring news that his comrade-in-arms had not suffered; he later sent a photographer to take a picture of the nondescript patch of brush where McPherson had fallen.

McPherson's death posed a problem, for his position had to be filled, and there were a number of aspirants. One of McPherson's corps commanders, John A. Logan, had taken over temporarily and had performed well in those first hours of his command as the Battle of Atlanta raged on. But as a man of civilian background and a prominent figure in the Democratic Party, Logan was what Sherman called a "political general" and for that reason unacceptable. Among Sherman's corps commanders there was another candidate, an eminent and eager one, Joseph Hooker. A professional soldier (USMA, 1837) he was well-known to Sherman, but not well-liked by him. Hooker was something of a discordant element, holding a command he considered beneath him; he was also an Easterner who resented what he called the "fancied superiority" of the Western army. Having commanded the Army of the Potomac until he was trounced by Robert E. Lee and Stonewall Jackson at the Battle of Chancellorsville, he felt himself by seniority and previous services the only logical successor.

Sherman's choice was the commander of the Fourth Corps, Oliver Otis Howard. He had many qualities that attracted Sherman. He had a West Point diploma (class of 1854), which Sherman considered the proper ticket of admission to high command. He had demonstrated loyalty and personal courage, qualities that Sherman prized. He was a zealous, practicing Christian, which Sherman was willing to overlook. Sherman also overlooked a less than brilliant record, which many attributed to Howard's errors of judgment and failure to carry out orders. When Howard's appointment was announced Logan the civilian proved to be a good soldier; he stayed on as corps commander despite his deep disappointment. Hooker promptly asked to be relieved, and Sherman gladly honored his request. "Fighting Joe" thus left the field, confiding to a friend that he was only sorry he hadn't broken his saber over Sherman's head.

After the Battle of Ezra Church the offensive punch had gone out of Hood's army. Sherman proceeded to put Atlanta under siege. The city was too strongly held to be taken by assault; for a time he hoped that his heavy bombardments would make the town too hot to hold on to; ultimately he produced its fall by severing the rail line connecting it to Macon, its only remaining link to the outside world. On September 7 he made his unhurried entrance into the battered city, five days after Hood had pulled out. The fall of Atlanta probably did more than any other event that summer to bolster Northern morale; and it may well have guaranteed Lincoln's reelection two months later.

The Atlanta campaign had been a hard-fought one, but despite the high stakes it had been a war between gentlemen. Soldiers of the two armies chatted back and forth along the picket lines, exchanging Yankee coffee for Confederate tobacco. Following battles there were often truces to evacuate the wounded and bury the dead. In the truce that followed the fighting at Kennesaw hundreds of soldiers from both sides gathered and communed. In one spot along the truce line a number of Union and Confederate generals came together to make or renew acquaintance and inquire after old friends; a bottle appeared and made the rounds. Though Sherman and Hood had sent each other some angry letters, they were able to arrange an exchange of prisoners.

But bad war also made its appearance. Grant and Sherman had speculated on the next step once Atlanta was taken; Grant suggested Sherman might seize all foodstuffs in the city and ration them out to rich and poor alike. But now Sherman found a simpler solution: on September 5 he issued an order expelling the population of Atlanta. They could go north or south—trains would be available to take them north—but they had to leave the city within five days. The new master of Atlanta did not want the useless mouths that he could not feed with his long and tenuous supply line, and anyway so deeply had he penetrated the Confederacy that he would need to convert the city into a purely military citadel; troops were soon laboring on a new defensive perimeter to protect Federal Atlanta.

There was great outcry in the Southern press and the measure was noted and commented on in the North and in Europe. The expulsion of the people of Atlanta is still treated by some historians as an act without parallel that broke new ground in violating the traditional laws and customs of war. But in fact the expulsion of populations was an option Union military leaders had already turned to. In July Grant had authorized his lieutenants to empty out part of the Shenandoah Valley: "I do not mean that houses should be burned, but all provisions and stock should be removed, and the people notified to move out." And in Missouri a year earlier Sherman's own foster brother, General Thomas Ewing, had driven the population out of three Missouri counties and part of a fourth—in all some 20,000 people; to make the move permanent, troops were then set to burning the abandoned houses. As for Sherman, he had given some thought to driving out the population of Memphis; moreover he had emptied out Iuka, Mississippi, the preceding October.

Whatever questions were raised in the North about the propriety of his unceremonious emptying of Atlanta, they were drowned out by the praise he received for taking the city. His campaign had already attracted wide attention, and next to Grant he was now most prominent in the constellation of Union generals. As early as May Phil Ewing described his foster brother as "on the path to glory"; Sherman noticed the change too. "I was in hopes I could remain unpopular," he wrote John in August, "but I see the newspapers begin to point out good points in my character." It was not just the press; at the beginning of the campaign he had begun to receive mail from unprecedented numbers of admirers, requests for his autograph, for locks of his hair.

Before the year was over he had adjusted to fame; now one of his aides handled fan mail, and in his spare time he put his signature on slips of paper destined for his admirers. With the fall of Atlanta the stream of mail became a torrent, and letters from ordinary citizens had to compete with messages from national figures. Lincoln thanked him in the name of the nation. Halleck labeled the Atlanta campaign "the most brilliant of the war," and McClellan wrote him that it would go down in history as "one of the most memorable ones of the world." For a man who had once wanted to shun the limelight he wore his laurels easily.

The taking of Atlanta was one thing, staying there was another. While Hood's army had been bested and was licking its wounds, the war to protect the Western and Atlantic went on. There in North Georgia bad war had flourished from the beginning, and as time passed Sherman's measures became more repressive. There the enemy was fast-moving rebel cavalry on raids, indigenous guerrillas and partisans, and to Sherman's way of thinking the population that aided them and cheered them on. That summer the rebel horsemen were a plague throughout Sherman's domain: in June General John Hunt Morgan raided into Kentucky and General Nathan Bedford Forrest, "that devil Forrest," trounced a Union force at Brice's Cross Roads, setting off alarm bells in northern Mississippi and Tennessee. In North Georgia, Wheeler's cavalry carried out a destructive raid, but the persistent enemy was the guerrilla. After a rash of attacks on the rail line in June, Sherman made North Georgia into a special circumscription which he called the District of the Etowah. He confided it to General James B. Steedman, to whom he gave sweeping powers. Steedman set to work clearing out a security zone extending three miles on each side of the Western and Atlantic Line. No persons could live within the zone (towns excepted) unless they could establish their loyalty to the Union; soon this highly restricted area embracing some 600 square miles contained only 291 "authorized" households.

Sherman had already extended his policy of punishing populations to include those living where Confederate cavalry was active, for he concluded that they tolerated and sustained such raiders, just as they accommodated the guerrillas who thrived in their midst. He urged his subordinates to proceed aggressively against Forrest, "devastating the land over which he has passed or may pass." The people of Mississippi and Tennessee would soon learn that Forrest would "bring ruin or misery on any country where he may pause or tarry."

He had exhausted his patience with Confederate partisans, whom he now classified as "wild beasts," not covered by the laws of war. Anyone caught damaging the railway, the telegraph line, or army stores, especially if in civilian dress, "should be disposed of finally and summarily." The partisans were fiendishly inventive: in the piles of wood used to fire the boilers of the locomotives they placed logs that had been hollowed out and filled with gunpowder. Under the tracks they buried "torpedoes," primitive land mines set off by a friction fuse when a train passed over them. To protect the rail line in the District of the Etowah that June Sherman issued a "ruling" that merits quotation in full:

"The use of torpedoes in blowing up our cars on the road after they are in our possession, is simply malicious. It cannot alter the great problem, but simply makes trouble. Now if torpedoes are found in possession of an enemy to your rear, you may cause them to be tested by wagon loads of prisoners, or if need be, by citizens implicated in their use. In like manner, if a torpedo is suspected on any part of the road, order the point to be tested by a carload of prisoners, or citizens implicated, drawn by a long rope. Of course the enemy cannot complain of his own traps."

That June Sherman also began implementing his idea of deporting undesirables and irreconcilables and repopulating the South. Action in this direction had been on his mind at least since April, when he wrote John about a program on the largest scale: "There is no doubt we have to repeople the country and the sooner we set about it the better. Some decision must be made to deed houses and land captured of the enemy. The whole population of Iowa & Wisconsin should be transferred at once to west Kentucky, Tennessee & Mississippi, and a few hundred thousand settlers should be pushed into south Tennessee."

The dispossession and deportation of inhabitants was based on a principle enunciated by him: "The civil power being insufficient to protect life and property *ex necessitate rei*, to prevent anarchy, 'which nature abhors,' the military steps in, and is rightful, constitutional, and lawful. Under this law everybody can be made to 'stay at home and mind his or her own business,' and if they don't do that, can be sent away where they won't keep their honest neighbors in fear of danger, robbery, and insult."

This decree was to be the justification for systematic deportation of the chronically disaffected in Kentucky, Tennessee, and North Georgia, outlined in a letter to General Stephen Burbridge, his proconsul in Kentucky, and distributed to other commands, including the District of the Etowah: "Your military commanders, provost-marshals and other agents may arrest all males and females who have encouraged or harbored guerrillas and robbers, and you may cause them to be collected in Louisville, and when you have enough, say 300 or 400, I will cause them to be sent down the Mississippi through their guerrilla gauntlet, and by a sailing ship send them to a land where they may take their negroes and make a colony with laws and a future of their own." He added this cautionary clause: "I wish you to be careful that no personalities are mixed up in this, nor does a full and generous love of country, 'of the South,' of their state or county, form a cause for banishment."

Sherman addressed a copy of this order to Secretary Stanton, along with further explanation. He proposed sending the exiles to Honduras, British and French Guiana, Santo Domingo, or Madagascar, though Baja California might do. Stanton gave his approval. In Kentucky, Burbridge made it known that Southern sympathizers could be "arrested and sent beyond the limits of the United States." At Sherman's urging he arrested Lucius Polk, a harmless old gentleman of distinguished family and brother of Confederate General Leonidas Polk; he was told to prepare for a long ocean voyage.

In the end it appears that no shipload of diehard rebels ever sailed to one of the exotic places Sherman had chosen for them. The general decided Lucius Polk could post bond and go home, but several hundred people were transported north of the Ohio River and ordered to stay there until war's end, including the entire female workforce of at least one Georgia textile mill.

Paradoxically, as Sherman was intensifying the war he was also involved in negotiations to end it. He had been in Atlanta scarcely a week when he made contact with three prominent Georgia Unionists: William King, Joshua Hill, and Judge Augustus Wright.

From their discussions came a plan to have Georgia change sides in the war: she would sever her ties with the Confederacy and renew those with the Union. For his part, Sherman pledged that his military presence would cease to be burdensome: "I will keep the men on the high roads and commons and pay for all the corn and meat we need and take." So feasible did the scheme appear that Sherman sent messages to Georgia Governor Joseph E. Brown and Confederate Vice President Alexander Stephens, who was then at his Georgia residence; it was well-known that both men were bitter enemies of Jefferson Davis.

Strictly speaking Sherman had no business discussing political matters with officials of an enemy government, and even less in initiating such talks without his own government's knowledge or consent. Washington did not hear of these initiatives until September 15, when Sherman wired Halleck that Governor Brown had sent the Georgia state troops home to harvest the state's corn and sorghum crops, adding, "I have reason to believe that he [Brown] and Stephens want to visit me, and I have sent them a hearty invitation." It was Lincoln himself who replied. If the president was troubled by his general's unauthorized foray into diplomacy, he made no mention of it, perhaps because it could bear rich fruit. He wired Sherman, "I feel great interest in the subjects of your dispatch mentioning corn and sorghum and contemplated visit to you." Sherman promised to keep the president fully informed on what could become "a magnificent stroke of policy."

It was not to be. The wily Alexander Stephens made an indirect and noncommittal response; Governor Brown, who had all the political acumen that Sherman lacked, saw the invitation to Atlanta as a bomb that could go off at any moment. He defused it by framing a reply that he immediately communicated to the press: since neither he nor Sherman was authorized by their governments to conduct negotiations, there was no point in their meeting. From that point on Sherman's plan hung fire; it was in fact destined to come to naught, though he could not see it that way. He kept up his contacts with the Unionist camp and late in October he even arranged for Judge Wright to travel to Washington and confer with Lincoln. But by then Sherman himself was preparing to travel; he was on the eve of a new campaign, one that would take him to the end of the war.

# 17

## THE GREAT MARCH

INITIALLY THE CONQUEROR of Atlanta gave every indication he would hold on to his prize. Having expelled the population and ringed the town with a new set of fortifications, he set the garrison to building winter quarters. But not long after he took up residence in Atlanta he had a visitor named Horace Porter. Sent by Grant, Colonel Porter had come straight from headquarters at City Point to consult on the next step in their operations. He had not met Sherman before, so he was careful to record his impressions: "I found him sitting on the porch of a comfortable house on Peachtree Street, in his shirt sleeves, without a hat, tilted back in a big chair reading a newspaper. He had white stockings and low slippers on his feet." Greetings were exchanged and Sherman's questions about the war in the East answered, and then the two men turned to the purpose of Porter's visit.

Now conversation was replaced by monologue. Porter fell silent as the Conqueror of Atlanta began "a marvelous talk about his march to the sea. His mind, of course, was full of it. He seemed the very personification of nervous energy." As Sherman talked on, Porter noticed that he rocked back and forth in his chair, his hands were at work shredding the newspaper they held, while his stockinged feet darted in and out of their slippers.

Grant and Sherman had given serious thought to the basic problem as far back as March: once in Atlanta, the army might not be able to sustain itself unless it either drew supplies from a new base somewhere on the Gulf or the Atlantic or itself moved to the coast. Following up on their March talks, Grant had sent Sherman a map with two blue lines penciled on it, one of them connecting Atlanta with Savannah and the other running from Atlanta in the direction of Mobile, which port, like Savannah, was then still in Confederate hands. As late as July, Grant told Halleck that Sherman would remain in Atlanta, where, with proper measures to husband his supplies, he would wait "until a permanent line can be opened from the South coast." Sherman was also giving thought to what he called "the next move on the chessboard." By September, when Porter arrived, he was proposing that Grant arrange for the capture of Savannah, so that his own force could advance there from Atlanta. "But the more I study the game," he wrote Grant, "the more I am convinced that it would be wrong for me to advance farther into Georgia without an objective beyond"; hence he might move east along the line Augusta-Columbia-Charleston. Another option discussed earlier now interested him less—a move in the direction of Columbus, Georgia, where his army could connect with the gulf by means of the Chattahoochee River. This line of operations would have the effect of further segmenting the South but would have less impact on the resolution of the conflict than shifting the Western army's strength to the eastern seaboard.

First, something had to be done about Hood. His army was nearby, encamped southwest of Atlanta. On September 20, the very day Horace Porter arrived, Sherman got the first inkling that Hood was about to take his army on campaign again. The Confederates were headed north, no doubt to threaten his rail line. Sherman had no choice but to start back north himself, leaving the Twentieth Corps in Atlanta. Hood was soon wrecking portions of the railroad, threatening garrison towns and taking some of them. Then he veered west. Sherman followed as far as Gaylesville, Alabama, where he halted his army on October 20.

On its way back through Georgia, Sherman's army was foraging heavily, with men and animals eating the region out in what Sherman described as a carnival atmosphere: "Our poor mules laugh at the fine corn-fields, and our soldiers riot on chestnuts, sweet potatoes, pigs, chickens, &c." The result, wrote one of Sherman's soldiers, was "worse than the army worm in the meadows of Indiana." When the inhabitants came to Sherman for relief he had a ready response: "Your friends have broken our railroads, which supplied us bountifully, and you cannot suppose our soldiers will suffer when there is abundance within reach."

During October heavy traffic moved over the telegraph wires linking Sherman's headquarters with Washington and Grant's headquarters at City Point; the messages they carried contained widely varying views on the next campaign. Halleck, consulted by both Grant and Sherman, came up with eight different reasons for heading for Mobile instead of Savannah. General Thomas gave Sherman his own objection to the proposed march: "I don't wish to be left in command of the defenses of Tennessee unless you and the authorities in Washington deem it absolutely necessary." If there had to be a great raid, Thomas suggested that it be done by Sherman's cavalry, which now had a new chief, James H. Wilson.

As for Lincoln, Stanton conveyed his view to Grant: "the President feels much solicitude in respect to Gen. Sherman's proposed movement and hopes that it will be maturely considered." Grant too had some misgivings, especially about leaving Hood for Thomas to deal with. "If there is any way of getting at Hood's army, I would prefer that, but I must trust to your own judgment." Sherman answered that he "infinitely" preferred to carry out his plan, to "move through Georgia, smashing things to the sea." Grant wired his consent.

By October 28 Sherman had waited long enough at Gaylesville. Since Hood was doubtless headed for Tennessee, he himself would leave sizable forces there under Generals Thomas and Schofield; with the rest of his troops he would head back toward Atlanta, making preparations for his own departure southward. The withdrawal, carried out systematically, was complicated by the November elections. All the soldiers had to have a chance to vote; on this the administration was adamant, and the balloting was arranged with some difficulty.

The withdrawal from North Georgia required elaborate preparation. The Western and Atlantic rail line and its accompanying telegraph line would be destroyed, but only at the last moment; in the interim Atlanta and all depots between there and Dalton would be emptied of stores, excess equipment, and the nonessentials that Sherman usually referred to as "trash." The flow of supplies was reversed; now the heavily laden trains were running toward Chattanooga and the "empties" toward Atlanta.

The army's concentration on Atlanta was accompanied by systematic destruction. When he was pursuing Hood, Sherman had adopted the practice of having houses set ablaze for signaling purposes, "making smokes," as he called it, to mark the progress of his columns. Then the army had used empty houses as warehouses, and now, as these were emptied of their contents in preparation for the departure, Sherman ordered these burned as well. Others went up in flames that October to settle accounts with guerrillas; General Corse slated the destruction by fire of three towns—Van Wert, Cedartown, and Buchanan—"for atrocities committed by gangs of thieves having their rendezvous there." And Rome burned, on Sherman's orders: "Destroy in the most effective manner, by fire or otherwise, all bridges, foundries, shops of all kinds or description, barracks, warehouses, and buildings especially adapted to military use, lumber or timber, also all cars off the track or material that cannot be moved." There were no orders to burn residences and Corse did not put them to the torch—but some of his men did. A Union officer who was there described these fires as "the work of rowdy soldiers." This pattern was repeated elsewhere: "official" destruction, carried out on orders, was supplemented with arson improvised by the rank and file.

On Monday, November 14, Sherman and his staff were witnesses to this latter phenomenon on the town square of Marietta. They stopped to watch Union soldiers who were at work manning a fire engine, putting out a blaze that had appeared in the courthouse. Soon flames appeared in other buildings and some of the general's aides took a hand in fighting them; a short time later fire burst forth again inside the courthouse. As the general and his party headed to a nearby house serving as their headquarters, they passed some soldiers. The general gestured toward them and said: "There are the men who do this. Set as many guards as you please, they will slip in and set fire. That court house was put out—no use—dare say the whole town will burn, at least the business part."

In Atlanta later that same day the general and his staff found men at work using battering rams to destroy the usual objectives: "storehouses, machine shops, mills, factories, etc." The work was directed by Sherman's engineer, Orlando Poe, who had been told to use fire only in the final phase of his work. But as early as the evening of November 11 fires broke

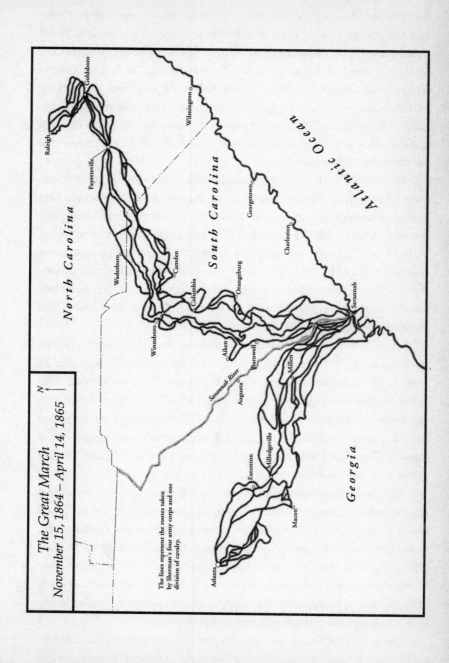

The Great March
November 15, 1864 – April 14, 1865

N

The lines represent the routes taken
by Sherman's four army corps and one
division of cavalry.

North Carolina

South Carolina

Georgia

Atlantic Ocean

Savannah River

Raleigh
Goldsboro
Wilmington
Fayetteville
Georgetown
Charleston
Wadesboro
Camden
Columbia
Orangeburg
Winnsboro
Savannah
Aiken
Barnwell
Millen
Augusta
Milledgeville
Eatonton
Macon
Atlanta

out here and there. By the following night these blazes were visible at considerable distance from the town; General Joseph Wheeler, who commanded the only Confederate force in the area at Jonesboro, wrote Hood on the thirteenth that the enemy was "burning things" in Atlanta. On the fifteenth Poe wrote in his diary, "much destruction of private property by unauthorized persons, to the great scandal of our army and marked detriment of its discipline." By then the bulk of the army was on the road south and east; by the evening of the sixteenth Sherman and all of his men had put the ruins of Atlanta behind them.

The expedition had been prepared by its commander in great detail, even to a digest of recent news items on the war and Lincoln's election that Sherman planned to present to Governor Brown in the course of a "friendly interview" at the state capital, Milledgeville—for he had not given up the idea of detaching Georgia from the Confederacy. The prominent Unionist Joshua Hill, with whom Sherman had kept contact, was already there, lobbying the Georgia legislature on behalf of peace. Unfortunately for Sherman, Governor Brown left Milledgeville two days before the invaders arrived.

The 62,000 men who marched or rode out of Atlanta were a picked force, eminently fit for an arduous campaign; the units had been culled of what their commander called "the sick, wounded and worthless." This select body was divided into two wings, each of which contained two corps (the right wing was General Howard's Army of the Tennessee, and the left wing, commanded by General Henry Slocum, was designated the Army of Georgia). The cavalry, commanded by General Kilpatrick, would follow its own route.

The expedition was superbly equipped. It even had a "government photographer," George N. Barnard, who captured little of the march itself in his photographs, but later made some striking images of Columbia, South Carolina, in its post-Sherman desolation. Two artists from *Harper's Weekly* accompanied the army and sketched scenes of action that Barnard's camera could not capture. Sherman had ordered new road maps of Georgia in a photographically reproduced pocket edition, run off in sufficient numbers for wide distribution to his command.

The army's ordnance specialists brought along rockets that could be used to signal from one column to the other at nighttime. During the day there was an easier way to spot the movement of a column—by its "smokes"; as in North Georgia, houses or barns would sometimes be set afire expressly for that purpose. One of Sherman's staff officers wrote, "I think I shall never see a distant column of smoke rising hereafter, but it will remind me of Sherman in Georgia and South Carolina."

Captain Poe was equipped with an odometer to determine the exact mileage of each day's march. Since the expedition would have to cross several rivers and could almost count on bridges being destroyed in its path, it would need pontoon trains. Sherman took two trains of recently introduced collapsible pontoons; these devices, made of wooden frames covered with canvas, had already proved their worth and could be carried on the standard army wagons.

Sherman had promised Governor Brown that if Georgia would drop out of the war he would move his army across the state without loss or damage to the inhabitants; in fact they would stand to profit from the Yankees' passage, selling produce to his commissaries. But now the army would simply take all that it needed, carrying in its wagons only enough rations for emergencies and drawing its sustenance almost entirely from the land it passed through. As an aid, the Department of the Interior had prepared special large-scale maps of the Southern states; hand-lettered over each county were figures for its population, livestock, and crop yields as recorded in the Census of 1860.

For both tactical and logistical reasons Sherman's force could not be moved cross-country in one body. His Special Field Orders, Nos. 119 and 120, issued November 8 and 9, 1864, laid down the rules for the movement. Where possible the army would march in four parallel columns, one for each army corps, with routes chosen so that the columns would be within easy supporting distance of one another—about ten miles on the average.

Special Field Order, No. 120 went into considerable detail on how the army would feed itself. The force was to "forage liberally on the country," a phrase that the soldiers took as a broad license, though the same order prohibited them from entering houses or committing any "trespass." Foraging parties were to "refrain from abusive or threatening language"; when they took food or forage they were to give a written statement of what they had taken and "leave with each family a reasonable portion for their maintenance." The destruction of houses, mills, cotton gins, and other structures could be carried out only by order of a corps commander. Where the army met no opposition there was to be no destruction of buildings or facilities not linked to the enemy's military effort.

Considering the patterns of foraging-pillaging and random, indiscriminate destruction that were already well-fixed among Sherman's soldiers, and that they were not about to renounce, there seems to be something almost chimerical about these two orders, which their author described as "clear, emphatic and well-digested." He included the complete texts of both orders in his *Memoirs*, insisting that "no account of that historic event is complete without them." But with a discreet phrase in the same paragraph he reconciled the strictures of Orders 119 and 120 with the grim realities of the Great March: the orders "were obeyed as well as any similar orders ever were, by an army operating wholly in an enemy's country, and dispersed, as we necessarily were, during the subsequent period of nearly six months." The commander had done his part; enforcement of his orders would fall to his subordinates.

A lack of serious opposition was virtually a necessity for the success of Sherman's enterprise, first because the stocks of munitions that it could take along were limited, perhaps enough for one serious battle (despite repeated orders to conserve ammunition the soldiers routinely shot the cattle, hogs, and chickens that would provide their next meal). Then too, if an enemy force should succeed in blocking the army's movement, food supplies would soon run out.

The most serious weakness of Sherman's force was in its draft animals. These had been for the most part kept in the Atlanta area, where fodder was in very short supply. The horses and mules were in bad shape at the beginning of an arduous campaign, and if they could not soon be replaced Sherman's artillery and the 2,500 wagons and 600 ambulances in his columns would be immobilized. Fortunately for the invaders the first several days out their appearance took the population completely by surprise; they found what animals they needed in stables and barn lots.

That Sherman was preparing an expedition toward the coast seems to have been common knowledge in the opposing army; by late October prisoners and scouts reported he would be heading to either Savannah or Mobile. Talk of the move at his headquarters was apparently quite open.

On the eve of his departure a New York newspaper published embarrassing details about the expedition; the leak was traced to one of his quartermasters, and Secretary Stanton was furious. Fortunately for Sherman, any foreknowledge the Confederates had would do them little good, for they had not the force to stop him. There was to be scarcely any fighting, save cavalry clashes with Wheeler, who shadowed the Yankees to Savannah. Georgia, and most of the Confederacy as well, was a hollow shell.

There was one imponderable, and an important one–the weather. The chief danger was in heavy rains that might bog the columns down in Georgia's red clay. Sherman apparently believed, as some farmers did, that a month's quota of rain often came in one wet spell, and when November opened with heavy showers it seemed to confirm his view. On November 8 he sent a message to commanders of all posts: "This is the rain I have been waiting for and as soon as it is over we will be off."

Once on the road he would have to provide for various contingencies and eventualities. Though the route he took was southeast, toward Savannah, he was not completely sure where or when he would reach "salt water," as he put it; to be on the safe side supplies would be waiting for him in depots at both Hilton Head and Pensacola.

The march through Georgia went smoothly and in generally good weather, though heavy rains in the period November 19–22 slowed the columns briefly; the Georgia "leg" of what Sherman would call the Great March ended at Savannah, where on December 13 Sherman's soldiers stormed and easily took Fort McAllister, which he regarded as the key to the city; eight days later came the peaceful and orderly occupation of Savannah, whose 10,000 defenders, commanded by General William J. Hardee, had evacuated the city the preceding night. The army thus reached the coast in excellent condition and spirits, and after connecting with Union naval forces there, was soon ready to take the road again.

By contrast the march across the Carolinas, especially South Carolina, was often arduous in the extreme. It began in the dead of winter and was attended by frequent heavy rains; moreover the army's line of march–laid out to keep the enemy guessing about his destination–lay across a particularly challenging stretch of land. It led over the floodplains of rivers and across innumerable creek bottoms. To keep wagons and caissons from sinking to their hubs it was frequently necessary to "corduroy" the roads, laying across them small trees, limbs, and fence rails. The men lived in mud, rain, and cold. One veteran of the march noted that there were but "three pleasant days" in the entire march from Columbia, South Carolina, to Goldsboro, North Carolina.

The lowland portions of South Carolina that the army crossed were poorer and more sparsely settled than Middle and South Georgia. Foragers had a harder time finding sufficient rations for the men and fodder for the animals; they had to go greater distances, up to fifteen miles from their columns, and this in the dead of winter when the hours of daylight were few. The soldiers slogging through the icy creek bottoms of the Palmetto State in those winter weeks of 1865 must have looked back on their march through Georgia as a summer lark. Then too, the atmosphere changed in a subtle but significant way when Sherman's men entered South Carolina. They were now in the cradle of secession, the home of the most rabid fire-eating rebels; the press and political leaders of South Carolina had promised they would meet the direst of fates if ever they trod her soil. So they were on the qui vive, anticipating violence and more easily moved to it themselves.

The collective experience of the Great March was the subject of so many memoirs and reminiscences that a historian of the expedition encounters an embarrassment of sources. Not so a biographer of the army's commander; if he appears at all in the many surviving diaries and memoirs his is a fleeting presence, more often seen than heard. And since the general was out of touch for weeks at a time, the stream of letters from him necessarily stopped in those periods. He carried a pocket diary for 1865, but what entries it contains are as a rule brief and insignificant. The fullest account he offers of this critical episode in his life is in his *Memoirs*, which are entertaining and informative, though far from the whole story.

But Sherman had his Boswell during the Great March—Henry Hitchcock. In 1864 Hitchcock was a thirty-five-year-old attorney and nephew of Ethan Allen Hitchcock, who arranged for Sherman to appoint Henry judge advocate on his staff with the rank of major. At the end of October Hitchcock reported to his command. Since there was little or nothing for him to do in his official position, Sherman made something of a private secretary of him, which meant that the two were thrown together almost constantly.

Hitchcock was an ideal observer and witness. With a quick and able pen and a lawyer's sense of relevant detail he committed his impressions and comments to his journal and to occasional letters addressed to his wife. Since all was new to him he wrote about everything, including—happily for the historian—matters that habitués to the war rarely bothered to mention. And he was fascinated by Sherman from the first, making him the chief subject of study.

After only forty-eight hours Hitchcock wrote his preliminary impression of Sherman, and it accords well with the judgments made by many others: "He impresses me as a man of power more than any man I remember. Not general intellectual power, not Websterian, but the sort of power which a flash of lightning suggests—as clear, as intense, and as rapid." From time to time he would return to his portrayal of Sherman, filling in details and occasionally making corrections.

Hitchcock also offers a privileged view of life in the general's official family. He recorded that Sherman never slept through the night and seemed particularly active between 3 A.M. and dawn. He noted the shortcomings of the general's mess: Sherman ate anything put before him by Manuel, his not very adept cook, and even nibbled hardtack without complaint. At headquarters the general dominated the conversation. He spoke of other generals with surprising candor and recounted endless anecdotes, including the one about the officer at Fort Corcoran he had threatened to shoot.

As the army advanced, its commander did not lead in the conventional sense; in fact one has the impression that he was carried along almost passively. For a time he and his party would be in one column, then another. Sometimes he was toward the rear of the line of march and sometimes he was in the van. Yet he did direct the army's movements. Much of his time seems to have been spent in coordinating movements, using couriers to communicate with the other columns, and deciding on routes and destinations for the next day's advance. These he fixed the preceding evening, based on the maps he had and on reports from the cavalry, which had explored the country ahead, especially the roads and their condition.

Somewhat to Hitchcock's disappointment there was relatively little combat; the Great March would not be accompanied by a great battle. The most important engagement, that at Bentonville, North Carolina, on

March 19–21, 1865, was fought against a hastily assembled hodgepodge of commands directed by Joseph E. Johnston. The Confederates struck General Slocum's force and checked its advance, but after some initial successes they were obliged to withdraw. Hitchcock noted that when Sherman met any sort of resistance—entrenched troops and artillery blocking a river crossing, for example—he preferred to compel the enemy to retreat by flanking him; the word in headquarters argot was "flinking." Hitchcock was an apt pupil. He wrote in his diary, "I begin to understand what a science war is in the hands of a master."

The most spectacular display of violence on the march, and the most memorable, was not a battle but the occupation of a city. Savannah, the grande dame of Georgia's towns, had received Sherman's army with good grace, and the Northern soldiers reciprocated with what witnesses on both sides acknowledged as generally good behavior; vandalism and pillage—never completely absent—were not much in evidence.

Columbia, South Carolina, was not so fortunate. The Northern soldiers' penchant for "settling scores with the Secesh" had increased in South Carolina, and some of them promised themselves a particularly spectacular show in the state's capital; yet Columbia only witnessed on a grand scale what was occurring in smaller towns. There on the evening of February 17 one could see the same scene as that enacted in nearby Winnsboro: soldiers in blue uniform rushing about setting fires while other soldiers, also in blue uniform, rushed about putting them out. The first Union troops entered a city where disorder and pillage were already in full swing, having been initiated by Wheeler's cavalry on their way out of town. Liquor was flowing freely. The troops first assigned as provost guard proved unequal to the task of restoring order and had to be replaced; some of them joined the revelers and the looters.

The raw materials for a vast conflagration were at hand: cotton bales in the streets, lint flying about, and a high wind blowing through block after block of wooden structures. Finally alcohol, that never-failing solvent for reason and restraint, was being consumed that night in large quantities—by white and black, soldier and civilian, Yankee and rebel alike.

When the flames finally died down much of Columbia lay in ashes. In coming years Sherman would spend considerable time trying to explain—in some way compatible with his *amour-propre*—just what happened; but his terse diary entry for February 18, 1865, probably tells the essential story: "Columbia burned fire high wind. Cotton in the streets fired by the enemy, and the great animosity of our men—just distress of people." The use of the word "just" is curious; by it Sherman presumably meant "merited."

Hitchcock was a shocked witness to the conflagration in Columbia. He described to his wife a scene he could never have imagined: "Our own officers shot our men down like dogs wherever they were found riotous or drunk." War, it seemed, and all that was connected with it, was filled with complexities and contradictions.

Hitchcock noticed other things that didn't seem to square with reason: as the army advanced toward Savannah it was wrecking the Georgia Central Railroad; indeed, the route of march followed the rail line expressly to ensure its complete destruction. Yet even a newcomer to war like Hitchcock could see that the job was not being thoroughly done; the men doing the work, from the Seventeenth Corps, didn't seem to be doing much more than burning the ties.

There were other things about this army that seemed "unsoldierly." It was attracting any number of blacks, captivated by the heady air of freedom the Yankees brought. Hitchcock followed the example of other officers and took on two servants, while other able-bodied black males found places in the army's pioneer corps; most of the ex-slaves were not that lucky, but the tatterdemalion mass followed along, at times threatening to encumber the columns; while officers tried to discourage these fellow travelers, the men in the ranks would often invite them to come along.

The army itself gradually took on a ragamuffin appearance. Uniforms were soiled and in many cases torn or frayed; here and there a soldier would replace regulation dress with something whimsical picked up along the way; he might discard a worn-out cap for a top hat brought back from a foraging expedition, and none of his officers would say anything. Sherman's force might be a first-rate fighting body, but it no longer had the appearance of one.

Then much of the foraging seemed to be done in blithe ignorance of Special Field Order, No. 120. Though foragers were forbidden from moving ahead of the army itself, the best pickings were there; when the cavalry units leading the advance entered a community they all too often came upon the army's "bummers" in scenes of drunken riot and plundering. And when the commandeered wagons and buggies rolled into camp, their cargoes were the most varied: "pumpkins, chickens, cabbages, guinea fowls, carrots, turkeys, onions, squashes, a shoat, sorghum, a looking glass, an Italian harp, sweetmeats, a peacock, a rocking chair, a gourd, a bass viol . . . and every other thing a lot of foot soldiers would take it into their heads to bring away."

Much of what had been brought in was abandoned at the campsite the next morning. The men as a rule refused to carry along much food, preferring to depend on what the foragers would bring in next time; the quartermasters supervised carefully what went into the army's wagons and no other vehicles were permitted to join the columns, so whatever a soldier had picked up he would have to transport himself. A man might decide to carry the pier glass he had "captured" for his wife all the way to Savannah, but his resolve would give out in a couple hundred yards and the mirror would end up in the ditch. The army's track was littered with such debris. Carl Schurz, who was along, recalled seeing a baby's cradle sitting abandoned along the roadside, a mile from the nearest habitation.

In some ways the indiscriminate destruction that attended the army's passage worked a hardship on the invaders themselves. Once in South Carolina a party of stragglers opened the sluice gates of a dam, flooding a ford downstream where a column was trying to cross. Soldiers set fire to granaries the cavalry needed to feed its mounts, and to mills that the army's commissaries needed to turn wheat into flour; pillagers ransacked courthouses and post offices before officers could examine their contents. Foragers might draw off a bucket of molasses, then pour out the remainder of the barrel "so the Rebs won't get it." Then another foraging party would conclude that the owner had poured out his own molasses to keep them from having any—and burn his house for him.

Sherman's subordinates issued stern reminders to supplement Special Field Order, No. 120. The relevant volume of the *Official Records* contains half a hundred issued at army, corps, and division level during the first month of the march, and rather fewer—a score or so—while the invaders passed through South Carolina (a Draconian directive of General Howard calling for pillagers to be shot on sight but was apparently not enforced). These orders—and no doubt others omitted from the *Official Records*—dealt over and over with the same problems: straggling, plundering, setting fires, and the unauthorized taking and using of horses and mules.

Here Sherman rarely intervened. General Wheeler sent word he would leave cotton for the Yankees if they would refrain from burning houses—and left behind 300 bales as a token of his sincerity. Sherman replied: "I hope you burn all cotton and spare us the trouble. We don't want it, and it has proven a curse to our country. All you don't burn I will. As to private houses occupied by peaceable families, my orders are to spare them, and I believe my orders are obeyed." Sherman directed his cavalry leader, General Kilpatrick, to spare occupied dwellings, but added: "If people vacate their houses, I don't think they should expect us to protect them." When the army moved from South to North Carolina early in March, Sherman told Slocum, and possibly others, that "a little moderation" would be politic.

Here and there, then, in various of its parts and workings, the vast and complex mechanism that was Sherman's army was not performing as it was supposed to. In the matter of discipline, where the failings were most obvious, it seemed to Hitchcock that the blame should be placed at the door of those on the spot when the transgressions were committed, the company and regimental officers. Charles Ewing claimed that the fault lay instead with the highest authorities in the Lincoln administration: they would not permit the army to apply the death penalty. Officers of the regular army, Sherman prominent among them, believed that the problem was the unwillingness of volunteer officers generally to act decisively in their commands; sitting on courts-martial, they diluted penalties even further, to the point that fear of punishment scarcely influenced the soldier at all.

But the more Hitchcock reflected on it, the more he became convinced that the source of many of the army's problems was its commander. Orlando Poe confided to him that he had gone to Sherman and offered to take his engineer regiment and wreck the Georgia Central the way it should be wrecked. But Sherman merely acknowledged that the men of the Seventeenth Corps "won't do it right." He had not given Poe orders to take over the work, and Poe was sure he wouldn't.

And Hitchcock himself was struck from time to time by the way this man of strong opinions, trenchant language, and vigorous action could show complete indifference before situations that seemed to cry out for remedy. One day as the general and his suite rode along they encountered a drunk soldier lying on the roadside. As they approached the man began to curse Sherman loudly and kept up with a stream of imprecations as they passed by. The general rode on by, seemingly oblivious, though he could not have failed to hear the choice oaths the drunk bawled at him; those in his suite, abashed by the scene, could think of nothing better to do than imitate their leader.

Sherman's passivity, his laissez-faire attitude in some matters was striking, notably concerning the depredations of his own troops. When an old woman called out to the general to help her save her chickens, which his soldiers were chasing and gathering up, he told her he could not stop. He explained to Hitchcock, who had witnessed the scene, "there's no help for it. I'll have to harden my heart to these things."

But this attitude of seeming indifference to the soldiers' excesses could encourage them in this very direction. General Jefferson C. Davis told Hitchcock that the men believed that Sherman actually wanted them to do such things, and he had heard soldiers say as much. It had not always been so; Sherman assured Hitchcock that during the first two years of war "no man could have done more than he against everything of the sort—has personally beat and kicked men out of the yards for merely going inside, etc., etc." But Hitchcock knew that he was no longer making the same effort, "indeed he admits as much." So Sherman, like many of his subordinates, had given up what had proved a hopeless cause.

At first Hitchcock believed that this slackness in matters of discipline was part of Sherman's general tendency to ignore a persistent problem, to let things slide: "He does not seem to me to *carry things out* in this respect." But eventually he came to the conclusion that this slackening of the reins was intentional, and a part of Sherman's policy to make war so distasteful to the Southern people that they would accept peace any way they could get it—thus in a way the general really did want his men to be fearsome and destructive—in their conversations Sherman seemed to be telling him as much. And in principle Hitchcock came to agree with him: "General Sherman is right—the only way to end this unhappy and dreadful conflict . . . is to make it *terrible beyond endurance*."

But here again the general was not consistent. When he and Hitchcock stopped to talk to a man named Vaun in Middle Georgia, Vaun told them he was worried over what might happen to his possessions. Sherman advised him "very kindly" to put as much as he could inside the house; he even got some sacks for Vaun to put his wheat in. When the general learned that his bummers had carried off all the food from the Confederate military hospital in Milledgeville, he called in its director for a chat and undertook to stock the hospital with food and medical supplies as well; he personally ordered an eight-man guard for the place. And there are other recorded instances of his kindness and consideration during the Great March.

And in his defense it must be said that had he been disposed to take up the struggle against plunder and arson that General Howard was then waging, to persist in the good fight for the "honorable, manly warfare" he extolled—even if it were exclusively for the well-being of the army itself—he would have had no better luck. It was possibly this consideration that led him to the "why fight it" attitude shared by many of his subordinates.

To do otherwise would be labor lost. These volunteer soldiers did indeed think for themselves, and they thought it no great crime to make life hard for the "Secesh," in fact to some it was no crime at all, regardless of what the Articles of War said. And even soldiers who abstained from robbery or vandalism were the silent accomplices of those who did such things. "Houses are fired under our very feet," Sherman complained to a fellow general, "& though hundreds know why & who did it, yet the commanding general cannot get a clue." Thus there was a sort of solidarity among the men that made the detection of the malefactors in their midst very difficult. It could also be dangerous. General Howard encountered a wall of hostility among the men when he tried to crack down; he sensed that if he persisted he could be faced with armed resistance from an "organized banditti." Sherman had already wrestled with mutinous regiments back in 1861; he had no desire to see such troubles reappear in mid-campaign.

However irksome he may have found the indiscipline in his troops, Sherman seems to have been in generally good humor on the Great March. Neither Hitchcock nor Poe noted in their diaries any episodes of rage or periods of brooding silence. One officer who was along said the general seemed as happy as a schoolboy on a holiday, and there is in fact considerable evidence to suggest that Sherman went through his last campaign on something of a high. On the first leg of the march he seemed confident he could reach Savannah, sure of his rendezvous there with the U.S. Navy, and untroubled by doubts that he could take the city; if he had any worries about the loose ends he had left behind him, these were laid to rest when news reached him in Savannah that General Thomas, with some prodding from Grant, had decisively beaten Hood at the Battle of Nashville. And this same confidence and optimism seem to have carried Sherman through the long, difficult trek across the Carolinas. The hard marching on the wretched roads was taking him to fame and glory, and he knew it.

He left Atlanta with the plaudits of the whole nation ringing in his ears. When he reached Savannah the surge of public acclaim was even greater than before. Once again he was deluged with letters, with requests for his signature and locks of his hair; the press now sang his praises and gifts flowed in, his known taste in bourbon and cigars catered to by countless admirers. It was enough to turn a man's head—and it was a narcissist's paradise.

And his reaction was that of the narcissist. That this adulation afforded him intense pleasure is eminently understandable, but his responses to it are significant. A letter to Halleck, whom he now addressed as "my dear friend," began with a complaint: "I deeply regret that I am threatened with that curse to all peace and comfort—popularity." To Grant he wrote as the dutiful subordinate, concerned that his recent successes might make Congress think that he was in competition with his superior, and a man driven only by ambition: "I know I feel none, and today will gladly surrender my position and influence to any other who is better able to wield the power." Yet in the same letter he was emboldened to offer his chief some advice: "As soon as possible, if I were in your place, I would break up the Army of the James, make the Richmond army one."

With his family he took a different tack. Here the letters are essentially devoid of those expressions of modesty or indifference that a man makes, for form's sake at least, when his cup runneth over—gestures such as he had made to Grant and Halleck. In his letters to family members, on the contrary, one has the impression that he could not have agreed more with those who were singing his praises. Indeed he found ways to join in the chorus, as in this letter to Thomas Ewing: "You have lived to see the little red-headed urchin not only handle a hundred thousand men, smoothly & easily, but fight them in tens and fifty thousand at a distance of hundreds of miles from his arsenals and sources of supply." And he went on to point out that such accomplishments had won him extraordinary trust and loyalty on the part of his men. Such was their devotion that without a moment's hesitation they would "march to certain death" for him. As for his nominal superiors, they too held him in greatest affection and esteem. He confided to Ellen, "Grant is almost childlike in his love for me."

Since Ellen understood little about war, he ticked off for her the most remarkable features of his operations in Georgia: "the quiet preparation I made before the Atlanta campaign, the rapid movement on Resaca, the crossing of the Chattahoochee without loss in the face of a skillful general with a good army, the movement on Jonesboro whereby Atlanta fell, and the resolution to divide my army with a part to take Savannah and the other to meet Hood in Tennessee." These achievements, he assured her, "are all clearly mine and will survive us both in history."

Sherman's tendency to enlarge upon his mandate as commander, sometimes into areas where he had no business, has been noted earlier, and appears to be closely related to his enhanced status and self-image and tendency to grandiosity. In the last twelve months of the war, when he stood second only to Grant, his correspondence reveals increasing evidence of initiatives in policy matters, with the launching of the deportation scheme within his Military Division and his peace overtures to Governor Brown being good examples. He rarely referred matters to Washington now, resolving questions that arose himself, including some of considerable importance. When McPherson wrote him about the disposition to be made of 170 bales of cotton—a matter that directly concerned the Quartermaster's Department and the U.S. Treasury—he replied with a sweeping pronouncement: "My decision is that all property, cotton, horses, mules, or any movable thing within the lines of the public enemy is lost, as much so as in a case of shipwreck or conflagration." And he considered such seizures his to dispose of (he sometimes made gifts of cotton to meritorious Unionists who had been despoiled).

These actions provoked no disagreements with Grant, for Grant did not believe in close supervision and let his lieutenant do largely as he pleased. Halleck was still supportive, but with Secretary Stanton there were several collisions. In the fall of 1864 Sherman lost his patience with various state functionaries sent to check on the needs and welfare of their state's volunteers. In September he sent Governor John Brough of Ohio a peremptory note saying, "don't send commissioners for sick, pay, or anything else to this army." Shortly thereafter, when he discovered an "Indiana State Sanitary Agency" operating out of an office in Atlanta, he shut the agency down, packing off its civilian personnel and turning its stock of supplies over to his quartermaster.

Governor Morton wrote a letter of protest to Stanton, who in turn wired the general for an explanation. Sherman replied that he had put the agency out of business because he "excluded all civilians," and Stanton might assure Governor Morton that Indiana had not been singled out. His action was "fair and uniform, and applies to all state agencies." And the letter to the secretary of war concluded on a light, almost flippant note: "Give Jeff. Davis my personal and official thanks for abolishing cotton and substituting corn and sweet potatoes in the South."

Then the Southern newspapers had taken to comparing the passage of his army to the barbarian invasion of Rome, and him to Alaric or Attila. Rather than being enraged, he was titillated by the simile—so much so that he took to using it himself. While in Savannah he sent a long, chatty letter to Mrs. A. A. Draper of Charleston in which he reminisced about Charlestonians he had known in the forties, about his own life since then, and how the world had changed: "But how strange! Is it not? That I who used to ride all night to dinner with Hardy or Sally Quash . . . should now be the Leader of the Vandal horde that has made its Mark from the Mississippi to the Atlantic." (Mrs. Draper never answered his letter, but contented herself with writing acid comments on the margins.)

In all of this ebullience, this soaring self-confidence, there is more than a hint of grandiosity. It would shortly lead him to commit the most spectacular blunder of his career.

The conflict was rapidly drawing to its conclusion now. On March 11 Sherman was in Fayetteville, North Carolina, where messengers from the Union forces in eastern North Carolina reached him. He and his army were once again in touch with the outside world. He drafted a long report to Grant, then pushed on toward Goldsboro. On the way occurred the collisions with Johnston's forces at Averasboro (March 16) and Bentonville (March 19–21).

At Goldsboro Sherman received supplies and reinforcements from the coast. Since he now had a rail line east he confided his command to

General Schofield and made a quick trip to City Point. Lincoln too had come to see Grant, so on the afternoon of March 27 and again about noon of the day following, the president, Grant, and Sherman met to talk about the coming end of the war.

Sherman briefly described the meetings as he could recall them a decade later. Fortunately Admiral Porter sat in on one of the meetings and wrote an account of it. According to Porter, Lincoln was anxious for the conflict to end as soon as possible and without any more of the terrible battles that had marked its course. Sherman was to assure North Carolina's Governor Zebulon Vance that once the Confederate forces laid down their arms his people would be welcomed back into the fold. According to Porter, Lincoln "wanted peace on almost any terms"; when Sherman told him that he could force Johnston to take about any conditions he wanted to, "the President was very decided about the matter, and insisted that the surrender of Johnston's army must be on any terms." The talks concluded and Sherman took his leave of the president; he would never see him again.

Thereafter events came in swift succession. The Confederate bastion of Petersburg fell to Grant's army on April 2; the following day Federal forces took possession of Richmond. On the seventh Grant and Lee exchanged notes about terms for the surrender of the latter's forces and two days later the two generals signed a simple surrender agreement. Lee's troops would be paroled and allowed to return to their homes upon laying down their arms and pledging not to take them up again; there were some stipulations regarding the disposal of the defeated army's weapons, equipment, and animals, but no "political" provisions, for Grant had received emphatic orders from Secretary Stanton that such matters were beyond his competence.

The news was some time in reaching Sherman, who set his army in motion on the tenth in the direction of Raleigh, with Johnston's force retreating before him. On the night of the eleventh word came of Lee's surrender; Sherman announced it the next morning, setting off a roaring celebration among his troops. Sherman himself had received peace feelers from Governor Vance on the tenth and had received his emissaries, but when Northern troops occupied Raleigh on the thirteenth Vance fled. The following day came a letter by flag of truce from Joe Johnston; it referred to "the results of the recent campaign in Virginia" and proposed a suspension of operations so that "civil authorities could make arrangements to terminate the existing war." Sherman resolved to accept, sending word to Grant and Stanton: "I will accept the same terms as Gen. Grant gave Gen. Lee, and be careful not to complicate any points of civil policy."

Sherman rode out to meet his opponent on the morning of the seventeenth; just before leaving he was handed a telegram announcing the assassination of the president. He swore the telegraph operator to secrecy, stuffed the telegram into his pocket, and went on. The two parties of horsemen met on a stretch of country road, and Sherman recognized Johnston from pictures he had seen of him. The two generals shook hands and introduced the officers accompanying them; then they retired to a nearby house whose owners, a couple named Bennett, agreed to vacate it while the generals talked.

Once they were alone together Sherman showed the telegram to Johnston, who pronounced it "the greatest possible calamity" and denied it was part of any Southern plot. Sherman said he was sure the Confederate Army was not involved. Then he proposed the same terms Grant had offered Lee; Johnston countered with a suggestion that they try to go further and "arrange the terms for a permanent peace." Both men were interested in a more comprehensive settlement, but for different reasons. The invitation to meet that Johnston sent to Sherman had actually been composed by Jefferson Davis. The Confederate president, now a fugitive in his own land, was playing the final card of a very weak hand; his reference to arrangements being made by "civil authorities" might at least allow him a voice in the extinction of his own regime.

What Sherman saw in a broader agreement was surrender of all remaining Confederate forces east and west of the Mississippi; if left to themselves, he was afraid that some of those commands, taken over by the "young bloods" and diehards, would slip into partisan warfare and banditry, keeping the South in turmoil for years. This had been on his mind; six days before he had written Ellen that Jefferson Davis "and at least 100,000 men in the South must die or be banished before we can think of peace. I know them like a book. They can't help it any more than Indians can their wild nature."

And there is no doubt that his mind was turning over a dazzling possibility: this would be a stroke far more brilliant than the detachment of Georgia from the Confederacy. And it would be "good war" at its best: the great conflict's final, dramatically charged scene with him the magnanimous conqueror, extending his hand to the vanquished. But for the moment he had urgent business back at headquarters: he had to break the news to his army that its commander in chief had fallen to an assassin's bullet; if the story got out in some distorted form the town of Raleigh might well go the way of Columbia. The two generals agreed to meet again the following afternoon, then Sherman hurried back to Raleigh. His soldiers were profoundly affected by the news of Lincoln's death, but committed no serious depredations.

When the two parties met the next day at the Bennett house, Johnston suggested that Confederate Secretary of War John C. Breckinridge, who had come along, might be helpful in drafting mutually agreeable terms. At first Sherman balked at the participation of a "political" figure in the talks, but Johnston pointed out that Breckinridge was also a Confederate general, and for the purposes of their meeting he could "sink" his cabinet status. After considerable discussion, Sherman agreed. It was Sherman who framed the agreement, to be applicable to all Confederate commands.

The Confederates were to agree to "cease from acts of war" and to obey state and federal authorities; the legislatures and officers of the various states would take oaths of loyalty and upon doing so would be recognized in their functions by the Federal executive; where there were rival state governments the U.S. Supreme Court would decide the issue. Federal courts would be reestablished, there was to be a general amnesty, with the people of the Southern states guaranteed the franchise and their "rights of person and property."

At Appomattox Lee's army had surrendered its weaponry on the spot, save for officers' sidearms, but in Sherman's set of terms the Southern units would keep their weapons and their organization until they reached their respective state capitals. Johnston had not asked for this concession, it was Sherman's idea: ever the friend of order, he reasoned that if news of the surrender sparked civil upheavals, Southern officials would need an armed force to suppress them.

Copies of the agreement had to be prepared for the signatures of the two generals, and while this was being done Sherman, Breckinridge, and Johnston rejoined their companions, who had been waiting in the yard. Sherman advised Breckinridge to leave the country and may have offered to help–it would be a more polite form of the deportation he had long advocated. It was nearly dark when the copies were finally ready. The principals signed them and the delegations parted.

The feeling was probably euphoric among the leaders on both sides (Sherman, Johnston, and Breckinridge had sipped bourbon while putting the agreement together). The accord would need approval by Presidents Johnson and Davis, but Sherman returned to his headquarters confident that he had brought an end to the war.

His immediate subordinates–Howard, Blair, Logan, and the others–also felt that way; Henry Hitchcock gathered as much, and General Slocum, who talked with them about it, knew that was the case. But Hitchcock, whom Sherman charged with carrying the agreement to Washington, left with the personal conviction he was carrying a dead letter. Carl Schurz, who had just joined Howard's staff and got the details from Slocum, was "astonished" by its provisions. He too saw that both Sherman and his subordinates had had their heads turned by the momentous nature of the event they were involved in. They did not or could not see that recognition of existing state and local governments in the South would confirm in power the most prominent partisans of rebellion; the provision confirming property rights was a veritable minefield. To Schurz "it required no extraordinary political foresight to predict the prompt rejection of the Sherman-Johnston agreement."

Sherman, now in a flurry of letter writing, was of another mind. He sent a copy of the accord addressed to Halleck, announcing that it would produce "peace from the Potomac to the Rio Grande," a happy phrase he would repeat several times in the next few days. He urged his friend not to let President Johnson change any of the terms, "for I have considered everything and believe that the Confederate armies once dispersed we can adjust all else fairly and well." He wrote Easton, his quartermaster: "I have no doubt that I have Thursday made terms with Johnston that will close the war and leave us only to march home." He was so sure of this that he shared the news with his troops and canceled a seaborne expedition to

Georgia, since "all the Confederate armies will be disbanded under a convention made between me & General Johnston."

It was Sherman's letters to Ellen that revealed most fully what was running through his mind as he awaited Henry Hitchcock's return. The day he signed the agreement with Johnston he explained it to her, adding, "I can hardly realize it, but I see no slip." And incredibly, he added,that the terms were "all in our favor." For those on the other side prospects would be bleak; he was not thinking of the Southern people here, but of Wade Hampton, Wirt Adams, Forrest, and the others who had gloried in the war and would now have no future; maybe he could help them get out of the country. Then he speculated on what he himself would do once his agreement was accepted: he would make a quick trip to Charleston and Savannah to arrange matters there, then he would have his army march to Frederick or Hagerstown, Maryland, where he would have "a magnificent pageant" before his men were all mustered out. He would bring her and the children down to see that final, grand review. Then, with peace restored, he would receive a military fiefdom at the hands of Grant: Meade would have the Atlantic slope, Halleck the Pacific, and he the great valley of the Mississippi, and somewhere there the Sherman family would at long last settle down–Chicago was a town with future greatness, but in winter too cold to live in. "How would Memphis suit you as a home?" Then he reined himself in: "But I am counting my chickens before they are hatched . . . "

He was anxiously awaiting Hitchcock's return from Washington; on the evening of the twenty-third he got a terse telegram from him, sent from Morehead City, saying he would be back the next day. There was no hint of the news he brought. About six the next morning Hitchcock appeared at Sherman's headquarters in the governor's "palace" in Raleigh. Sherman was up but not dressed; when he went to meet Hitchcock he found his emissary was not alone; with him was Grant. Sherman may have known at that moment that the news they brought was not good, that his agreement with Johnston had been rejected. Grant explained that was indeed the case. Sherman would have to break the news to Johnston, then negotiate a new accord–and Grant had been sent to see that this time it was done correctly.

Sherman was now aware that he had incurred official displeasure; soon he would make another discovery: he and his agreement with Johnston were the target of denunciations by both public figures and a frenzied press. The shock of Lincoln's assassination, the bizarre circumstances surrounding it, and the wild and lurid stories they engendered, had accustomed the public to swallow the most implausible stories. Now it was reported that Sherman had lapsed once more into lunacy, or that he had been working hand in glove with Jefferson Davis to smuggle out of the country a great horde of gold from the Confederate treasury, or that he was angling to replace Andrew Johnson in the White House.

On April 26, while this agitation was sweeping the country, Sherman and Johnston signed another agreement, similar to that Grant had made with Lee. Under Sherman's signature Grant later placed his own, along with the word "approved." The most mortal conflict in the nation's history was over. So, quite possibly, was the career of Major General William Tecumseh Sherman.

# 18

## THE FIRST
## YEARS OF PEACE

SHERMAN WAS UNDERSTANDABLY worried about how he stood with his superiors in Washington, especially President Johnson and Secretary of War Stanton. He knew that Johnson had been unhappy over his refusal to send forces from Kentucky to support Tennessee loyalists during those dark days in the autumn of 1861; and then Johnson would have preferred George H. Thomas as Grant's successor in the Military Division of the Mississippi. Sherman had also had differences with Stanton, and at the moment Stanton seemed to be the dominant figure in the new president's entourage. So he addressed the secretary of war a carefully reasoned letter, admitting his error–he even used the word "folly"–and stressing his loyalty and good intentions.

But Sherman was isolated there in the executive mansion in Raleigh. Grant had departed on the twenty-seventh, apparently without saying much about the possible fallout from the abortive agreement of the eighteenth. There was as yet no direct telegraphic link with Washington or the North, where the general's blunder at the negotiating table had in fact become the talk of the hour, competing for public attention with the hunt for Lincoln's assassins. The tide of opinion was running strongly against Sherman. The press was outspoken on the issue, with the *New York Times* portraying the first accord as "Sherman's surrender to Johnston." A

Philadelphia publisher later told the general, "My paper was the only one in Philadelphia that did not denounce you." President Johnson was continually receiving reports on the public mood in various parts of the Union, and these too indicated strong condemnation of Sherman's action. "Poor Sherman has at last verified your opinion of him," wrote a correspondent in Tennessee, "and after a long and gallant defence of the country, has I fear spoiled it all. Even the soldiers here are loud in his abuse. He must have been crazy."

Not until April 28, when a four-day-old copy of the *Times* came into his hands, did Sherman learn of the firestorm he had unwittingly ignited. Secretary Stanton seemed to be feeding the flames. A letter he had written to General John Dix had gotten into journalists' hands, so that the country at large could share the secretary's fears: Sherman's truce might aid Jefferson Davis in his flight from Richmond and make it easier for the rebels to evacuate the gold from their treasury—which they might offer to share with Sherman in exchange for safe passage. Moreover a friend of long standing was subscribing to these views: Halleck. Stanton had unceremoniously evicted him from his post and exiled him to a command in Richmond; once there Halleck ordered his subordinates to disregard orders coming from Sherman.

The general reacted with rage. Carl Schurz saw it at white heat one evening in the executive mansion in Raleigh: "About a dozen or so of generals were assembled in a large, bare room. They were all in a disturbed state of mind at the turn affairs had taken, and had come to get from Sherman the latest news. They sat or stood around in rather mute expectation. But Sherman was not mute. He paced up and down the room like a caged lion, and, without addressing anybody in particular, unbosomed himself with an eloquence of furious invective which for a while made us all stare. He lashed the secretary of war as a mean, scheming vindictive politician, who made it his business to rob military men of the credit earned by exposing their lives in the service of their country. He berated the people who blamed him for what he had done as a mass of fools, not worth fighting for, who did not know when a thing was well-done. He railed at the press, which had altogether too much freedom; which had become an engine of vilification; which should be bridled by severe laws, so that fellows who wielded too loose a pen might be put behind bars—and so on, and so on."

It was a still-seething Sherman who issued orders to his troops to begin the move northward by easy stages to Richmond, where he would rejoin them. He would have to travel in another direction, taking ship to Charleston, Savannah, and other points along the coast, arranging for their garrisoning. The trip was a quick one; by May 10 he caught up with his troops in the former capital of the Confederacy. There Halleck addressed him in cordial tones, assuring him of his friendship, inviting him to stay in quarters set aside for him, and indicating his desire to review part of his host as it passed through. Sherman's response was anything but cordial: he would not be needing accommodations and his troops would not pass in review; moreover he advised Halleck to stay out of sight—perdu, as he put it—during their passage; if the veterans of the Western army spotted him they might do something regrettable.

Nor was that all. He wrote out a report on the final month's operations, and it was like no report he had written before. It was not so much a chronology of marches and encampments as a wide-ranging defense and justification for his conduct. There was a reproof for Stanton for making "official matter" (the secretary's letter to General Dix) available to the press and affording reading matter for "every bar-room loafer in New York." Halleck too got severe handling: "It may be that General Halleck's troops can outmarch mine, but there is nothing in their history to show it."

The report was a gesture that could well jeopardize his career, but at that point he had probably decided he would never again be given a position of importance in the army anyway. Stanton and Halleck were powerful enemies; President Johnson had little reason to favor him. The only man who could defend him effectively was Grant, but he had not heard from his friend since they parted in Raleigh two weeks previously, and this long silence could mean that Grant too was distancing himself. On May 10 Sherman wrote him a bleak letter: "I regard my military career as ended, save and except so far as to put my army into your hands."

But in the next few days his outlook changed. He received a letter from Grant saying that despite their differing views on that first accord with Johnston, he valued Sherman's services and would continue to count on them. Then press attacks were fading before the general euphoria over victory and peace washing over the country. A friend close to President Johnson sent word that "already the reaction in your favor is manifest."

Sherman was thus at his most peppery on May 14 when he informed General Rawlins that "Vandal Sherman is encamped half way between the Long Bridge & Alexandria to the west of the Road, where his friends, if any, can find him. Though in disgrace, he is untamed and unconquered."

By the time he reached Washington he had probably dropped the idea of leaving the service. In the army he enjoyed a status and position not easily found in civilian life. His salary, about $500 a month, would be sufficient for his family's needs; moreover efforts were under way to raise sums of money to present to him as a tangible form of appreciation for his services—in Ohio they were seeking a hundred thousand dollars; these efforts could be jeopardized if he quit the uniform. Then too, according to Sherman's view of things, the war was not yet over; having whipped the Southern secessionists, the army would soon have to deal with the fractious mobs of the great cities. In the New York draft riots of 1863 he had seen the onset of what he called the "second stage" of the revolution—a fight he did not want to miss.

On May 22 Sherman gave testimony before the congressional Committee on the Conduct of the War. There, in a nimble volte-face, he managed to put a completely different face on his role in the negotiations with Johnston: he never had any intention of framing a surrender agreement himself. He had only drawn up an agreement filled with "glittering generalities," counting on his superiors in Washington to recast the draft he sent them and fix the surrender terms. He would embroider on this theme, telling historian John Draper: "I wrote out those terms myself as carelessly and hastily as I write this letter, in the firm belief that they would never see the light, other than after the Cabinet had remodelled them and sent me specific 'orders.' I know that I never aspired to the position of 'Pacificator,' or dreamed of pointing out the wisest and best way to lead our people from a state of active war to one of peace."

He could not yet bring himself to make a conciliatory gesture to Stanton, but on May 24 he wrote the assistant adjutant general in Washington with an obvious if indirect message for his superiors in the capital: "I will be willing to shape my official and private conduct to suit their wishes." Four days later he undertook further fence-mending: he wrote Grant to say that while he would leave the army "if not wanted," should he remain in uniform he would "serve the President not only with fidelity but with zeal"; he added that he had absolutely no presidential aspirations, despite the wild stories carried in the newspapers. And he had a favor to ask of his friend: "I would like Mr. Johnson to read this letter . . . "

He now felt sufficiently emboldened to do something he had been thinking of for some time. Back in March he had described his army to Ellen as "ragged, dirty and sassy," and he confessed to her: "I would like to march this army through New York just as it is today, with its pack mules, cattle, niggers and bummers." Just why he wanted to do this one can only guess; perhaps to flaunt before Eastern cosmopolites the distinctive Western "style" in soldiering, or perhaps to show their majesties the mob that in this body of warriors they would meet their match. Now a final chance to parade his army was at hand—not in New York but in Washington. The Grand Review was to be carried out in two phases: on May 23 the Army of the Potomac would parade through the city and on the following day the Western army.

By May 16 he had resolved he would personally lead his army down Pennsylvania Avenue just as it was, "in the rough," as he put it; he wired Ellen to come and bring the children (she had to leave the children in school, but she would be there to savor her husband's triumph). The serried ranks of Eastern soldiers presented a properly ordered martial display, but Sherman's veterans offered a far more animated and colorful spectacle and stole the show. After passing the dignitaries on the reviewing stand, Sherman left the marchers and joined the president's party. There he shook the hand of every man save the secretary of war; as to whether Stanton extended his hand witnesses disagreed.

In his letter to the assistant adjutant general, Sherman had expressed a desire to be posted somewhere in the West, and Grant could now confirm that he would have the Military Division of the Mississippi (soon to be called the Division of the Missouri), one of five great circumscriptions to be commanded by major generals in the regular army.

He had known that he would not have a Southern command, since his conservative views about the region were well-known to the Congress, which did not share them. It was just as well, if one may judge by a speech he made in Arkansas. The state was briefly in his jurisdiction and he made a trip to Little Rock in December 1865, where he sounded a familiar theme. For the state's multitude of problems he had a single remedy:

"What is right or wrong I do not know and do not think you know, but we all understand what is law. Inquire as to what is lawful, that is plain and simple. Act lawfully, and you will do right." If they didn't know what was legal, Arkansans had only to address themselves to lawyers, of which there was an ample number.

Sherman's headquarters would be in St. Louis, where he needed to be settled by September so the Sherman children could start school. But for the moment he was going to take leave and enjoy a traveling vacation with his wife. Their first major stop was New York City, and it must have been a profoundly gratifying experience, for it demonstrated that despite the recent hue and cry, Sherman the soldier remained broadly popular. The Board of Aldermen saluted him with pardonable exaggeration as "the victor of a hundred battles," and proclaimed the couple the city's guests. The Shermans' presence was the occasion for an endless round of ceremonies and festivities: at ten o'clock on Saturday morning, June 3, there was an "informal levee" at a large residence on East Twenty-third Street. There the Shermans met a host of dignitaries, but these were supplemented by "unbidden guests" who poured through the mansion's open doors for most of the day.

The general slipped away at two in the afternoon but made the mistake of riding in an open carriage. He was shadowed by a *Times* reporter who recorded that everywhere people in the street recognized his "familiar and strongly marked features"; as a result the progress of the carriage was slowed and occasionally stopped by crowds of well-wishers. He had intended to call on some friends, but when he reached their homes he was prevented from leaving the carriage by the press of the crowd; he could only send in his card. At five he escaped the throng to a dinner at the Union Club where the mayor made a speech of welcome and Sherman responded; a crowd had gathered outside, so the general had to go to the window and say a few words "to appease their earnest clamors."

He had become a celebrity, and he was destined to remain one for the twenty-five years of life that remained to him. This status would be to him an endless source of both pain and pleasure. He never enjoyed what the French call the *bain de foule*, immersion in the swarming, sweating, overfamiliar crowd, the masses frantic to shake his hand, to pat him on the back, or simply to touch him. Under these assaults he would sometimes grow violently angry, cursing and pushing his tormentors away. But on other occasions, when he looked down a banquet table at the line of admiring faces fixed on his own, or when he rose and made his way to the rostrum of a packed hall, striding into a swelling thunder of applause–then the pleasure he knew must have been of the keenest kind; over the next quarter-century he would never tire of it–indeed he would make it a central element in his life.

The Shermans' trip ended with another popular outpouring, one that the general probably relished less, a welcome to Lancaster delivered in the form of a long and sonorous speech. By then it was the end of June and time was short; he needed to settle his family in St. Louis and take up his new duties. In the past transplanting the Sherman menage had meant heavy expenses, but this time the financial burden would be lightened by popular subscriptions. In Ohio the "Sherman Testimonial Fund" was trying to raise its target sum of $100,000 from the nickels and dimes of the masses (an idea that Grant seems to have floated, through the agency of his father); but receipts were so disappointing that the sponsors issued an appeal to "patriotic citizens possessed of pecuniary ability." Sherman's worst fear was that the award might come in the form of a house in Cincinnati: "I cannot imagine any contingency of life that would keep me in Cincinnati longer than a day." He need not have worried. In November 1865 Ohio's governor presented the general with the modest sum of $9,696.10 in cash and bonds, along with the observation that the fallout from Sherman's first ill-advised agreement with Johnston had been enough to "mar the success of the enterprise." The episode gave Sherman another grudge to hold against his native state.

But St. Louis, on the other hand, came through handsomely. On August 15 a group of prominent men of the city, including his old friend Henry Turner, presented Sherman with $30,000, "with the wish that you will with it purchase a home in our midst." He did so immediately. Ellen said the house he bought at 912 Garrison Avenue pleased her as much as if the architect had consulted her. It had a multitude of rooms, three of them parlors, spread over four levels. The Shermans were soon installed there, attended by a cook and four maids (a nurse would be added to the staff when Ellen gave birth to the last of their progeny, Philemon Tecumseh Sherman, on January 9, 1867).

Sherman's home life in the late sixties, what there was of it, can be reconstructed thanks to Ellen's diaries and letters. They make it clear that, as before, when the two were together their lives meshed, but not perfectly. An old pattern repeated itself: the couple received countless invitations, but usually Sherman went out alone. Ellen noted, apparently without any resentment, "Cump goes to so many parties he cannot keep well." She spent most of her time supervising the household staff and attending to the children's needs, her religious devotions, and her correspondence. Tom was giving less trouble now, though his mother occasionally noted that he came home from school "weary and nervous"; for the time being one of his sisters had replaced him as the family's problem child: "Rachel is rude and outbreaking as a boy," Ellen recorded, "there is Sherman enough in her."

Ellen carefully watched over her children during their various illnesses and her diaries and letters reveal much preoccupation with her own health. "I shall adopt the whiskey and salt water bath," she wrote her father, who had apparently suggested it for some undisclosed ailment. She was troubled by a weight problem that would continue until her death; a small woman in stature, she complained about "the 145 pounds that oppress me."

Sherman brought many of his problems home with him and was still in the habit of pouring out his worries to her; as always, she listened with a sympathetic ear. Frequently he told her he felt like walking away from all the vexations of his job and joining one of the Indian tribes, to wander the prairies, as unconcerned as a Sioux brave about what the next day might bring.

But Sherman had little patience with his wife when she in turn aired her complaints. One evening in March 1868 he had some particularly harsh words, which she related in a letter to her father. It was a Friday, and Ellen had forgotten to buy something for her dinner other than fish, which disagreed with her. She made the mistake of eating the fish and a second mistake in telling her husband it had made her ill. "Cump says I have been complaining all my life and never was sick an hour," she wrote Thomas Ewing. Then, she continued, Sherman had told her that "it is pure imagination when I complain & that if I chose I could always be well." Then Ellen did a rare thing–she fired a salvo of her own: "I think he missed his calling when he took a civilized wife, as nature made him for the spouse of a squaw ... When he retires to the tribe in which he may lose himself, should they give him any power or he attempt to force his views into practice, he will kill off by severity and want of all sorts of kindness and comforts the unfortunate doomed tribe."

Sherman's most frequent complaint when he was home was about money. His allowance for quarters was $48 a month, which was not enough to rent the most decrepit shack in St. Louis, much less cover the monthly expenses on Garrison Avenue. On January 11, 1869, Ellen wrote her father that her husband was "on a rampage" over taxes; he demanded that she put some of her property on the market to ease the tax burden. "I demurred somewhat, but afterwards consented, as *an emigration to South America* would have been the desperate suggestion following a further refusal."

Sherman doted on the children, teasing them at the same time. When in their innocence Rachel and Elly tried to teach him the Articles of the Creed, he would recite instead the Articles of War. The arrival of Philemon had something of the same effect on him as the birth of Willie. Ellen remarked something curious when her husband was with the baby: "he attempts to alter his voice in speaking to him when he thinks no one is by and to assume the tender and persuasive, but he makes a horrible failure of it." She acknowledged that her youngest child ruled the household, "his father being his most obsequious slave."

In the house on Garrison Avenue Sherman was destined to be more a visitor than a resident. The administration called him to Washington for one task or another, revising army regulations for example. In the winter of 1866–67, when the French were ending their intervention in Mexico and President Johnson was anxious to see a smooth transfer of power to the government of Benito Juarez, he sent Sherman and an official envoy to Vera Cruz by steamer (he had wanted to send Grant, with whom he was having differences, but Sherman had agreed to take his friend's place). Unfortunately Juarez remained in the interior, where Sherman had no intention of going to seek him out. He ended his mission, declaring that "these Mexicans are as unable as children to appreciate the value of time."

Then his responsibilities in the Division of the Missouri made the greatest demand on his time; from spring till fall he might be away for weeks at a time. He plunged immediately into the affairs of his new command. It was vast and varied, embracing over a million square miles, a full third of the continental United States; it had endless stretches of mountain, forest, grassland, and desert, with its heart what are now the two tiers of states south of the Canadian border, running from the Mississippi to the Rockies. It was for the most part a region he had never seen, and with his passion for visiting new country he would explore it many times.

This Western command, which he would hold until 1869, and his tenure as commanding general of the army from 1869 to 1884, have customarily been treated by most of his biographers as a kind of postscript to his military career. Historians of the West do not always agree; indeed his Western tenure has been saluted by some as "the climax of his career," his "harvest years." This argument is not an easy one to sustain.

To begin with, the Division of the Missouri presented Sherman with challenges he was not particularly well-equipped to meet. In his new situation there was only one element familiar to him, the regulars he commanded; he had to contend with the baffling and inscrutable red man, and with a no less baffling civilian agency, the Department of the Interior and its Office of Indian Affairs, with whom he was obliged to share competence; then there were the influential Eastern philanthropic groups determined to save the Indians, and the clamorous Western settlers no less intent on having them disappear from the landscape; and then there was the final authority, the Congress, that did not seem to know its own mind or provide the wherewithal to fulfill its own intentions. To have been successful in his tenure Sherman would have needed both political acumen and diplomatic skill, and he was not endowed with either.

When he was at home he laid out his worries to Ellen, who in turn related them to her father by letter. On September 25, 1868, she wrote that Sherman was "very much harassed" by problems in his division: "He cannot get the money from the Treasury that Congress gave him orders to disburse. The Western people are telegraphing fearful accounts of Indian outrages & imploring help, whilst the Radical hounds are denouncing him for his aggressive movements."

Though he came to know his division well through frequent travel, he was not able to appreciate all of its potential; there were parts that he never saw at their full value. He believed that vast stretches of the high

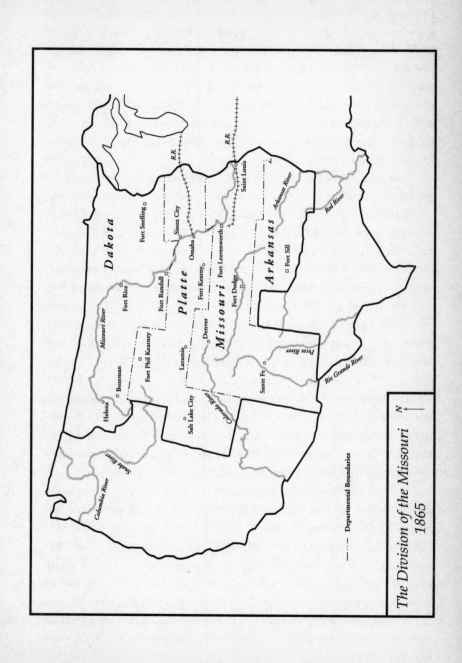

The Division of the Missouri
1865

N

----- Departmental Boundaries

plains would never support more than the thinnest sprinkling of herders: a country without trees, a country in which a major general had to help gather dried buffalo chips so his cook could heat his morning coffee–this was a country without promise. Then to his disgust his division contained part of the Southwest. He had never agreed with its acquisition from Mexico, and in later years the mere mention of Arizona and New Mexico could set him off: "I want to see this country lay hold of Mexico again," he would declaim, "and thrash her till she promises to take those d——d territories back again."

Sherman's command embraced the most rapidly growing portion of the country–and the only portion visited by war, now that the South was under military occupation. Here the warfare was intermittent and of low intensity. Between 1865 and 1890 there were 948 engagements between the United States Army and various tribes of Indians, most of these clashes involving very small numbers of combatants; even the most celebrated of these engagements, Custer's disastrous encounter at the Little Big Horn, cost only part of a single regiment of U.S. Cavalry. But the violence in the West always played well in the East. If three prospectors were scalped at a water hole in Utah, the journalists–still Sherman's bane–would see that New York and Philadelphia read about it over morning coffee; in Congress there would be loud complaints that the army was not doing its job.

The fighting was only part of a more complex struggle between two seemingly incompatible ways of life. In the contest between the settlers and the Indians the army was of course committed to the cause of the former, though there were times when Sherman's nominal allies gave him more trouble than his enemies. There was a dynamism in Western expansion that escaped control or regulation. The frontier was not a steadily advancing line. Its pace was pell-mell, the pattern of exploration and settlement diffuse and irregular, spreading west from the Mississippi and now east as well, moving down from the slopes of the Rockies. These wayfarers often took routes through dangerous country where the U.S. Cavalry could not offer protection. They might or might not observe the army's recommendations that they travel in convoys of at least twenty wagons and no fewer than thirty armed men. If they chanced upon Indians whose intentions they were unsure of, they were inclined to reach for their Winchesters.

Frustrating as Sherman's contacts with the settlers often were, the Indians presented him with a far greater challenge. Like any other officer serving in the West, he had to take the measure of his adversary, to determine not only the Indian's worth as a fighting man, but also his status and his value as a human being. Much depended on this latter point, including the degree of thoroughness with which the army made war on these people and the discrimination that it might or might not make between enemy warriors and noncombatants. Some studies have suggested that the majority of officers serving in the West in this period were inclined to think that while the red man was decidedly different, his unusual ways—most of them considered backward and some of them deplorable—were the result of his adaptation to a harsh environment and his cultural isolation. This being the case, the Indians might well be capable of further adaptation. In practical terms this meant the heathen warrior and hunter of the plains could become a peaceable, Bible-reading, tax-paying farmer or rancher. And since he could be "salvaged" for civilization, he should not be exterminated.

Some of the generals who fought in the Indian wars could not conceal a certain fascination with their opponents, and one or two became their frank admirers and outspoken defenders. General George Crook even found a kind word for the Apache: "Were he a Greek or Roman, we should read with pride and enthusiasm of his determination to die rather than to suffer wrong." Lecturing to West Point cadets, Crook explained that the Apache brave was moving along the same road their own ancestors had taken: "He is cruel in war, treacherous at times, and not over cleanly. But so were our forefathers."

Sherman had wrestled with this matter at least as early as his tour of duty in Florida. While he confessed his admiration for the way the Seminoles continued the struggle against all odds, he had been mystified by their comportment in other situations, especially when they were taken captives or fell into the army's hands when wounded; even their children suffered the gravest wounds without a whimper. Was this remarkable behavior in some way an acquired stoicism, a learned response imposed by their culture? Or was it simply part of a different nature, the appropriate reaction in a race or breed that lacked the sensitivity, the feeling of the white man, and other of his traits as well?

Lieutenant Sherman did not know, but ultimately General Sherman came up with the answer. Races *were* different in nature, in their mental, moral, and physical capacities and in the level of culture they could attain if left to their own devices. His views were common enough, indeed they were shared by any number of his colleagues in the Indian-fighting army. General William B. Hazen thought in Darwinian terms: the Indians were "in the way of the evolutions of progress" and would simply have to go. As for Sherman's chief subordinate and good friend Philip Sheridan, he saw war on the plains in black and white. According to him the war against the Indian was "a clearcut struggle of civilization against barbarism."

As for Sherman, his notions seem to have jelled over the years as he remarked and reflected on the behavior of the Seminoles, the Californios, Pimas and Yumas of California, the Mexicans, and the blacks of the South, fitting all by 1860 into a racist matrix that he expanded as other wars and other ethnic groups came to his attention. Thus in 1882, when he was asked by a Canadian reporter if the Egyptians would be able to drive the British out of their country by force, his answer was "I think not." The Egyptian, he explained, had sufficient physical strength, but did not have the dauntless courage of "the Anglo-Saxon or the Northern nations," and the general threw in the comment that the Turks couldn't beat the British either—for the same reasons.

In the case of the Western Indians, his studied opinion was that they were doomed to defeat and extinction, or at least near-extinction. They could not evolve because of limitations nature had placed on them. Notably they were incapable of centrally directed common action to some worthwhile end, for they were "democratic in the pure sense." He argued that "no chief has authority. He has influence but the moment he crosses the popular prejudice he goes under." The red man could never follow purely peaceful pursuits because he was addicted to war. "Now war with the young is always popular, because it is so much easier to steal than to work. And to kill occasionally makes them popular with the women." As for becoming civilized, it was beyond their capacity; their attempts in that direction were to him "simply ridiculous." He was convinced that their primitive culture was the only one attainable by them. As a matter of fact he used them as an example to demonstrate a similar incapacity in blacks: "The Indians give a fair illustration of the fate of negroes if they are released from the control of whites."

But what if racial distinctions disappeared through intermarriage? Would this not alter the picture? Yes, but only for worse. Sherman called this "amalgamation"; he considered Mexico the prime example of it, and the result disastrous: the blending of races had produced "general equality," which led inevitably to "Mexican anarchy." Amalgamation also had deleterious genetic effects, and here he regarded New Mexico as a prime example of this phenomenon. He described its inhabitants as "mongrels," and "a mixed breed of Mexican, indian and negro, inferior to either race if pure."

As for solving the Indian problem, he conceded that he would probably have to wait out the idealists and reformers. He agreed that the red man should be given a chance to find a productive place in a society run by the "dominant race," but he was not sanguine about the outcome. He believed that he would be obliged to stand aside, to watch the civilizing schemes fail one after the other. When they were done, the army would settle the matter. Briefly, after Grant's election in 1868, he thought the Office of Indian Affairs might be abolished and the army given a free hand—but this did not happen. In any event dealing with 300,000 Indians—that was his estimate of their number—would be a mammoth task for the minuscule force placed at his disposal. In 1865 he had about 25,000 effectives for use in the West, and that number would gradually decline as Congress whittled down the army's budget.

But Sherman soon discovered that several factors were strengthening his hand in the contest with the Indians. One was the continuing arrival of immigrants in large numbers, ensuring the vitality and viability of the settlements that received them. The immigration was in turn encouraged by the rapidly developing rail lines, especially the two parallel lines pushing toward the Rockies at the rate of a mile of track a day.

The railroad also aided the army, reducing shipping costs for its supplies and giving it increased mobility. Moreover Sherman hoped to turn the territory between the rail lines into an Indian-free buffer zone separating northern and southern tribes. To eliminate the temptation that the area held for Indian hunting parties, he spoke of wiping out the great buffalo herds that grazed there, perhaps inviting European sportsmen to help with the slaughter.

By the late 1860s Sherman and Sheridan had refined the formula for effective campaigning against the plains Indians: go after them in the dead of winter, when their movements were hampered by snow and their undernourished ponies lacked the stamina of the cavalry's grain-fed mounts. Even if they succeeded in fleeing their lodges, the Indians left behind the stores of food they had collected to carry them through the winter. When these were destroyed, starvation would soon bring them to heel.

Sherman saw for the Indians one of two fates, extinction or confinement to lands set aside for them. "Sooner or later," he confided to General Sheridan, "these Sioux will have to be wiped out or made to stay just where they are put." And it made eminent good sense to him to put them in areas where white men would not or could not live. One gains the impression from some of the general's remarks that he felt their numbers and combative spirit would have to be beaten down by war until they were more tractable and more manageable in mass. "The more I can kill this year," he wrote John in the fall of 1868, "the less will have to be killed next war." Once reduced to confinement within the lands set aside for them, their preserves would be something like those later set aside for the buffalo; both Indian and bison would continue to live in their natural state, though the former, with their freedom to hunt restricted and their ability to take up agriculture next to nil, would probably have to be maintained "as a species of paupers."

It should be said that the propounder of this grim and heartless scenario was not always relentless and pitiless in its execution. He was often angered by the behavior of the settlers, whom he suspected of provoking the Indians so as to draw the army into the affair, with the purpose either of simply getting the Indians killed off or making money on grain and hay sold to the troops brought in. Once when there was friction between the Utes and the settlers, Sherman came down on the side of the impoverished Utes: "They are scattered, and not hostile further than a necessity compels them to steal occasionally a cow or a sheep to appease hunger. Of course the rangers, Americans and New Mexicans, want the troops to kill them all . . . I will not permit them to be warred on . . . "

In the summer of 1868 Sherman and Samuel F. Tappan, both of them members of a peace commission named to negotiate with the Indians, paid a visit to the Navajos. After much fighting, the tribe had been forcibly removed from their homeland four years before and resettled at Bosque Redondo, in southeastern New Mexico. The place had no good water supply, the soil was too poor to grow crops, and the Navajos were living in abject misery, barely subsisting on an alien diet of army rations. Sherman was disgusted with the situation he found; Tappan compared the conditions at Bosque Redondo with those in the infamous Confederate prison camp at Andersonville. After several days of talks with Barboncito, the Navajo spokesman, Sherman decided to move them all back where they had come from, though their reservation would be only a part of the territory they had once roamed. Somewhat grandiloquently he told them: "My children, I will send you back to your homes." Despite considerable opposition he kept his word; within a month the Navajos were back home.

In the 1870s, as commanding general, Sherman presided over the army's efforts either to destroy the Indians or to force them onto reservations, and by the end of the decade the job was for the most part completed. Speaking somewhat expansively during the 1882 newspaper interview cited earlier, he described the transformation that had taken place: "As near as I can estimate there were in 1865 about one and a half-million buffaloes on the plains between the Missouri River and the Rocky Mountains. All are now gone, killed for their meat, their hide and their bones. This seems like desecration, cruelty and murder. Yet they have been replaced by twice as many neat cattle. At that date there were also about 165,000 Pawnees, Sioux, Cheyennes, Kiowahs and Arapahos who depended on these buffaloes for their yearly food. They too are gone, and have been replaced by twice or thrice as many white men and women who have made the earth blossom as the rose, and who can be counted, taxed and governed by the laws of nature and civilization. This change has been salutary and will go on to the end."

Sherman's work in the Division of the Missouri was frequently interrupted by other assignments. Despite the remoteness of his Western fiefdom, in the late 1860s Sherman was drawn into the political broils of Washington. Increasingly the calls to Washington were rooted in politics. President Johnson and the Congress were at odds over the "reconstruction" of the Southern states. Johnson, a Southerner himself, had come to favor a moderate approach, essentially returning political control to the whites who had previously held it; the Congress, increasingly under the influence of Republican Radicals, was bent on creating a new order that would give Southern blacks—and the Republican Party—a significant role in Southern political life.

As the contest between the president and his congressional enemies became more envenomed, the army was inevitably drawn in; Federal forces occupying the South were the key to what happened there, and those who controlled the army were thus important as potential allies or enemies. By the end of 1865 President Johnson had identified Secretary of War Stanton as an enemy; then he became distrustful of Grant as well. He began thinking of bringing the nation's second most distinguished soldier to Washington as an ally.

While the president's favorable view of Sherman was a recent development, it rested on solid reasoning: here was a man who would have no traffic with the congressional Radicals. He had made no secret of his conservative political views, which included restoring much of the old regime in the South. In his view the Southern people had received their punishment for rebellion in the war itself, and he had done his part in administering that punishment. But now it was time for them to once again become master in their own house.

Sherman sent clear signals that he was on the side of the president in the standoff with Congress. At the beginning of 1866 he wrote Johnson a polite note of apology for not "paying respects" during a quick trip to Washington, adding that his business had taken him to Capitol Hill, which he described as the home of "noisy innovations of mere theorists and experimental legislators." That fall Johnson had him summoned to Washington and dangled in front of him the prospect of taking Stanton's place. Sherman was cold to the idea; he wanted no civilian's role and doubted if he could take the post without resigning his commission, which he would not do. Instead he went on the fruitless mission to Vera Cruz in the place of Grant.

Sherman was clearly not angling for an appointment in Washington. He had long regarded the national capital as a sink of chicanery and corruption. "In the West we made progress from the start," he explained to the president, "because there was no political capital close enough to poison our minds and kindle into life that craving, itching for fame which has killed more good men than bullets." And now Washington was the seat of a government embarked on a dangerous new course, further debasing the barely competent electorate of the country. Giving the vote to freed slaves he saw as only the beginning of a disastrous extension of the franchise: "We must go through the whole series," was his grim prediction, "women, minors, convicts, etc."

President Johnson kept him in mind. In the late summer of 1867 the president suspended Stanton from his functions and asked Sherman to take over the War Department ad interim. When the general declined, Johnson confided the post to Grant, who had no difficulty in sitting at Stanton's desk and writing orders to himself as commanding general. And still Johnson persisted; early in 1868 he proposed creating for Sherman a new military circumscription with headquarters in Washington; again the general found a way to say no: "For eleven years I have been tossed about so much that I do really want to rest, study, and make the acquaintance of my family."

With 1868 came the impeachment crisis; Johnson survived the critical vote but his political stock was exhausted. For the presidential elections of that year the Republicans had found an attractive and eminently available candidate in Ulysses S. Grant.

The Democrats were hard put to match the Republican ticket with a martial hero of the same stature. James Buchanan and other Democratic leaders saw Sherman as their chief hope, but he rejected all overtures; he would continue to do so, whether the feelers came from Republicans or Democrats, for the rest of his life.

In November the voters handed Grant a victory as sweeping as Vicksburg; he gave up what had been a spectacular military career for a political one that even the most charitable historian would classify as lackluster. One consequence of Grant's election was that Sherman got a promotion. He had succeeded Grant as lieutenant general in 1866, and now he inherited his rank of full general. Sherman also assumed direction of the army as its commanding general, which would require his presence in Washington.

He briefly considered turning down his promotion because of the move and the cost involved. As commanding general he would have social obligations. In addition to being a political Gomorrah, Washington was an expensive place to live; even with an increase in salary, the move to the capital could be financially prohibitive. Fortunately a group of wealthy and public-spirited businessmen—mostly New Yorkers this time— offered to buy from Grant the fine house on I Street that had been given to him and present it to the new commanding general (moreover the furniture would be thrown in). The Shermans graciously accepted. Ellen began preparations for the move, which would open a new chapter in the couple's life.

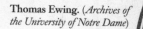

Thomas Ewing. (*Archives of the University of Notre Dame*)

Maria Boyle Ewing, with her granddaughter Minnie Sherman. The photo was taken early in the period 1853–57, when Minnie stayed with her grandparents in Lancaster during her parents' move to California. (*Archives of the University of Notre Dame*)

Main Street, Lancaster, Ohio, about the time of the Civil War. The Ewing and Sherman homes are on the left along the incline at the far end of the street. (*Ohio Historical Society*)

One of Sherman's teachers while he was at West Point: Dennis Hart Mahan, professor of military and civil engineering and the "Art of War." (*U.S. Army Military History Institute*)

Another of Sherman's mentors while he was at West Point: Major Richard Delafield, who became superintendent of the academy in 1838. (*Massachusetts Commandery, Military Order of the Loyal Legion of the United States and U.S. Army Military History Institute*)

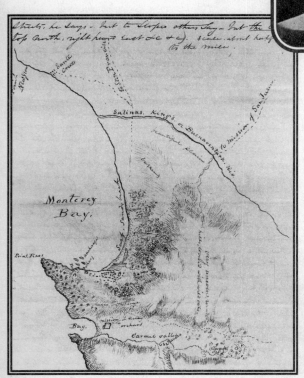

At the academy Sherman honed his skills as an artist; he made this sketch of Monterey Bay for Ellen Ewing in 1847. (*Archives of the University of Notre Dame*)

Colonel Richard B. Mason, Sherman's commanding officer during his first stay in California. (*California Historical Society*)

The Customs House, Monterey. (*Pat Hathaway Collection, Monterey, California*)

A panorama of San Francisco in the mid-1850s. Sherman lived in the house with the fenced yard in the right foreground. (*California Historical Society, George A. Berton Collection, FN–19615*)

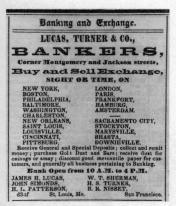

Lucas and Turner advertisement that ran in the *Daily Alta California* in 1855 and 1856. (*California Historical Society, North Baker Research Library, FN–31665*)

James H. Lucas of Saint Louis, principal investor in Lucas, Turner, and Company. (*Missouri Historical Society*)

Sherman's bank in 1953; it lost its third story in the San Francisco earthquake and fire of 1906. (*California Historical Society, FN–2738913*)

**George Mason Graham.**
(*Walter L. Fleming Collection, Louisiana and Lower Mississippi Valley Collections, LSU Libraries, Louisiana State University, Baton Rouge*)

**David French Boyd.**
*(LSU Photography Collection,
RG# A5000, Louisiana State
University Archives, LSU
Libraries, Baton Rouge)*

**The Louisiana State Seminary
of Learning and Military
Academy, Pineville, Louisiana.**
*(LSU Photography Collection,
RG# A5000, Louisiana State
University Archives, LSU
Libraries, Baton Rouge)*

Ellen Ewing Sherman and four of her children. (*Ohio Historical Society*)

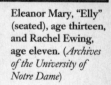

Eleanor Mary, "Elly" (seated), age thirteen, and Rachel Ewing, age eleven. (*Archives of the University of Notre Dame*)

Mary Elisabeth–"Lizzie"– at age nine. (*Archives of the University of Notre Dame*)

William Tecumseh Sherman Jr., called Willie. This much-worn photograph of Sherman's favorite son has written on the back, in Ellen Sherman's hand, "This picture is blessed." (*Archives of the University of Notre Dame*)

Officers of the Sixty-ninth New York Infantry posed for a group portrait on the ramparts of Fort Corcoran. (*National Archives*)

### Gen. William T. Sherman Insane.

The painful intelligence reaches us in such form that we are not at liberty to discredit it, that Gen. W. T. SHERMAN, late commander of the Department of the Cumberland, is *insane*. It appears that he was at times when commanding in Kentucky, stark mad. We learn that he at one time telegraphed to the War Department three times in one day for permission to evacuate Kentucky, and retreat into Indiana. He also, on several occasions, frightened the leading Union men of Louisville almost out of their wits, by the most astounding representations of the overwhelming force of BUCKNER, and the assertion that Louisville could not be defended. The retreat from Cumberland Gap was one of his mad freaks. When relieved of the command in Kentucky, he was sent to Missouri and placed at the head of a brigade at Sedalia, where the shocking fact that he was a madman, was developed, by orders that his subordinates knew to be preposterous and refused to obey. He has, of course, been relieved altogether from command. The harsh criticisms which have been lavished upon this gentleman, provoked by his strange conduct, will now give way to feelings of the deepest sympathy for him in his great calamity. It seems Providential that the country has not to mourn the loss of an army through the loss of the mind of a general into whose hands were committed the vast reponsibility of the command in Kentucky.

The article on Sherman's "insanity," published in the Cincinnati *Commercial* of December 11, 1861. (*Ohio Historical Society*)

Shiloh Church. (*Ohio Historical Society*)

Sherman and his staff, in a photograph taken at Memphis, September 1862 *(left to right):* Lieutenant John T. Taylor, aide-de-camp; Major J. H. Hammond, assistant adjutant general, Captain L. M. Dayton, aide-de-camp; Major Ezra Taylor, chief of artillery; Captain J. Condit Smith, divisional quartermaster; General Sherman; Colonel Thomas Kilby Smith, aide-de-camp; Captain Joseph W. Shirk, USN, naval attaché; Major W. H. Hartshorne, divisional surgeon; Colonel W. H. H. Taylor, aide-de-camp; Captain James McCoy, aide-de-camp; Major W. D. Sanger, aide-de-camp. (*Photographic History of the Civil War*)

The Vicksburg campaign gave Sherman his first opportunity to work closely with naval forces. He made this sketch of the gunboat *Cincinnati* in September 1863. (*Archives of the University of Notre Dame*)

*Gun Boat "Cincinnati"*
*Sept 30. 1863.*

This highly flammable wooden trestlework at Whiteside, Tennessee, just outside Chattanooga, illustrates the fragility of Sherman's supply line in 1864. As he said, to cut the line required only "one man with a match." (*National Archives*)

Sherman stressed complete destruction of enemy rail lines, with fire as the chief agent, including heating and twisting the rails longitudinally, so that they could not be straightened unless they were sent back to a rolling mill. But his soldiers preferred the easier way demonstrated here: When sufficiently heated, the rails would bend of their own weight. (*National Archives*)

A section of Missionary Ridge, outside Chattanooga, Tennessee. (*National Archives*)

A portion of Sherman's invasion map of Georgia; hand-lettered over each county are demographic and economic statistics taken from the census of 1860. These include the number of whites, slaves, free blacks, and men of military age as well as yields of corn, oats, wheat, and so on. (*National Archives*)

The Sherman residence at 912 Garrison Avenue, Saint Louis. (*Archives of the University of Notre Dame*)

Sherman was a "writing" general, one whose daily output could run to fifty letters and notes; yet it is rare to find him portrayed as he is here. His taste ran to stern, formal portraits. (*Cumberland Gallery Collection*)

Thomas Ewing Sherman and his younger brother, Philemon Tecumseh Sherman, 1875. (*Archives of the University of Notre Dame*)

The Sherman clan at John Sherman's home in Mansfield, Ohio, August 1886. Standing (*from left to right*): Charles W. Moulton; Frances "Fanny" Sherman Moulton; Elisabeth Sherman Reese; ____ Babcock, secretary of John Sherman; Cecilia Stuart Sherman; her husband, John Sherman; Lampson Sherman; General William T. Sherman; Sarah Sherman; Hoyt Sherman; Ellen Ewing Sherman; Kate Willock. Seated (*from left to right*): Mary Sherman, Philemon Tecumseh Sherman, Minnie Moulton, and Addie Sherman. (*Archives of the University of Notre Dame*)

"Latest Battle of the Books." This cartoon represents Sherman doing battle with Adam Badeau, Charles A. Dana, and others of Grant's partisans over the question of which general, Grant or Sherman, authored the March to the Sea. From *Frank Leslie's Weekly Newspaper*, June 5, 1875. (*Ohio Historical Society*)

Among the people who figured in
Sherman's postwar military career were
William W. Belknap (*upper left*), Emory
Upton, and this delegation of Jicarilla
Apache whose *carte de visite* the general
kept in his photo album. (*National Archives;
Massachusetts Commandery, Order of the Loyal
Legion and the U.S. Army Military History
Institute; Ohio Historical Society*)

Augustus Saint-Gaudens's statue of Sherman, New York City. (*New York Historical Society*)

Union veterans gathered for the dedication of the Sherman monument in Washington, D.C., in 1903. (*Collection of Bob Zeller*)

# 19

## COMMANDING
## GENERAL

$S$ HERMAN SERVED AS commanding general of the United States Army until 1884, making it his last and longest assignment. Though it put him at the pinnacle of the army's command pyramid, it cannot be regarded as the zenith of his career. First because the post carried with it far less power and prestige than that of present-day chief of staff; and secondly because it thrust him into a world of high politics and bureaucratic ways in which he could never feel at home or move with the confidence and mastery he had known in the field; he and the army suffered for it. And significantly, while in the war he had rarely been away from the front, in this peacetime billet he was tempted to absent himself from his command when the occasion presented itself.

Then too, he found himself in something of a legal and administrative jungle, holding an office whose powers and functions had never been well-defined. It had been created in 1821, with the centralization of the army's various services or "bureaus," those of the adjutant general, commissary general, and so on, which collectively acquired the somewhat misleading name "general staff." Thus from 1821 the army had three central elements: secretary of war, commanding general, and general staff. Though they had now functioned together for almost a half century, their roles and relationships had never been fixed. They rested on what one commanding general, John M. Schofield, called a "heterogeneous mass of regulations, customs and special legislation." This haphazard machinery worked, but it did not work well or smoothly.

There was a long-standing line-staff controversy. Almost inevitably a fracture had developed between those in the "line" elements of the army (infantry, cavalry, and artillery) in the field, and those in Washington who ran the staff departments. The staff or bureau heads, whether in charge of subsistence, pay, or other functions, enjoyed permanent assignment to their specialty and statutory authority to run their departments. As a consequence they tended to think of themselves as collaborators with the commanding general rather than his subordinates—a relationship likely to produce stresses and strains in peacetime and sharp crises when the nation was at war. By 1869, when Sherman became commanding general, there were ten of these staff bureaus, with the head of each solidly entrenched in his prerogatives. Pushed to its extreme, this independence of the staff departments meant that if a captain in Montana wanted his commissary to issue double rations to his company, the order might have to pass by way of Washington.

Staff officers were better paid than line officers. They lived comfortably in Washington, while the great bulk of the army, its regiments of cavalry and infantry, were stationed in the depressed and war-ravaged South, or were living in remote Western outposts with Cheyennes and Paiutes for neighbors. And somehow promotions seemed to come through more rapidly for officers processing bills of lading in Washington than for those chasing bands of Sioux over the prairies.

At high command level the most serious point of friction was between the secretary of war and the commanding general, or general in chief, as he was sometimes called. Here jurisdictional disputes or "turf wars" were frequent; they had led Winfield Scott to distance himself from the capital and exercise the functions of commanding general from New York. At the commencement of the Civil War General McClellan and Secretary of War Cameron barely communicated; after much experimentation, by 1864 the war was being managed by a four-man civilian-military team: Lincoln, who fixed broad policy; Stanton, who oversaw the administrative machinery and advised on operations; Halleck, who coordinated the military effort; and Grant, who directed the army in the field. With the end of the war the old order returned and with it the old problems, though Secretary of War Stanton was for the time being the commanding presence.

As line officers, Sherman and Grant had the same concerns, and had talked about the need for a change in the army's command structure, particularly in the relationship between the commanding general and the secretary of war. Grant wrote Stanton in 1866 that in his view: "the general in chief stands between the President and the Army in all official matters and the Secretary of War is between the Army (through the General in Chief) and the President." One year later he wrote Sherman: "I am in hopes of getting the command of the army back again where it belongs, and if I do, there should always, for some years at least, be some one present to exercise it lest it revert again to the Sec'y of War."

Now, with Grant moving into the White House, with Sherman taking over as commanding general and General Schofield doing duty for the moment as secretary of war, changes could be expected. In fact, there was something of a military "coup" planned, whose purpose was to reduce both the power of the secretary of war and the privileges and pretensions of the staff departments. Grant was sworn in on March 4, 1869; the following day he formally designated Sherman commanding general and Schofield issued General Order, No. 11, reorganizing the War Department so that both staff and line came under Sherman's direct control. Sherman in turn issued General Order, No. 12, which provided that all the paper flowing down from the president and secretary of war would necessarily pass across his desk. On the sixth the new commanding general sent word to his subordinates, heads of staff departments included, that they were to meet with him on the eighth. Just what happened at the meeting is not recorded, but apparently the new order of things found grudging acceptance.

With the new system in place and functioning, Schofield gave up his job in the War Department and returned to regular duty. Grant was still in the process of filling various offices; he had intended to reward his Chief of Staff John Rawlins with a post in the Southwest, since Rawlins had lung trouble and the dry air would do him good (he was in fact dying of tuberculosis). Rawlins learned of the proposed appointment, was unhappy about it, and let Grant know he wanted the War Department instead. Grant gave it to him. Though moribund, Rawlins demonstrated remarkable energy and resolve. He immediately set to work to undo all that had been done to limit his authority, especially General Order, No. 11, which in effect put the War Department under Sherman.

Sherman knew Rawlins exercised great influence over Grant, so when he learned the recent changes were in jeopardy he dashed off a note to the president not to do anything until the two of them could talk: "it would put me in a most unpleasant dilemma because the Army and country would infer your want of confidence"; as for General Order, No. 11, he reminded his chief, "the order was yours and was long contemplated." He was too late; that same day, March 26, Grant ordered that the sections of General Order, No. 11 dealing with the reorganization of the War Department be stricken. When Sherman heard what had happened he was hard put to explain why his friend had thus canceled out a reform which he had favored. He told friends that Grant had had his wits temporarily addled by the stridencies of the press and clamor of office seekers; when he had time to think through the consequences of his action he might well change his mind. Most contemporaries saw the episode as confirming the great influence Rawlins seemed to have with his chief.

It may be that Sherman himself, or at least the way he comported himself, was in part the cause for the failed "coup." Schofield thought he "entertained views of his authority which were too broad"; General James H. Wilson, who was then in Washington, was struck by the flamboyant way Sherman announced that bureau chiefs would now form "his" general staff; Wilson also heard that Radicals in Congress were speaking of recent changes at the War Department as "odoriferous of military dictatorship." In Congress, where Sherman was regarded as something of a loose cannon, there were in fact some rumblings of concern about the increased influence of the military in the councils of government: "With the independent and unpredictable Sherman as commander, many Congressmen feared that the Army might not readily submit to civilian direction." If Sherman detected these rumblings he paid them no heed, but Grant seems to have understood their tenor and their importance and to have adjusted his policy accordingly—even if it meant clipping the wings of his old friend.

Rawlins's tenure at the War Department was a brief one. By July he was too ill to continue his work; Sherman filled in for him. He died in

September 1869, with Sherman at his bedside. His successor was William Worth Belknap, an Iowa lawyer who had fought at Shiloh and risen to command a division under Sherman, who had taken him under his wing. It was Sherman who had proposed Belknap's name, among others, to succeed Rawlins, so he no doubt anticipated cordial relations with him. But if the commanding general counted on restoring his own control over the staff, he ran into a stone wall. The new secretary saw immediately what Rawlins had seen: increasing Sherman's authority would diminish his own.

Relations between Belknap and Sherman deteriorated rapidly. The assertive and hard-driving secretary was now the aggressor, issuing orders to the army and reassigning personnel through the adjutant general's office, bypassing Sherman. Moreover he was intruding in areas customarily reserved to the commanding general, including the running of West Point, about which Sherman was particularly sensitive. Appeals to the president brought no satisfaction. Grant replied that when he was in Sherman's place he thought all orders should pass through his office too, but now he realized that it was not that simple.

Now not only was Sherman abandoned by his old commander in the White House, but one of his wartime subordinates had reappeared as a powerful enemy in Congress. In January 1870 John A. Logan, chairman of the House Military Affairs Committee, introduced a bill calling for massive reductions in the army and in its appropriations. At bottom Logan wanted to change the entire military establishment, getting rid of the regular army and relying on the militia and the volunteer soldier who had after all won the Civil War. But in defending his bill in the House of Representatives he also aimed at Sherman and the post he held. He alluded to the possibility of a commanding general carrying out a coup d'etat; he indicated that Sherman's salary and allowances, a princely $19,000 a year, might well be cut. Sherman, who was sitting in the House gallery that day, walked out in a rage.

Logan's bill passed the House. Sherman hoped it would be turned back in the Senate, which ended by adopting a much-altered version. The compromise legislation drawn up by conference committee made modest and relatively painless cuts in the army, but Sherman's own pay dropped by a considerable sum. And Sherman had other problems than those in

Washington. Some 10,000 troops were in the Southern states, assigned to tasks for which they were ill-suited and their officers without adequate training and guidance. Mostly they were doing police duty, chasing nightriders and barn burners, and trying to maintain order and legality in elections. In the West they had their hands full with the Indians—and with the Indian Bureau, for antagonism arose again between the War and Interior Departments; in January 1870 troops attacked and killed a hundred Piegan Indians under conditions that led to a congressional investigation.

Sherman was feeling a good bit of frustration, and the pressures on him were considerable. He sensed "a growing jealousy of Genl Grant and his cabinet, who think I do not blow their trumpet loud enough." Then too, his time was being "frittered away in little details," while Belknap handled the matters of appointment, promotion, and discipline, which were of paramount importance in a peacetime army; finally, steps were being taken that troubled him deeply—admitting blacks to West Point, for example.

In the fall of 1871 he did something he had never done in war; he temporarily abandoned the field to the enemy. In the spring of 1867 he had come close to accepting an expense-paid passage to Europe and had even obtained White House approval. Now Admiral James Alden invited him to accompany the Atlantic Squadron for a cruise to the Mediterranean, leaving in November; with official approval he accepted, taking with him his aide, Joseph Audenreid, and Grant's son, Fred.

Their itinerary took them through the Mediterranean, to Turkey and Russia, and on a tour of European capitals. What is known of the trip comes chiefly from travel notes kept by Audenreid, who seems to have been preoccupied with sanitary conditions along their route. But the journal also reveals Sherman as an enthusiastic and indefatigable tourist. In his fifty-first year he climbed 300 feet up the pyramid of Cheops (Audenreid and young Grant, who went on to the top, needed several days to recover). He met an endless procession of generals and field marshals and called on assorted heads of state from Queen Victoria to Sultan Abdul Aziz. With the eye of a specialist he walked over old battlefields, visited fortresses, and inspected troops; he concluded that the siege of Sebastopol had been a small affair compared with the contest for Vicksburg, and in general he saw little to emulate or copy in the military systems of Europe—

confirming what Sheridan had told him after his own tour in 1870. In all Sherman was gone ten months.

Little was changed in Washington when he returned. With each session of Congress there were the usual assaults on the army and the battle over the budget. As the years passed, Southern states began to send to the Congress men who had worn Confederate gray, and Sherman complained that these legislators were disinclined to do anything for the bluecoat army. In the periodic budget battles the administration and the army's partisans fought mostly to hold the line against further cuts. Sherman's relations with Belknap did not improve. He complained to a correspondent that he did not get to see important orders to individual officers and units until they were published in the newspapers; officers were court-martialed and he learned about it through the grapevine. And now even Grant was ignoring him: "The President . . . generally announces his appointments before I am consulted," Sherman wrote an applicant for a position, "& before even I have a knowledge of the vacancy." He confided to a friend, "I never treated a corporal thus."

In 1874 Sherman decided to execute another withdrawal. He asked permission to follow the precedent set by Winfield Scott and transfer his headquarters from Washington to St. Louis. Grant gave his approval and Sherman made the move, though any number of his associates, including Sheridan, advised strongly against it. With reliable telegraphic communications between the two cities the commanding general could argue that he would not be that remote, yet at the same time the distance would liberate him from the baneful influences of the capital, where he said he "would have to bend to the personal influences of Senators and members of Congress who naturally wish rather to serve their friends than to watch over the economy and good of the service."

With these few words Sherman revealed the chasm that separated his notions of statecraft from that vital if sometimes sordid interplay of interests that makes representative government work. By withdrawing from the capital he reduced his chances of getting for the army the legislation that it needed—though probably he did not reduce those chances by much. Even in Washington he had generally refused to play the political game, to seek out the chairman of an important committee, or explain the cavalry's pressing needs to a senator he had invited to lunch.

Actually he did not even have to seek these people out; he saw them constantly at dinners and receptions, and rubbed shoulders with them frequently at official functions. But he would not lobby. "I cannot and will not condescend to importune members to obtain any special end, no matter how desireable . . ." He had always been scrupulous, almost fussily so, about staying on his side of the line separating the professional soldier from the politician. "By my office I am above party," he proclaimed, and he would not descend into the pit. Neither would he hesitate to lose his temper with the legislators if sufficiently provoked. When a Democratic Senator named D. H. Armstrong wrote him about moving a young cavalry officer from his post in the Far West and placing him in the Commissary Department at Washington, he made this response: "I have no more right . . . to transfer Lieutenant Thompson to the Commissary Department than you have to transfer Mr. Hayes, Postmaster of Saint Louis, to be Collector of Internal Revenue." He had recently tried to get a law permitting him to move officers back and forth between line and staff, "but you Democrats listened to the howl of the nest of staff officers here in Washington and voted it down."

His correspondence for the seventies offers scant evidence of his "working" the Congress. As a rule he supplied the legislators with what they asked for by way of information and opinion, or sent formal, comprehensive programs for enactment and looked on in rage as they were picked to pieces. When a critical vote on an army appropriations bill was coming up in the House in the spring of 1874 he announced that he was going out of town, so he would not be there "to influence the opinion of any single member."

Sherman did have some allies; in Congress there were a number of veterans, men like James A. Garfield, who would lend their support; Stephen Hurlbut, whom Sherman had never cared for as a subordinate during the war, became a valuable ally when he entered Congress in 1872. Then there was another valuable friend and supporter of pro-army legislation in William C. Church, editor of the *Army and Navy Journal.* Sherman carried on an extended correspondence with Church, got the *Journal*'s endorsement for a number of measures, and even used it once to poll army personnel on their preferences in weaponry.

Sherman's removal to St. Louis, which lasted for nineteen months, could not but encourage Belknap to ignore him. By 1876 the secretary seems to have been leaving the commanding general out of the loop completely. Sherman sent him only about one letter a month, and rarely got an answer. "How can I be responsible for the discipline or instruction of the Army," Sherman asked a correspondent in March 1876, "if orders are made without my knowledge or advice?"

But the problem of Belknap was being resolved even as Sherman wrote. The Democrats had taken control of the House of Representatives and were systematically investigating corruption in the administration. The leads ran in many directions, and one of them went to the door of the secretary of war. Belknap, it developed, was receiving payments from men who had been given lucrative contracts to operate trading posts in Indian country. The evidence was damning. Belknap resigned in disgrace in 1876; Grant, himself under something of a cloud, departed one year later at the end of his second term.

Sherman gave up his self-imposed exile and returned to Washington in the spring of 1876. His reaction to Belknap's fall was unusual. He said that even when their relations were at their worst, Belknap had been a "manly" opponent. After the secretary's fall he wrote his old friend Ord, "I think this whole thing is more damning to Congress than to Belknap, yet the latter must be the scape goat."

Belknap was succeeded by a series of men who were less imperious and more willing to meet the commanding general halfway. Sherman's relations with the five secretaries of war who occupied the position during his remaining years were generally satisfactory, though the role and powers of the commanding general remained cloudy until the passage of the Dick Act of 1903, which abolished the position and replaced it with that of chief of staff. The military occupation of the South ended in the late 1870s, relieving the army of that thankless chore; there would be fierce battles over the army's budget, but Sherman was able to keep its strength above what he called the "bedrock" figure of 25,000, which meant there was only one American soldier for each 3,000 inhabitants. Even in the West, Sherman claimed, to find the army one would need a search warrant.

The years of Sherman's command fall in a period that a military historian later defined as the "Army's Dark Ages," an era of niggardly appropriations and general neglect on the part of succeeding administrations and sessions of Congress—but also of conservatism and stand-pat policy on the part of the nation's military leaders, engendered by the self-satisfaction flowing from the victory of 1865; the army seemed to rest for decades on laurels won in the sixties. On certain aspects of army policy Sherman himself clearly had that mindset. The war had not converted him into a believer in the citizen-soldier. His idea of a good postwar military establishment was an army of 100,000 regulars; what he got was an army a quarter that size in which the volunteers only gradually gave way to regulars. West Pointers had to vie with "outsiders" for commissions; even so they were given preference and held eighty percent of the junior officers' slots by 1890.

The man's essential conservatism accorded well with a number of time-honored army traditions. Like most in his profession, he held to certain customs that were not compatible with the principles of good administration. He readily acknowledged that there was no justification for maintaining the Division of the Atlantic other than to provide an honorable niche for George Meade—but the army took care of its own. He did not like to oust a man from a comfortable billet he had worked years to obtain, and had thus in a sense earned. He revered the rule of seniority: "To pass over a worthy man to select a junior, overslaughing, is in the Army and Navy the most mortifying blow an officer can receive—worse than death."

Though the "revolution" Sherman spoke of never came about, he was sensitive to any tremor felt in the social structure. When labor violence rose in the 1870s, Sherman saw in it another reason to keep the army up to strength. He continued to speak of a violent reckoning between haves and have-nots, one not too different from that envisaged by his contemporary, Karl Marx—though they were looking at the phenomenon from opposite sides of the barricade.

To Sherman it seemed perfectly natural for the army to take the side of a copper mine owner in his quarrel with striking miners. He knew that for fighting city mobs one needed regulars; they could always be counted on not to go over to the crowd, whereas conscripts and militiamen might defect. Unfortunately, with all the army had to do, it could not instantly respond to urban unrest. To a worried city official who wrote in, the general emphasized the need to maintain at least three fortified strongpoints within a large city; it would be wise to ban the sale of arms to "suspicious parties" and to "watch the embryo Communists and break them up before they acquire too much head."

The academy was still being criticized as a citadel of privilege, but it was one thing Sherman would lobby for; he visited it often and apparently took pleasure in always finding it the same. He stayed in touch with Dennis Hart Mahan, the academy's most venerated teacher and in a sense the high priest of a West Point "cult" until his tragic death in 1871 (shortly after being informed that he would have to retire from the academy, Mahan either fell or stepped from a boat and drowned in the Hudson).

Sherman's first reaction to change at his alma mater was often negative. In 1881 Captain E. S. Godfrey, the academy's instructor in horsemanship, proposed adding exercises and gymnastics to his program. The superintendent did not believe the existing course needed any changes, having been taught the same way to the general satisfaction for a half-century, but he sent the proposal on to Sherman, along with a notation that this was the second time Godfrey had proposed changes. Sherman wrote back that horsemanship was taught for service in the army, "and not in the circus." Moreover "young officers are apt to construe innovation as improvement. The instruction laid down in cavalry tactics has made good soldiers in the past and will in the future." Finally, the captain's actions were "indelicate and disrespectful." There was a clear implication in Sherman's letter that Godfrey would soon be moving on to duty elsewhere.

Sherman opposed racial integration of the academy; it was not the place to try "the experiment of social equality." When a black cadet was discovered one morning in his room, bound with rope and with cuts about his head, Sherman readily accepted the story that the cadet had staged the incident to win sympathy because he was having scholastic problems: "The Darkee is in the wood pile yet," he wrote a friend, "this one is cunning and has managed to raise a great smoke to cover his danger at an approaching examination." As for the public clamor over the elitism of the academy, Sherman regarded it as habitual and of no consequence: "We have such outbursts against West Point every ten or fifteen years."

He was nonetheless the agent of change at West Point. Despite the pleadings of Mahan, he endorsed the removal of the academy from the control of the engineers. He wanted "a school for soldiers, not technocrats," and he saw to the installation of the able soldier Schofield as superintendent. As the academy lost its primacy as an engineering school, he favored a broadening of the program of study, but he preferred expanding the military offerings rather than the humanities. Cadets could do what Cadet Sherman had done, explore literature and history in their free time.

To Sherman's way of thinking great commanders were made, not born. They did need one innate quality, a drive or impulsion which he called "motive power," but that alone would not make men excel in war. The motive power had to be directed by knowledge, and that came from instruction, such as that dispensed at West Point. This conviction led Sherman to develop a lively interest in a new form of military school that had recently developed in Europe, one that offered postgraduate work for officers in various arms. The French, who were prominent in setting up such schools, designated them *écoles d'application*. Thanks in large part to Sherman's advocacy of such institutions, the army's engineers, cavalry, and infantry had their schools of application by 1881, and a preexisting artillery school had been upgraded to an equal standard. Sherman was strongly supportive of them but quick to criticize anything he did not like.

In such practical matters as armament and equipment the commanding general was usually a friend of innovation. He liked the breech-loading, rapid-fire Krupp guns that had helped Prussia and her allies defeat the French so handily in 1871, and he wanted similar weaponry for the U.S. artillery. He felt the gaudy, impractical uniforms of earlier wars had wisely been abandoned, and none too soon. What was needed was utilitarian clothing that would not restrict the movements of the soldier or hamper him from using his breech-loader with the greatest freedom. The bayonet should be abandoned as useless, and the noncommissioned officer's sword as well. The soldier should carry the best rifle available, a Colt's revolver and a utilitarian knife. The cavalryman would find little use for his saber: "with breech-loading arms the bold sabreur must disappear and the horse [be] used chiefly to bring the cavalry to the point of danger, there to fight as other men do—on foot." The cavalrymen did not agree; the sword also had its value for esprit and its psychological impact, even in drill and parade. To them there was nothing like the flash of sabers in the sunlight.

Then for a decade and a half Sherman was the patron and collaborator of an extraordinarily gifted young officer named Emory Upton. A West Point graduate of 1861, Upton had a brilliant record in the field; at war's end he held the brevet rank of major general—at age twenty-five. He had also demonstrated remarkable ability to analyze and resolve tactical problems; it was he who authored a new tactical manual incorporating the lessons of the war. He had greatly impressed Sherman, though the two had minor differences over certain tactical questions. Their surviving correspondence indicates that the two men developed a working relationship in which Upton did the spade work and writing that Sherman then read and critiqued, though recent research suggests that Sherman's input was greater than generally thought.

As early as 1873 Sherman proposed to Belknap that Upton be sent abroad on a study tour of foreign military institutions. Sherman was particularly interested in having him study how war was waged in the Middle and Near East, believing that the region had climate and topography much like the American West. Since Congress would never vote a special appropriation for such a trip, he suggested sending Upton to California on temporary duty, then ordering him to return to the East Coast "via India, Afghanistan, Persia and Circassia." Belknap gave conditional approval but warned "it might be difficult to arrange the matter of mileage."

It apparently was, for not until 1876 did Upton and two other officers make the tour. Though Sherman had written ahead to British, Turkish, and Russian authorities to open doors for them in the Middle East, the American officers' stay there was brief. Their interest was inexorably drawn to Europe. Upton was greatly impressed by the German army, which had recently been formed around a Prussian core. It had resoundingly defeated France in 1871, and the more closely Upton examined it, the more he admired it. He returned home believing that the best course for the United States was to reorganize its army along German lines.

Sherman did not believe in adopting the military system of Germany or any other country; he had come to the conclusion that a nation's army reflected its geographical setting and its national character. Huge armies based on compulsory military service had no place in the United States, first of all because there was no need for them: "War with us seems an impossibility, because we have no neighbors capable of grappling with us; all our enemies are within . . ." The only potential danger he saw was from the country's imperialist urge to expand southward to absorb the Antilles, which he referred to as "the wild, barbaric Spanish colonies." Annexation of them might "disturb the equilibrium of races and place the turbulent in majority."

Sherman's solution to the nation's military needs was the "expansible" army, an idea that dated from John C. Calhoun's time, and that Sherman and Upton refined. In time of war each regiment would be capable of rapidly incorporating enough men to make the whole army swell to nearly 200,000 men. This, he decided, was "quite as large a force as we will ever need." In fact the 200,000 figure was exceeded only eight years after Sherman's death, when army strength reached 210,000 during the Spanish-American War; in 1918, the army would have two and a half million men under arms, and in 1945 it would contain over eight million—but such are the perils of prophecy.

While the commanding general thus occupied himself with the army's future, as befitted his role, in private—one might even say in secret—he was spending countless hours communing with the nation's recent military past. Given the man's preoccupation with his "image," it is hardly surprising that well before the war ended he was giving thought to how history might view his role in the great struggle—and he was doing something about it.

In 1863 he had reestablished contact with his old California friend Samuel M. Bowman, who had begun writing articles about him in the *United States Service Magazine.* "Whether agreeable to you or not," Bowman informed him, "the people from Maine to California are in love with you." The two began collaboration on a book-length biography for a Cincinnati publisher. Though the extant correspondence is too fragmentary to follow their joint effort in detail, Bowman did the writing and Sherman supplied the documentation and apparently went over Bowman's text. The results were predictable: their account of the Sherman-Johnston negotiations presented such sinister portraits of Stanton and Halleck that the publisher insisted on a rewrite. And one wonders whether the general actually read and passed on these fulsome lines Bowman put in the book about his command style: "His written orders are luminous of the inspiration of his own matchless genius; and when his directives to subordinates in command are given orally, they are absolutely irresistable . . . "

*Sherman and His Campaigns* appeared in the fall of 1865, with Bowman and R. B. Irwin listed as authors. A brief prefatory letter from Sherman stated simply that Bowman had had access to his order and letter books and should thus have at hand "all authentic facts." The book had some success; the two partners corresponded about a second, expanded edition in 1866–67, but fortunately Sherman decided instead to tell his own story. He probably began work soon afterward; the Ohio Historical Society has a manuscript he wrote on his prewar years bearing the date 1868. He had a complete draft by early 1872, when young Tom Sherman recorded in his diary that he had finished numbering the pages of "Papa's autobiography." Yet not until the end of 1874 did the general go shopping for a publisher.

For Sherman approached the publication of his memoirs the same way he had approached Confederate Atlanta–with care and caution. He concealed the very fact that he was writing them from all except a small circle; he did not even tell his brother John about the manuscript until it was on its way to the publisher. He got a confidential reading by historian John Draper, who was enthusiastic; he contacted another, George Bancroft, who also encouraged him. He consulted William Church and detached the final chapter on "Military Lessons of the War" as a trial balloon in the *Army and Navy Journal*, where it was well-received. There are indications he sounded out others as well. After considerable negotiations he signed a contract with D. Appleton. The general imposed some unusual terms: he paid for and kept control of the plates from which the book was printed, so that he could destroy them whenever he wished.

The two volumes of *Memoirs* were the major publishing event of the spring of 1875. Briefly put, the reception was very favorable. Most reviewers saluted it as a major achievement, even when they found flaws here and there (readers in Ohio were distressed to find no mention of Sherman's early years there, for he began his narrative in 1846). Today informed opinion has not changed. His memoirs, along with those of Grant, figure among a handful of classic works on the war. As with everything Sherman wrote, the style is vigorous and engaging; the expositions are clear and the scope the broadest compatible with reader interest.

Though Sherman often misjudged the political and intellectual competence of the American public, he had a good idea of what it liked to read: he wrote over a hundred crisp and vivid pages on California in the gold rush days, then used only thirty or so as a bridge to bring the reader up to the war. Once there he understood what has fascinated the public since the *Iliad*–the spectacle of great armies in the field, the ring and clash of battle, yet he deftly wove in strategic and logistical threads to make a richer, more textured narrative. And as Russell Weigley pointed out, the *Memoirs* are "almost unique among such literary efforts in their thoughtfulness about the future of war."

Sherman awaited the reaction of the press and public with more impatience than most authors. He was particularly concerned about how Grant might take his work, so he sent proof sheets to Grant's close friend George William Childs, asking him if he had been just to his old comrade-in-arms. Apparently Childs's answer was yes, and later, when Grant had read the book, he said it had done him full justice.

The nation's newspapers had almost all noticed the *Memoirs* by the end of May 1875; often their reviews provoked a stream of letters to the editor. Then magazines and reviews took up the *Memoirs* in turn, so the literary cauldron boiled all summer. Though Sherman had assured Bancroft that he had reworked any passages that might have provoked controversy, in this case as in so many others, he had completely miscalculated the reaction he was about to provoke.

At the outset most of the heated discussion about the book concerned Sherman's treatment of other public figures, many of whom were still alive. Even Grant criticized him for categorizing such men as Frank Blair and John A. Logan as political generals. Sherman's portrayal of another "amateur," John A. McClernand, was particularly unflattering. Where the general considered that subordinates had signally failed him, he cited chapter and verse. The success of his Meridian campaign was limited because "General W. Sooy Smith did not fulfill his orders, which were clear and specific, as contained in my letter of instructions to him of January 27, at Memphis, and my personal explanations to him at the same time." Smith wrote Sherman to protest, but to no avail; the general replied: "You seem to think that I should not have written my memoirs, but yours."

In the case of the failed assault at Chickasaw Bayou, Sherman was equally emphatic: "I have always felt that it was due to the failure of General G. W. Morgan to obey his orders or to fulfill his promise made in person." Morgan did not accept the condemnation in silence; at a veterans' meeting that August he rose and "spoke half an hour, taking occasion to attack Gen. Sherman's book savagely." And Morgan soon took up the pen in his own defense.

Friends warned Sherman that he could expect to have volleys of criticism over his treatment of Shiloh, and he received a lot from people ready to testify that he and Grant had indeed been surprised. One particularly relentless critic was Thomas Worthington, ex-colonel of the Forty-sixth Ohio. He roamed the halls of Congress and the streets of Washington, telling all who would listen that it was he who had tried to warn an obtuse and unbelieving Sherman that an attack was imminent.

In his description of the fighting around Chattanooga Sherman succeeded in giving offense to an entire army. Explaining the situation of Thomas's Army of the Cumberland, he wrote that Grant had told him that "the men of Thomas's army had been so demoralized by the battle of Chickamauga that he feared they could not be got out of their trenches to assume an offensive"; were this not bad enough, he went on to say that Grant wanted Sherman's Army of the Tennessee to set an example, "to take the offensive *first*; after which, he had no doubt the Cumberland army would fight well."

While Sherman's proper task as a memoirist was to recount his own role, some critics felt he did so with entirely too much enthusiasm. This was the case with a reviewer for the *St. Louis Globe Democrat*: "Until the Sherman romance was published we labored under the impression that the successful prosecution of the war was the result of the skill, valor and intelligence of a very great number of people; but a careful perusal of that remarkable work almost convinced us that the suppression of the rebellion was the product of one man's supernatural genius in overcoming the blunders of everybody else."

Such passages stung Sherman to the quick, and there was worse to come. Henry Van Ness Boynton, the Washington correspondent of the *Cincinnati Gazette*, began publishing an extended series of articles on the *Memoirs* that were also printed in one of the Washington dailies. Boynton could not be dismissed as a peevish scribbler treating matters he knew nothing about. He had led an Iowa regiment in the war, had been seriously wounded on Missionary Ridge, and had received a brevet as brigadier general on the personal recommendation of George H. Thomas.

Moreover Boynton was well-connected in Washington; he obtained permission to consult the army's war records, which was exceptional (probably through the influence of one of Grant's associates, Orville E. Babcock). Thus Boynton could document himself on various episodes even more thoroughly than Sherman, since he could read not only the general's letters and reports, but also those of Grant, Thomas, and others (Sherman suspected Boynton was being encouraged and paid out of the army's contingency funds by enemies in the staff departments).

Then, to compound the mischief, Boynton turned the series of articles into a book entitled *Sherman's Historical Raid: The Memoirs in Light of the Record.* In its preface he described Sherman's memoirs as "intensely egotistical, unreliable, and cruelly unjust to nearly all his distinguished associates." The title of each of Boynton's chapters announced some blunder, some misdeed: "Shiloh–the Question of Surprise," "Chickasaw Bayou–Plunging an Army through Deep Swamps against Impregnable Bluffs," "Kennesaw–Ungenerous Treatment of Thomas," and so on. Scattered through the book were documents from the War Department archives that tended to substantiate Boynton's various charges. Impressive as the whole compilation appears, it was artfully constructed to show Sherman in the worst light, just as in the *Memoirs* Sherman had presented himself, consciously or not, from the most flattering angle. As in all such disputes, somewhere between these two poles one must look for the truth.

The impact of Boynton's charges was such that Sherman's friends and advisers–his brother-in-law Charles W. Moulton, Samuel Bowman, Henry Hitchcock, General Jacob D. Cox, and others–had to ask themselves if some response was not called for; yet if Sherman were to reply to Boynton, it would only dignify his charges and help the circulation of *Sherman's Historical Raid.* This may have been why Sherman wrote his brother John at the end of the year that he had "the blues bad."

Bowman wanted to stand firm: "No truer memoirs have ever been written since the New Testament." The general and others of his allies seriously considered rushing out a second "revised" edition of the *Memoirs* with some corrections in the text and an explanatory preface. But in the end they decided that the best response would be a brief volume directed at the errors in Boynton's work. It would not carry Sherman's name, but that of Moulton.

Early in 1876 Moulton rushed out a short book with a long title, which does not seem to have had much impact; when the press did take notice of it, the comments were often unflattering. The *Baltimore Gazette* noted its appearance with these words: "As Colonel Moulton is a brother-in-law of General Sherman, and won his spurs in glorious service as a quartermaster, his reliability and qualifications will not be questioned." Long after the main battle over the *Memoirs* was over, Sherman and his lieutenants would continue to skirmish with Boynton and other critics.

A second edition appeared in 1885. It included a word from the general for all those who had found fault with the earlier edition: "Critics must differ from the author, to manifest their superiority." But Sherman also said, "where I have found material error I have corrected." There were some cosmetic changes as well, usually through the introduction of more charitable adjectives. Sherman also added a fast-paced chapter on his life from 1820 to 1846 and a final chapter on the postwar years; in its last paragraph he made the customary apology for things he had done or left undone, but on the whole, he said he was content. The same can be said of legions of his readers over the last century and a quarter.

# 20

## THE BANQUET YEARS

I N HIS LAST YEARS AS commanding general, Sherman seems to have mellowed considerably; perhaps in writing his memoirs he unburdened himself of much of the bile within him. He settled old quarrels, with John A. Logan, for example, and occasionally found kind words for the nation's press. When he wrote to Murat Halstead, who had helped spread the story about the "insane" Sherman back in 1861, he signed his letter, "yours in friendship." He was also cordial in his letters to Whitelaw Reid of the *New York Tribune*. Reid, in turn, was quick to apologize if an unflattering reference to the general slipped into his paper while he was absent and to promise a "corrective."

Though still possessing remarkable energy as he entered his sixties, Sherman does not seem to have been expending a great deal of it on his work at the War Department during those final years of his tenure in Washington. His outgoing letters as commanding general are a trickle compared with the Niagara of his wartime correspondence. His calls for reforms and increased appropriations in the army are essentially confined to his portion of the secretary of war's annual report, to which Congress paid scant attention.

He habitually arrived at the War Department between nine and ten, having come downtown on the horse-drawn streetcar where he enjoyed the general attention; from time to time, with elaborate gallantry, he would offer his seat to a lady–all the more gallantly if she were attractive. He would remain at the War Department for perhaps six hours, with a break for lunch.

His headquarters consisted of a simple suite of rooms filled for the most part with map cases and book-lined shelves. While clerks toiled silently in the background, for the balance of the morning he would sit at a littered desk, handling his mail. He and his aides usually took lunch in the office, gathered around a table; the fare consisted of sandwiches, and in the general's case some bourbon to wash them down with. Most of his aides had been with him for some time–Audenreid had joined him at Jackson, Mississippi, nearly two decades before. The atmosphere was casual, with Sherman still dominating the conversation, and as ever in such a group, emphatic and uninhibited to the point of indiscretion.

In this setting the commanding general appears as an affable, unhurried man, an accomplished raconteur playing to an appreciative audience. And back at his desk he found the time to occupy himself with what seem the most trivial matters. The president of Columbia College wrote in, seeking a military man's judgment on the feasibility of the Israelites' miraculous flight across the Red Sea as described in Exodus. It was the commanding general who answered his query in precise, mathematical terms: assuming that each man, woman, or child in the act of walking took up a space of two feet by three, and each ox three feet by nine, then two million Israelites with a million head of cattle "could be compressed into a space half a mile broad and six miles long." If they could maintain a speed of one to two miles per hour they could easily pass through a half-mile wide defile in the space of six to nine hours. "Of course the Israelites were not organized or compact as an army," the general noted, "but rather resembled a stampeded crowd"; even so, he felt that with pharaoh's host at their heels they would have moved along at ample speed.

He received sundry visitors in the afternoon, including such "drop-ins" as the postwar mayor of Savannah, who happened to be in the city and stopped by for a lengthy chat. Sherman was generally gone by four o'clock. He was in no hurry to go home, but would walk about the town and was often seen along the street in conversation, "bestowing his benign blessing as a resident war hero on officials and visitors to the capital." He rejoined his family in time for dinner.

As Sherman was mellowing in the Washington years, Ellen was giving way to old age and a variety of infirmities. Her weight crept up and her stamina gradually declined, though in 1871 she found the energy to join briefly Admiral John Dalgren's wife and other prominent women in an antisuffragette movement—no doubt with Sherman's hearty approval. The purpose of the organization, proclaimed by its journal, *The True Woman*, was to lead "the great conservative movement of women against the fantastic doctrines of the long-haired and socialistic agitators for female suffrage."

But Ellen gradually dropped almost all outside activities save religious ones. While the Shermans entertained fairly frequently in the early 1870s, the daughters often replaced their mother in the role of hostess; Minnie had been mobilized for this purpose as early as 1867. Ellen ceased going out socially; she found it fatiguing, though her husband persisted in laying her withdrawal from the banquet and drawing room circuit to sheer indolence. And while she had once proclaimed that her home was her kingdom, Ellen now increasingly relied on her daughter Lizzie for the everyday operation of the household; eventually she even ceded to her daughter her traditional place at the end of the table.

The family group gathered around the table had shrunk as the older children departed in one direction or another. While Cumpy, the youngest, remained a fixture in the house, Minnie, the oldest and the first to marry, departed in 1874. Her marriage to Lieutenant Thomas Fitch of the United States Navy was an affair of considerable pomp, attended by high dignitaries of state and rivaling the wedding of her mother in the same city twenty-four years before. (Elly would marry Alexander Thackara in 1880.)

In the 1870s Elly and Rachel were enrolled in a nearby convent boarding school and were home on weekends. In the early part of the decade Tom was also living with his parents. He kept an extensive diary for the year 1872, and it permits a privileged look at this fifteen-year-old destined to lead a troubled and ultimately tragic existence. He appears in the diary pages as a dutiful, serious-minded child. Like his mother, he took his religious obligations in earnest. He carefully noted and observed the various holy days and made comment on sermons he heard; thus on February 28 he recorded, "Fr. Clarke gave us a very fine sermon on 'Hell' in the morning." He wrote more about his studies than most fifteen-year-olds would. He was not without friends, yet he seems to have spent much time in solitude, studying in his room or "reading in the parlor." In the diary entries there is little evidence of boyish pranks and escapades, and almost no effort at humor. Only once is there a flash of temper, when he is assigned to write a poem about General Lee at Appomattox: "Not a good subject for one who will not praise a *rebel* and a *traitor*."

Tom's delight was to work with his hands, and particularly to build things of wood; this may be why his father wrote a friend—no doubt in rueful vein—"I design him for a carpenter." But Tom acceded to his father's wishes that he become a lawyer, which would put him in good position to supervise the business affairs of his mother and sisters once Sherman himself was gone—a consideration that figured increasingly in the general's thoughts. He would leave his widow and daughters each some property, but a man would have to manage such things; now Tom would be able to assume that burden. He began grooming his son for his family obligations, familiarizing him with various properties and their management.

Things went as Sherman planned through the spring of 1878. Tom had completed his studies at Yale and entered law school in St. Louis, finishing his work there that spring; his father had already made arrangements for him to enter the St. Louis law firm of Henry Hitchcock, virtually assuring him a successful career. Then one day in May catastrophe struck in the form of a letter from son to father: Tom announced that he would not take up the practice of law and would not assume the role of guardian of the family's interests. He was instead answering a higher call: he would become a Jesuit priest. To spare the feelings of "the kindest and tenderest of fathers" he was going abroad and would discreetly enroll in a European seminary.

Coming out of the blue, the letter was a considerable blow to Sherman. His response to his son must have been as severe as his shock and disappointment; its very violence may explain why it has not survived in the Sherman family papers. But in his anger and his anguish the general wrote letters to others. He told John, "I have barely been able to bear this all the past week." To an old friend he wrote, "he is absolutely lost to me." He labeled his son a "deserter" who had abandoned his own family in its hour of need, and in doing so he painted the situation of the Sherman clan in overly somber colors: "I will need his help in the near future and his sisters need him NOW. They are young ladies and need the office of a BROTHER and not that of a PRIEST." He was bitter not only at his son, but also at a church that had robbed him of his son. "I know the Catholic clergy will boast of their conquest," he told his brother; he even speculated that they were "laying their ropes" to catch him. Later there would be a reconciliation of sorts between father and son, but Sherman's bitterness against the Catholic Church lingered and for a time he visited his anger on Ellen as well.

These worries might have weighed more heavily on the general had he spent his evenings brooding over them. But his evenings were otherwise occupied. If he wanted to get away from family problems or the frustrations of his office, he could always distance himself from them—literally. He had always been a traveler, and in the seventies both the opportunities and the necessity to travel were manifold. He spent considerable time away from Washington on trips that had some official purpose but were also important social occasions. There were the yearly trips to West Point, where the commanding general's appearance on the plain was greeted by the firing of cannon and considerable time was spent communing with old comrades. There were the Western inspection trips, which were a bit like the progress of a medieval king, greeted at each stop by his subjects, with ceremonies and diversions appropriate to the occasion.

Then the general had been offered honorary membership in a long list of fraternal and cultural associations. He seems to have almost invariably accepted; he even agreed to become an honorary member of the Marietta, Georgia, Riflemen, but with the proviso that "always they use their rifles and organization in support of good local government and always prove faithful to the nation." Many groups that claimed him as a member also sought him as a speaker, and here again he probably accepted more invitations than he should have, for periodically he would voice the same lament: "I am overtaxed with invitations here and there and everywhere . . . drawn hither and thither . . ." And yet the honors, the plaudits and the praise that came with these invitations and appearances were irresistible to him, for they fed a hunger that could never be assuaged.

Though Sherman lived long enough to have his voice recorded by a primitive phonographic device, his speeches are known only from their texts and newspaper accounts. This is unfortunate, since the general came to be regarded as one of the country's foremost after-dinner speakers; Chauncey Depew, himself an orator of legendary ability, said Sherman was a peerless performer on the banquet circuit.

Nowhere in the surviving texts of Sherman's speeches does the term "War is Hell" appear; on the other hand he could and did evoke the great struggle of the sixties in all its martial splendor. Having saluted the minuscule postwar army that Congress had authorized to contend with the Indians, he reminded his listeners, an assembly of Knights of Saint Patrick, of what once had been: "Many of us ... well remember when we had real armies, with their hundreds of thousands of men in serried ranks in all the majesty of glorious war, with a holy cause and a foe worthy of our steel."

The general would on occasion acknowledge that he was "somewhat glib of tongue" and felt obliged from time to time to "say a few words" to his old soldiers. Veterans' organizations had begun to spring up immediately after the war. The largest of them was the Grand Army of the Republic; the various armies had their "societies," with mostly fraternal and social activities, and there were annual reunions of regiments and even companies.

These organizations provided Sherman with his most loyal and enthusiastic audiences; for speaker and audience each meeting was a nostalgia trip—and for the speaker an ego trip as well.

Of all the fraternal groups with whom he communed, the most important was without doubt the Society of the Army of the Tennessee. It had been founded on April 14, 1865, by a group of officers meeting in the Senate chamber of the North Carolina state capitol, with its chief animator Frank Blair; initially Sherman seems to have stayed on the periphery, but at the 1869 meeting the nominating committee put his name forward for the presidency of the society. He protested that he had commanded several armies, and to accept the post would offend veterans of other societies. "I ask you to relieve me. However, if you insist, I can but submit to your wishes, though I assure you I would prefer it otherwise." Elected thereupon, he graciously accepted the presidency, but warned his fellow veterans that he would "press the same plea" the next year. He remained president until his death twenty-one years later.

The society's meetings took place in a hall swathed with bunting and decorated in a martial theme. They offered much florid oratory, martial music, and a Lucullan banquet that could run to eight courses and three hours' duration; then there was the endless hobnobbing and reminiscing in barrooms and hotel lobbies. Sherman liked the organization's gentlemanly camaraderie (only former officers could belong) and the communing with old friends—and manifestly he liked to run it. In the chair he was a benevolent, paternal autocrat. If someone slated to respond to a toast was absent, he would fill the spot himself or dragoon someone else, overriding the substitute's protests with the assurance that "any damn fool" could respond to that toast. At the Chicago meeting in 1879 he told the conventioneers: "Let the applause be short and emphatic ... don't drawl out into a long gaggle or a noise."

In 1879 Sherman communed with the past in another way when he visited the South. This Southern tour, made at the beginning of the year, was ostensibly to inspect government property, but like all his travels now, it received full press coverage and involved a number of personal appearances, as he moved through Tennessee and the lower South (his itinerary did not include South Carolina). He was well-received, even in Atlanta, where he was lodged in the splendor of the Kimball House, the town's newest and most ornate hotel; it was there that he received the local press. He was not assailed with howls and catcalls, though he might have been, had the public learned that he felt he deserved some credit for the town's renaissance, having brought it national publicity by his 1864 visit, as well as ridding it of old, unsightly structures and making room for new ones.

Despite the hollowness of the official functions that required his attendance in Washington, he took to the social life of the capital. Despite Ellen's ill health the Shermans still received in the 1870s. For a time they sponsored "Ohio sociables" in their home, an idea one suspects was Ellen's rather than her husband's. Then too, any army officer coming to Washington or even just passing through would follow the tradition of coming by to leave his card—and quite likely receive in turn an invitation to dinner. What is more, during Sherman's Washington years a considerable contingent from the high command of the vanished Confederacy also came calling, a number of them having taken seats in Congress. Former opponents Joe Johnston and Joseph Wheeler dined at the Shermans'; Cumpy particularly remembered the visits of a one-legged man with sad eyes–John Bell Hood. Sherman's old adversary had fallen on hard times and was trying to sell his collection of military papers to the government; Sherman tried to help him arrange the sale (in the interim the papers were kept by Lizzie Sherman).

There was now less contact between the Sherman household and the Ewing clan than before the war, and this change no doubt had an adverse effect on Ellen. Maria Ewing had died in 1864; Thomas Ewing was much in evidence when the Shermans first moved to Washington, but his death in 1871 had been a severe blow to his daughter. Though Hugh and Charles had served under Sherman for much of the war, the bonds of friendship had slackened with the coming of peace; Sherman suspected that the coolness of his brothers-in-law sprang from their belief that he could have done more to foster their military careers.

Perhaps one night out of two Sherman would go out, most often by himself, but sometimes escorting one or more of his daughters.

He had taken up socializing again when he settled in St. Louis, and in Washington this aspect of his life seems to have intensified; it would not stop until virtually the eve of his death. He became a steady patron of the Washington theater; when he was in town he was drawn into an unending succession of dinners, receptions, cotillions, garden parties, and Saturday morning gentlemen-only breakfasts. This vibrant social scene was essentially new to Washington and a byproduct of the war. The conflict had transformed the nation's capital socially; previously the prominent families had been Southern and their circle somewhat exclusive–but the war had ended their reign. The new elite had, understandably enough, a military cast, being presided over by a sequence of former Union generals in the White House. The voices heard in the fashionable drawing rooms and banquet halls no longer had the soft, slow cadence of the South; it had been replaced by the Yankee twang and the energetically turned syllables of the Middle West.

As the nation's second most distinguished soldier–and far more sociable than its most distinguished soldier–Sherman had automatic entrée to the homes of this new elite, and he was soon moving through them with perfect ease. He was somewhat wary at first; after all, most of those with whom he mingled had as their calling the dirty business of politics. But his friend Admiral Porter told him that Washington would grow on him: "Like everyone else, you will learn to like it." And as far as Washington society was concerned Porter was right.

Clark Carr, who encountered the general frequently in the salons and drawing rooms of the capital, recalled him as a "social lion" of the era. "Every hostess wished him to attend her receptions and balls, and it may be said truthfully that almost none were disappointed." It was often the case that there were several social functions taking place at the same time, and Sherman would have received invitations to them all. According to Carr the general would arrive–early or late–go directly to greet his host and hostess, linger to chat here and there, then head to the next affair; Carr claimed that over an afternoon and evening Sherman could put in as many as a dozen appearances.

Others who moved in these same social circles remembered how the general enlivened any group he joined. In his diary future President James A. Garfield frequently recorded encounters with him, and invariably with pleasure: "I am always charmed with the vigor and heartiness of General Sherman." A Washingtonian named Marian Adams once found herself at a dinner for fourteen, marooned between "a mild and dull New Yorker named Jay" and the taciturn secretary of state, Frederick T. Freyling-huysen. "At last in desperation I spoke across the table to General Sherman and he became very lively over his 'march to the sea' and repeated it with knives and forks on the tablecloth. Finally he swept the rebel army off the table with a pudding knife, much to the amusement of his audience."

The rich tapestry of the general's social life in its last decades contains a thread that only came to light long after his death; it consists of a series of discreet liaisons with women. Several names are known, though the list of them would no doubt be incomplete; nor is it always possible to fix the exact nature of the relationships, the evidence being essentially epistolary and anything but graphic. What, for example, might one make of this let-ter of 1880 that Sherman addressed to Mrs. Charlotte Hall, one of twenty he is known to have sent her between 1877 and 1890: "If you are ready next Saturday I will take you to Van Courtland Park early in the day so that you can be back before sunset. Destroy this."

In the case of sculptress Vinnie Ream there was what has been called a "quasi-bohemian alliance" beginning in 1873, when Sherman was in his fifty-third year, and continuing intermittently for a decade and a half, during which time Vinnie married Richard Hoxie. Sherman's letters reveal him in the role of lover, confidant, and ultimately fatherly counselor. Then in the summer of 1880 Sherman's longtime aide Joseph Audenreid suddenly died; soon thereafter the general formed with Audenreid's widow, Mary, an attachment whose nature seems to have been essentially sensual—and with Mary taking much of the initiative in the affair, at least judging from Sherman's side of their correspondence. This liaison lasted to the end of the decade. It has been argued that with Mary Audenreid, as with Vinnie Ream Hoxie, "Sherman found a source of pleasure and intimacy that he never established with Ellen Ewing Sherman."

The general's interest in the other sex sometimes manifested itself in another way in the postwar years, and most obviously in the last fifteen years of his life; and it too had certain sexual overtones: any man in his sixties who describes a nineteen-year old beauty as "full of life as a ripe peach" does not have his mind on her vivacity. When he met an attractive girl or young woman the general would rely on his age and his fame to permit him a certain familiarity, which usually took the form of a kiss planted on her cheek. This gradually became a habit; some women were flattered, or at least accepted it as a compliment; some endured it, and others did their best to avoid encountering the general. The wife of General Hazen was one of those who did not appreciate his attentions; she said there was a studied cunning and a feigned absentmindedness in his approach; he might suddenly plant a kiss, then apologize, saying he had mistaken his victim for someone else.

By the mid-eighties these habits were being mentioned in the newspapers, which presented them as the harmless eccentricities of a much-revered figure from the past, a living relic—to be treated with respect, but also with a touch of humor. A reporter who covered the general's visit to a St. Louis circus in 1884 described his encounter with Miss Mattie Jackson, a bareback rider: "He patted her cheeks and tipped her chin as if he were going to administer one of his remarkable kisses, but he didn't. He merely remarked 'you are a nice little girl, a very nice little girl.'" In the sideshow he gave "close scrutiny" to the Egyptian princess, and asked her:

"Were you born like that or do you frizz up your hair?" She replied that she was born like that.

In 1884 Sherman's long association with the United States Army—"the great family of the army," as he called it—came to an end. On February 8 of that year President Chester A. Arthur issued the formal order for retirement, an order that was read that day at evening parade to the officers and men in every army installation in the country. Sherman's life in retirement formally began that day—it was his sixty-fourth birthday, and he had left to him exactly seven years and six days of life.

The year 1884 brought him particularly strong appeals to become a candidate in the presidential election. James G. Blaine informed him that his nomination by the Republican Party was "more than possible"; he had only to say the word. His son Tom received the message essentially offering him the nomination and carried it into the library of the St. Louis house, where the general had company. Without abandoning his guests, without even removing the cigar from his mouth, he scanned the message and wrote out the reply for Tom to take to the waiting messenger: "I will not accept if nominated and will not serve if elected."

He thus remained firm in the resolution he had made in California in 1856 to shun all public office. One motive undoubtedly was the preservation of his privacy, the avoidance of that searching and potentially embarrassing scrutiny that a man in politics was subjected to. But he certainly thought about the possibility of having presidential power, for occasionally, when he was particularly exasperated, he would tell intimates how he would use that power. On occasion the newspapers could still stir his anger. He once wrote a confidant, "if power should ever settle into my hands, their number and unbridled license would cease immediately." Infuriated by the rates charged by the Baltimore and Ohio Railroad, he confided to John, "the first place I would visit with exemplary punishment would be the city of Baltimore." The administration of President Sherman would have been a memorable one.

After the general's retirement, he and Ellen had decided to spend their remaining years in their house in St. Louis and then be laid to rest in one of the city's cemeteries. Sherman had already made the arrangements. He had decided that Willie should sleep on the banks of the great Mississippi; the boy's body had been moved to St. Louis's Calvary Cemetery, along with the body of another infant son, Charles Celestine Sherman, whom the general had never seen (the child had been born in June 1864, when his father was driving on Atlanta; Sherman learned of his death from a Savannah newspaper six months later).

For most people retirement meant a certain repose, a decline of activity, but not for Sherman. The house on Garrison Avenue was for him less a haven than a point of departure for frequent trips to speaking engagements or meetings of clubs and associations to which he belonged, to veterans' reunions and the like. At the GAR's Grand Encampment of 1886 he led the "Western" forces in a fight over election of the next grand commander and the selection of a site for the next national meeting; he reported to a friend, "we succeeded in defeating New York and the East in securing Fairchild as our next leader and St. Louis as our next place of meeting." And he boasted that during the heated debates, "when I rose to my feet to speak the tumult subsided and perfect order and silence prevailed."

He was once again spending much of his time communing with the past, and now when he was at home in St. Louis he was again writing about it. There was the new edition of his *Memoirs*, but now he was also writing articles on various aspects of the war. But this busy retirement was interrupted by a move: as it turned out St. Louis would not be the last earthly abode of the Shermans. With Cumpy enrolled at Yale–to prepare for the career in law that Tom had abandoned–his father and mother resolved to go East (Minnie and her husband had settled in Pennsylvania and that was another reason). Sherman wrote John that Ellen would insist on moving nearer to her youngest child, and he could not afford to maintain a household there and in St. Louis as well. In the fall of 1886 they and their two unmarried daughters, Rachel and Lizzie, made the move to New York City, where for two years they made a temporary home in a Fifth Avenue hotel.

The sixteen-year-old cadet who had wanted to spend his life in the wilds of the Rocky Mountains was now, fifty years later, living in the nation's largest and most cosmopolitan city; curiously, he still talked occasionally about retiring to that hermit's cave in the Rockies, but now it was idle whimsy. In fact he was immediately drawn into the vibrant, animated world about him. It was Washington all over again: doors opened to him everywhere, he figured among the guests of honor at endless banquets and benefits, he was touted as the new and most distinguished member of any number of clubs and societies. He made acquaintance with the city's theaters and was delighted by what he found. At home he was occupied with his reading, numerous callers whom he received in his "office," and a steady stream of correspondence.

In the fall of 1888 he bought a house at 75 West Seventy-first Street. By then Ellen's condition was deteriorating rapidly; she had the shortness of breath and swollen feet characteristic of circulatory disease. On November 28 Sherman was reading in his office when Ellen's nurse called down to him that his wife was dying. He bounded up two flights of stairs calling "Ellen, wait for me . . ." But when he reached her room she had breathed her last.

The blow was telling, but with Rachel and Lizzie to console and encourage him, he gradually emerged from his depression and ventured out into the city again, once more ready to sample what life had to offer. On the occasion of his seventieth birthday his daughters went to great lengths arranging the festivities; a reporter from the *New York World* came to interview him and wrote a glowing account of the spry septuagenarian who rode the trolley and took the el, on occasion offering his seat to ladies. And Sherman went through the motions of protesting the attention he drew when he went out in public: "I begin to feel like a freak in a dime show."

But the ranks were thinning around him now, and he noted the death of each contemporary. When Sheridan, eleven years younger, died in 1888, he was especially affected; he took to referring to himself as "the last of the Mohicans." There were episodes of depression. Nine months before his death he wrote an old friend: "I am the last of the three great generals of our Civil War. I ought to be allowed rest, but there is no rest for me . . . I think it would be a merciful act if after a war Congress should enact that the successful generals should be shot to close the volume."

There is some evidence that as he approached his seventies he had brief episodes of mental confusion. A story circulated that in Augustus Saint-Gaudens's atelier he was introduced to Robert Louis Stevenson but took him for one of his "boys," and launched into war stories. Then in June 1889 he went to West Point for graduation exercises, as he had done for many years; afterward he sat on the veranda of his hotel, where an admiring crowd gathered about him, hanging on his every word. He talked and answered questions for many hours.

"Sherman was never in a happier mood than that day," recalled the father of one of the graduates who was present. At some point the subject of Atlanta came up. The general explained that he had planned to take Hood's entire army by coming at it from four directions at once; Hood was able to escape only because one of Sherman's generals was slow to advance. At this point someone in the crowd called out to know who that tardy general was. "General Sherman took a puff of his cigar, looked slowly around, and then pointed to a gentleman sitting at the further end of the porch, and scarcely lowering his voice, said: 'That's the man, confound him!'"

Such lapses must have been few and brief. Most accounts of Sherman in those last months describe an eminently lucid man. That was certainly his condition when on February 3, 1891 he penned a brief letter to his brother John, one of the last he would ever write: "Dear Brother—I am drifting along in the old rut—in good strength, attending about four dinners out per week at public or private houses, and generally wind up for gossip at the Union League Club . . . "

The following evening he went out in bad weather to join a group at the Casino Theater, and the next morning he awoke with the symptoms of a cold; yet he got up, handled some paperwork, and went to a wedding—despite the protests of Rachel and Lizzie. On the sixth Cumpy wrote in his diary, "Papa unwell." On the seventh the general's face became red and swollen; on the eighth, his seventy-first birthday, the family called in physicians, who diagnosed facial erysipelas. Modern medical opinion holds that the general was being visited by what earlier generations called "the old man's friend"—pneumonia. Telegrams went out to Minnie and Elly, and a cable to Tom, who was in England. Sherman's condition gradually worsened; he seemed to drift in and out of consciousness. The children summoned a priest, who administered Extreme Unction while their father lay insensate. The death watch began.

In his moments of consciousness the old warrior knew he was dying. Most men entrust their epitaph to posterity, but Sherman had composed his own; as family members bent over to hear him he repeated it insistently: "Faithful and honorable." "Faithful and honorable." On February 14, at 1:50 in the afternoon, the general took one short breath, then there was a sigh. A nurse bent over him, then whispered: "He is dead."

# 21

## To Posterity

SHERMAN HAD STIPULATED that his funeral be a simple military one. It was military but not simple, for the nation would not have it that way. Over the preceding thirty years he had fixed himself in the public mind and memory, and now that he was gone there was much to remember: that lone voice from Louisiana, proclaiming loyalty to the Union even as it seemed to be foundering . . . the spectacle of his "insanity," then his redemption at Shiloh . . . the resounding fall of Atlanta, that put his name on everyone's lips . . . the enduring suspense and dazzling success of the Great March . . .

The funeral service in New York City was a private one, but the procession that took his body to the funeral train was watched by tens of thousands. They packed the sidewalks and the windows along Fifth Avenue, and even climbed into trees and clung to lampposts; there were the dirges, the muffled drums, the tolling of bells as the cortege crept by: successive bodies of troops, military bands, dignitaries in carriages—and at the supreme moment the flag-draped casket borne on a caisson, and behind it the riderless horse.

The general made his final journey back to his "Great West," past silent crowds that had gathered at stations along the way. Once the burial ceremonies in St. Louis's Calvary Cemetery were over, the members of

Sherman's family resumed their lives. Rachel took a husband before the year was out, and like her sisters already married, she raised a family; only Lizzie was destined to live out her life a spinster. Thomas continued in his clerical role, but was increasingly troubled in mind and spirit. In 1911 he suffered a nervous breakdown; he lived an increasingly cloistered life and died in 1933. Philemon Tecumseh—Cumpy—became a lawyer and spent his life in New York City, dying there on December 6, 1941. He never married, so that of Sherman's four sons, none passed on his name.

While his progeny gradually disappeared from public notice, the general entered into that second existence reserved to the famous. In the immediate aftermath of his passing his name was on people's lips everywhere—and now it was to be conferred on streets and squares. This process had already begun west of the Mississippi, where many of the settlers were veterans of Sherman's armies: there was a Sherman Tunnel, a Sherman Pass, a Sherman Hill; Sherman County, Kansas contained Lincoln, Grant, Logan, McPherson, and Shermanville Townships.

The general was further immortalized in bronze. The Society of the Army of the Tennessee conducted a fund drive for a large statue of their leader on horseback, along with lesser figures, to be erected in downtown Washington. Unveiled in 1903, it occupies the approximate place where Sherman had watched his army march down Pennsylvania Avenue on May 24, 1865.

The brilliant, eccentric Augustus Saint-Gaudens set to work on a magnificent equestrian statue to grace New York's Central Park. Saint-Gaudens knew his man: his Sherman seems in a hurry, with horse and rider caught in vigorous, purposeful movement; for Sherman's mount the sculptor took as model a celebrated thoroughbred of the day. Preceding the general is a winged victory with features borrowed from Sherman's niece, the beautiful Elizabeth Sherman Cameron; under one of the horse's hooves Saint-Gaudens put a crushed pine bough, which he said represented the state of Georgia.

In the short term the general's passing provoked a flood of glowing reminiscences about the man and his times. Newspapers printed interviews with old soldiers who had served under him. The meetings of various veterans' organizations featured speakers who had known him; old comrades related anecdotes that were often embellished for purposes of publication. There was a new edition of the *Memoirs*, and various other tributes and testimonials.

But the general's death also had the effect of liberating historians who had felt obliged to tread carefully around such subjects as the first day's fighting at Shiloh and the burning of Columbia as long as the revered old soldier lingered among the living. Now revisionism set in. In April 1891 the *National Tribune*, the voice of the Grand Army of the Republic, published an article by General Aquilla Wiley about "Sherman's inaction" at Missionary Ridge. Then the following August the *Atlantic Monthly* published a reflective essay on Sherman by John C. Ropes that provoked a lively war of words.

In thirteen pages Ropes ran through the general's wartime career, paying him a number of compliments, but it was not these that caught the readers' attention. Ropes passed over the troubled period of Sherman's command in Kentucky, but lingered over Shiloh, noting that while Sherman and Grant both denied they were surprised there, "the world has never been able to take their statements seriously." Similarly, he noted that in their accounts of the struggle for Missionary Ridge, the two generals "allowed their personal feelings to color, if not distort, the narrative."

To Ropes, George H. Thomas had shown himself by mid-1863 a better commander than Sherman, who "was seen to be a careful, energetic and trustworthy corps commander. But that is all."

Had justice been done, Thomas, not Grant, would have been given command of the Military Division of the Mississippi; Sherman succeeded to that command chiefly because he was "Grant's favorite officer." As for the Atlanta campaign, had Sherman followed Thomas's advice in the operations around Resaca, "he might have ended the campaign with a sudden and brilliant victory." In his subsequent operations that summer Sherman "sometimes lost more men than he need have lost." While Ropes saluted the Great March as a magnificent military *tour de force*, he also quoted a number of the general's statements regarding "his often-announced policy of devastation." Here, Ropes said, Sherman went beyond the attainment of purely military ends, applying devastation as a "punishment for political conduct," which it was neither his business nor—by the laws of war—his right to inflict.

With this article Ropes drew the battle lines for a war that would continue for years. The reaction of the general's partisans was rapid and vigorous. One of them, writing in a Buffalo, New York, newspaper, professed himself shocked by Ropes's "extraordinary temerity," but went on to reassure his readers: "The severe and distorted article cannot damage the military fame of General Sherman," safely preserved as it was "in the hearts of thousands of his surviving and admiring comrades." Ropes came in for some personal attacks. He was referred to as "a Boston literator," a "literary chiffonier." The *Portland Oregonian* denounced him as "a closet soldier, valiant with his pen and mighty in magazine strategy." Ropes's own failure to serve in the war was thrown in his face.

Ropes's article was particularly well-received in two places, and one of them was the South. The *Richmond Times Dispatch* lauded it as "calm, impartial and discriminatory." The *Charleston News and Courier* was only too happy to find support for its own views on Sherman "from an unexpected quarter." The *Atlanta Constitution* proclaimed: "The north is approaching the period when she will be willing to hear the truth of history."

The other place where Ropes's article was read with manifest pleasure was among the veterans of the Army of the Cumberland. General Henry M. Cist, corresponding secretary of the Society of the Army of the Cumberland and former member of General Thomas's staff, could not help pouring out his own feelings in a letter to Ropes, writing effusively of General Thomas, of his qualities of mind and heart, and of his greatness as a soldier.

Then Cist turned his attention to Sherman and he did not mince words: "The other was a military sham, a popinjay soldier who could do no good hard honest work on the battlefield. If you will take Sherman's memoirs and read through you will find that he never fought a greater command than a corps that he did not get whipped. He always is ready with the excuse that someone else—a subordinate—failed to do what was needed at the right time."

But by 1891 these were becoming old soldiers' quarrels, losing their immediacy as the years passed and the generation that had fought the war dwindled away. With the new century came a new era; the nation now had the status of a great power, and with the new status came new wars, new heroes; people had other preoccupations and concerns, from which they would briefly turn their thoughts on Memorial Day.

In the popular mind and memory, that vast cultural milieu that acquired its knowledge of the nation's history in bits and snippets, by random reading and tales handed down, the image of Sherman inevitably blurred and faded; schoolchildren confused Sherman with Sheridan, something their parents would never have done. Here and there sharper memory of the man survived, and also veneration: in the Society of the Army of the Tennessee, which gradually shrank to a roomful of frail octogenarians; in Lancaster, Ohio, where his boyhood home was transformed into a museum; and in St. Louis, where every February 8 for over four decades the local unit of the Grand Army of the Republic decorated the general's grave in Calvary Cemetery. Then in 1934 the Union veterans' ranks had so dwindled that they turned their responsibility over to the American Legion.

A stronger memory of the man lingered generally in the Middle West, which had supplied his soldiers; there veterans' anecdotes would be passed on, and tales about the greatest adventure of their lives, following "Uncle Billy" through Dixie. In the South too the memory of his passage remained vivid, cultivated by a people determined never to forget. Occasionally the bitter legacy found public expression and made the nation's newspapers: in 1906 Father Thomas Sherman found himself at the center of an uproar in Georgia when it was known that he would accompany a group of army officers on a tour of the old battlefields of 1864; when he gave up the project an Atlanta newspaper proclaimed "Sherman in full retreat." In 1911 South Carolina's Governor Cole Blease refused to pay for a new history textbook for the state's schoolchildren because it did not specifically state that Sherman had burned Columbia; the book's author thereupon made suitable changes in the text.

In 1937 the Postal Service unwittingly spawned a controversy when it announced plans to issue a stamp bearing Sherman's portrait (Grant and Sheridan would also appear on the stamp, but it was Sherman's image that gave offense south of the Mason-Dixon Line). The hue and cry was raised by the United Daughters of the Confederacy and taken up by the last frail residents of Confederate veterans' homes, soon joined by legislators in a number of Southern states.

At the same time a related controversy developed over issuing a stamp in the same series bearing portraits of Robert E. Lee and Stonewall

Jackson. The Illinois-based Society for Correct Civil War Information, run by two daughters of one of Sherman's veterans, sounded the alarm. The society's newsletter urged its members to bombard their legislators with protests over the "traitor stamp." Both stamps were eventually issued and enjoyed lively sales. Newspapers often treated the whole episode in humorous vein, seeing it as a tempest in a teapot involving eccentrics and little old ladies.

The thirties also saw some increase in a broader spectrum of interest in the Civil War, fueled by several popular works. Sherman was the subject of a major biography for the first time in several decades. The author, Lloyd Lewis, a journalist by trade, knew how to tell a lively tale. He rarely resisted a choice anecdote, and his readers happily followed him through 650 pages. Though the author has been faulted on his research–he seems to have overlooked at least one major trove of Sherman material–nearly seven decades later his *Sherman* is still an engaging read.

But even more important in reacquainting America with its Civil War has been the enduring popularity of Margaret Mitchell's novel, *Gone With the Wind*, which appeared in 1936 and was followed three years later by the film version. The impact was considerable, resurrecting a whole era from the nation's past. The spell it cast has been an enduring one, prompting a historian to observe recently that "more Americans learn about the Civil War from Margaret Mitchell than from any other author." In both book and film Sherman was a tangible, looming presence, and the handiwork of his men on their way through Georgia was given prominent play.

Then the second half of the century, with the centennial of the conflict, and more recently the Ken Burns television series, has done much to make the "Irrepressible Conflict" America's favorite war. Buoyed by the strong tide of interest in the war that has flowed for the past four decades, Sherman is alive and well in this popular venue, with his memoirs available in paperback and three new biographies in the 1990s. But here perceptions of the man are formed not only through reading and reflection, but also by the dictates of the heart–and sometimes the promptings of the viscera. He is especially well-known to Civil War buffs, to Civil War round tables, to war gamers, reenactors, local historians, to the Sons of Sherman's March to the Sea, the Sons of Confederate Veterans, the subscribers to *Cump and Company: a Newsletter for Friends and Fanciers of*

*General William T. Sherman*, and an infinitely larger number of unaffiliated Americans who just like to read about "the war."

Were all these enthusiasts to meet, what would they have to say about this man? Actually they do meet—on the Internet, among other places. One has only to enter the Mason-Dixon chat room and type in the name "W. T. Sherman" and the reactions come, from "Stonewall," "Billy Yank," "Belle," and others. There is a distinct cleavage of opinion, and not surprisingly it tends to follow sectional lines. A recent visitor to the chat room reported that the exchanges were often spirited, but they frequently concluded with the message "no hard feelings," or "I really admire the way you handled yourself."

In the past few years the annual ceremonies at Sherman's tomb in Calvary Cemetery have been assumed by blue-uniformed men of the William T. Sherman–Billy Yank Camp No. 65 of the Sons of Union Veterans of the Civil War. This past year they had as their guests representatives of the Sons of Confederate Veterans. The fellowship was genuine. Here, as in the Mason-Dixon chat room, the participants demonstrated a desire for fruitful dialogue that their forebears could well have used in the great crisis of 1860. On Sherman, however, no consensus is yet on the horizon; his fame and infamy go marching on. But then the end of this whole war over the war is nowhere in sight—the participants are enjoying it too much.

But the Civil War aficionados march to their own drummer. In the American public more generally Sherman is probably remembered primarily as Sherman the Destroyer. In the 1880s Georgian Henry Grady could count on a hearty laugh from his audience—and from Sherman, who was in that audience—when he referred to the general as "a kind of careless man with fire." Today, when opponents of a crosstown freeway claim that roadbuilders will go through their neighborhood like Sherman went through Georgia, everybody knows what they mean.

The general, who held popular opinion to be opinion "of the crudest species," would have preferred to trust his fair fame to more informed observers, and in truth those observers have generally given him fair treatment. In the nation's universities, in professional military circles, he has long been the object of curiosity and of inquiry conducted in dispassionate, analytical fashion.

This was the case with researchers in the "Leavenworth" schools, whose creation the general had fostered. By the 1890s officers there were already projecting "on-site" studies of the Atlanta campaign. Major Eben Swift, who planned a study tour in Georgia, used it to highlight Sherman's essential caution and conservatism in the conduct of operations; Swift believed that the general's solutions to the problems he encountered on the road to Atlanta "provided excellent lessons in safe leadership."

Sherman had frequently used the term "strategy" as a synonym for "maneuver." In his *Principles of Strategy*, another army officer, Lieutenant John Bigelow, adopted this same definition and argued that Sherman was a master of this approach, his endless and intricate maneuvering having taken him from Dalton to the gates of Atlanta without having to fight a single major battle. For Bigelow, as for Eben Smith, Sherman's approach was that of a master chess player—studied rather than intuitive, and supremely cerebral.

Today Sherman's name still animates discussions in the classrooms at West Point and figures prominently in the pages of professional journals such as *Civil War History* and the *Journal of Military History*. Despite the mountain of manuscripts the general left behind him—more likely because of them—the man is not always easy to figure out, or even to follow. In his later years he would blithely contradict himself, and few had the temerity to point this out to him. He would tell one audience that the Civil War was perhaps the worst conflict in history—and assure another group of listeners that it was the best.

So two historians can defend diametrically opposed views of the man, each armed with quotations supplied by the general himself. The kaleidoscopic quality he presents to researchers is well-illustrated in the subtitles of biographies written over the years, where he has been called *Soldier, Realist, American; Fighting Prophet; Merchant of Terror; Advocate of Peace;* and *Champion of the Union*.

Just what was his peculiar genius, those qualities of mind that made him excel? Dennis Hart Mahan described his former pupil's mind in this fashion: "Brilliant, with the scintillations of genius in which Alexander, Hannibal, Caesar and Napoleon abound—eager, impetuous, restless, he always worked with a will." Any student of Sherman's career would agree with two of the three adjectives Mahan used, but while Sherman may have been impetuous as a cadet, he was not so as a field commander. He made war essentially the way he had run his bank–avoiding unwarranted risks. As a rule he calculated each step he took, weighing the possible consequences based on the best evidence, making investments of men and materiel only when a favorable return seemed the likely outcome.

Had caution alone governed his actions in such a hazardous game as war, he might have been led to the exclusively defensive posture, even to simple immobilism. But the man was driven by his desire to excel, to have "the esteem of men"; with him fame really was the spur. Subordinates recognized and welcomed his drive and his clear sense of purpose from the very first. General L. B. Parsons, who supervised the rail system supplying Sherman's armies, had admired him from the beginning: "General Sherman is a trump and makes things move," he reported to Halleck in December 1862. "I like his business mode of doing things, his promptness and decision."

Had the general been called on to indicate the campaign that offered his best claim to military glory, it would no doubt have been the one he conducted in Georgia in the spring and summer of 1864. As a conventional military operation the Atlanta campaign marks the zenith of Sherman's generalship. Its successful direction required the exceptional ability to store within one head and maintain for instant retrieval a mind-boggling mass of information on the strength and disposition of four armies (his three and that of the enemy), to factor in data on terrain, road and rail conditions, logistics, and other matters, calculating probabilities where information was sketchy or nonexistent–all to regulate the movement of armies so as to maintain momentum and to produce maximum effect. This was work for a rare intellect. It was also the kind of work that only a specialist could understand and appreciate; to the casual observer and the armchair strategist, waiting for a decisive battle that would not come, Sherman seemed to be taking his own good time, as the war in Georgia ground on.

In the aftermath of World War I, with its disastrous "battlefield constipation," the drive, intensity, and seemingly unfettered nature of some of Sherman's later operations–above all the Great March–attracted increasing attention among theorists and practitioners of war. In 1925 General C. P. Summerall told an audience that to Sherman "the most merciful way to conduct war was to bring it speedily to an end. Victory was the only consideration, and no suffering, hardship, or loss to combatants or non-combatants was allowed to swerve him from the means and the methods that he considered the most efficacious."

In that same year a young British officer named B. H. Liddell Hart took up this vision of a shorter if more intensive war in a slim volume entitled *Paris, or the Future of War*. In it he described the outbreak of a new European conflict that did not repeat the bloody stalemate of 1914–18. Instead, one of the belligerents launched a massive air attack that sowed high explosives, incendiary bombs, and poison gas over the enemy capital, paralyzing all governmental functions, producing wild panic in the population—and forcing the enemy to capitulate in a matter of days, even before its forces at the front had been defeated.

This new path to victory Liddell Hart called "the indirect approach," and in 1929 he published a biography of Sherman saluting him as a practitioner of this new strategy. According to Liddell Hart, when faced with a stalemate, as Sherman was once he had taken Atlanta, "a Great Captain will take even the most hazardous indirect approach—if necessary, over mountains, deserts or swamps, with only a fraction of his force, even cutting himself loose from his communications." The Great March was a brilliant expression of this strategy. It enabled Sherman to abandon the futile pursuit of Hood and strike at a vital but previously ignored objective, the Southern people; for according to Liddell Hart, Sherman saw or at least sensed that "the resisting power of a modern democracy depends more on the strength of the popular will than on the strength of its armies." So in the last months of the war Sherman's chief purpose was to bring the war home to the population of the South. He was in a sense making war on the Southern mind.

The U.S. Army's current doctrinal manual, FM 100–5, incorporates the principle of the indirect approach; the army's own volume on American military history stresses the modernity of the Great March: "Sherman's campaign, like Sheridan's in the Shenandoah, anticipated the economic warfare and strategic aerial bombardments of the twentieth century." A highly regarded and often cited history of the nation's military strategy and policy asserts that Sherman "not only carried on war against the enemy's resources more extensively and systematically than anyone else had done, but he developed also a deliberate strategy of terror directed against the enemy people's minds."

Sherman's own treatment of this part of his military record is interesting. In his memoirs he gave the Great March full coverage—his readers would have expected no less. But in his writings and speeches for a more sophisticated audience, he preferred to talk about other, more conventional campaigns in which he faced a foe worthy of his steel. Though not one to quarrel with a public that sang his praises over the destructive course through Georgia and the Carolinas, privately he acknowledged that as a military operation the Great March was no innovation in warfare; indeed anyone who had read Xenophon's *Anabasis* knew as much.

Writing to Philemon Ewing from Savannah in January 1865 he ridiculed the clamor over the March to the Sea and announced tongue-in-cheek that he was about to "cut loose to attempt another one of those Grand Schemes of War that make me stand out as a Grand Innovator. Of course, I know better, that I have done nothing wonderful or new, but only in our risks proportioned to the ability to provide for them." After the war he lamented to another friend that he would always be known as the general who conducted the March to the Sea. "I have never considered it the most meritorious work I was permitted to take a hand in during the war ... The battles and campaigns it fell my lot to conduct previously were, I think, better tests of a soldier's abilities."

There is no doubt that at the time Sherman hoped the Great March would have something of a moral effect on Southern civilians, but he had a specific plan in mind, prodding Georgia into dropping out of the Confederacy. With Atlanta burning behind him, he headed for Milledgeville, still pursuing the same will-o'-the-wisp: meeting Governor Brown in Georgia's capital and negotiating Georgia out of the war. The canny Governor Brown had left Milledgeville well before Sherman arrived; yet two months later the general had not given up, for he wrote Stanton from Savannah that "with a little judicious handling and by a little respect being paid to their prejudices we can create a schism in Jeff Davis' dominions."

But nothing came of these hopes either, even though he had by now used fire and sword from one end of Georgia to the other. Perhaps someone more adept than Sherman in using the stick and carrot approach might have been more successful, though it is doubtful. In any event in the next leg of his march, South Carolina, the general used the stick only. In the end the Great March delivered no message other than that of naked military power and capacity for destruction. Sherman's admirers say it was supposed to "kill the enemy's courage," to use the Clausewitzian term, to produce demoralization, defeatism, and wholesale abandonment of the "Cause," thus sealing the fate of the Confederacy—and the history books say it did.

It's too bad that Jefferson Davis's government did not do opinion sampling during its brief life, for the results would have been interesting. Such soundings have been made among civilian populations that have undergone ordeals as severe as the Great March in more modern times, and they provide compelling evidence that on his own turf the civilian remains tough, resourceful, and dedicated, even under very adverse circumstances. Could this have been the case with those Southerners in the path of Sherman's host? With what sketchy evidence there is for the Civil War era—mostly letters and diary entries—historian Gary Gallagher has recently argued with considerable effect that the Southern population had more determination, more "staying power" in the face of adversity and imminent defeat than is generally believed.

If one may thus debate the degree to which the Great March produced demoralization and disaffection, there is no doubt about the degree of hatred it engendered. Had a triumphant Confederacy held war crimes trials, William Tecumseh Sherman would have been the first man indicted. If nothing else his habitual verbal violence, his talk of making Georgia "howl," of inciting "fear and dread," and the like, would have condemned him.

After an exceedingly rocky start in 1861, Sherman ended by being enshrined in the nation's pantheon of heroes, taking a place in the Civil War contingent second only to Grant. But even in that assembly his image carried a faint blemish; some of his colleagues in the Union Army had on occasion disapproved his methods. Buell said his later campaigns were attended by "barbarities," and Hooker, highly critical, said he made war "like a brigand." Among Army leaders of the First World War, Southern-born Robert Lee Bullard, who commanded the U.S. Second Army in France in 1918, acknowledged that of all the Federal generals Sherman had been the most effective, but in saying that he made it clear that he did not regard Sherman was his "ideal."

Then when the *Dictionary of American Biography* was being prepared in the 1930s, the entry for Sherman gave problems. Its author, the career soldier Colonel Oliver Lyman Spaulding Jr., had trouble covering his subject with the number of words allotted, so editor Dumas Malone gave him an additional 500 words. Malone urged Spaulding to "buttress" his statements. Sherman's last campaign was particularly troublesome, especially the passage of his army through South Carolina: "The question is so complicated," Spaulding complained, "that a brief statement is hard to draft"; yet when he went into detail he seemed to be favoring one side. Malone too was concerned about fairness; he wanted Spaulding to read some South Carolina sources about the burning of Columbia, which he considered very important. "From many points of view," he acknowledged, "this will be a controversial article."

Of course it was the Great March that was most controversial and indeed remains so. To many it has become the paradigm of military aggression directed at civilians, invoked countless times as one war has succeeded another: Sherman's incursion into Georgia and the Carolinas has been compared to the Kaiser's invasion of Belgium in 1914, German atrocities in the Second World War, and the American intervention in Vietnam. Often such works are colored by an animus against one of the parties compared, and even when they are not their value is questionable. Each war has its *spécificité*, as the French say, its own particular dialogue of violence, its greater or lesser degree of distinction between combatant and noncombatant.

The Civil War began as a "good war," a gentlemanly affair in which the operations of the contending armies were to be carried out almost in vacuo, with the civil populations simple spectators. It could not remain so for several reasons, the challenge of logistics being one, the powerful emotions engendered by the war another; the prospect of advantage which lured generals to "bend the rules" was still another.

While General Sherman may be said to have applied the hard hand of war with more alacrity and with more enthusiasm than most others, he was swimming with the tide in a general evolution of policy that would bring to enemy civilians in the army's path increasing stress, privation, and loss. Yet he and others who led the armies at the end of the conflict would still claim that they had not gone beyond the customary and accepted rules of war of their era.

In the specific case of the Great March, Sherman's Special Field Orders, Nos. 119 and 120, governing the treatment of civilians and their property, were in accord with the laws of war. The plain fact is that such laws, whether they be those laid down in Emmerich de Vattel's *Law of Nations*, in Halleck's *Elements of Military Art and Science*, or in the War Department's own General Orders, No. 100, permitted the military to do any number of things that to most of us do not seem either right or fair. In one of its basic provisions General Orders, No. 100 states that "the unarmed citizen is to be spared in his person, property, and honor as much as the exigencies of war will permit"–with Sherman and other field commanders left to determine those exigencies.

What is troubling here is the door of opportunity that such latitude might open for someone of Sherman's particular mindset. Whether that particularity arose from empathic deficiencies related to narcissism, or had some other origin, Sherman seems not to have been sentient and sensitive in the same way most of us are.

Had he had the time, would he have launched his scheme for repeopling the South, packing off boatloads of disgruntled Georgians and Mississippians to Madagascar? Possibly. But the question is idle; for a man–and particularly this most garrulous of men–has to be judged on what he did, and not on what he talked about doing. Even the historian E. Merton Coulter, whose pro-Southern sentiments were so pronounced that he has been described as "neo-Confederate," conceded that Sherman made dire threats that he never carried out.

Sherman's record in the Civil War, while far from spotless, does not stand out for illegal acts or exceptionally repressive measures. A British specialist in the laws of war, James Molony Spaight, concluded that he was guilty of a single violation of the War Department's own ground rules in General Orders, No. 100: he had failed to notify the authorities in Atlanta before he began bombarding the city on July 20, 1864.

There is, moreover, evidence of a different kind to support the assertion that Sherman's record is not exceptional; it is in the National Archives where it has been gathering dust for nearly a century. In 1902 the secretary of war asked his staff for copies of "Union and Confederate orders relating to the exercise of extreme repressive measures in the conduct of the Civil War." The staff pored over countless orders issued during the war, weighed their severity, and made their selection.

They compiled a list of what they considered "the most notable cases." Sherman's name is there, and four of his orders are cited (more than for anyone else), including the one in which he authorized the use of enemy civilians as guinea pigs when torpedoes were suspected on the rail lines. But he has distinguished company on the list, some of whom are also cited more than once: Union Generals Grant, Halleck, Schofield, Sheridan, Pope, McClellan, Butler, Frémont, and Thomas Ewing, as well as several other generals and officers of lower rank—and on the Southern side General Jubal A. Early, Secretary of War James A. Seddon, and President Jefferson Davis.

Finally, America's most distinguished military figures of succeeding generations—men whose opinions Sherman would respect—have given him an impressive stamp of aproval. John J. Pershing called him "an outstanding genius and the beau ideal soldier." George S. Patton, another admirer, told Liddell Hart that in the 1930s he had dedicated a whole month's leave to following Sherman's track through the South. General Norman Schwarzkopf revealed in his memoirs that when he was directing the Gulf War he kept before him on his desk a Sherman quotation: "War is the remedy our enemies have chosen. And I say let us give them all they want."

So in the end Sherman has come to enjoy in death all that he sought in life: he has survived in history, as he had hoped to; he has that "fair fame" he spoke of, and that fame is periodically nourished and perpetuated by praise from other distinguished warriors. In whatever Valhalla it may dwell, that turbulent, driven spirit should finally know peace.

# NOTES

## Abbreviations Used in Notes and Bibliography

AC–Army Corps

AAG–Assistant Adjutant General

AG–Adjutant General

AGO–Adjutant General's Office

AL–Abraham Lincoln

BU–Mugar Memorial Library, Boston University

CalHS–California Historical Society

CE–Charles Ewing

Coll.–Collection

DHS–Delaware Historical Society

DDP–David Dixon Porter

DFB–David French Boyd

Ed.–Edition

E–Entry

EE–Ellen Ewing

EES–Ellen Ewing Sherman

EMS–Edwin M. Stanton

Filson–Filson Club Historical Society

GBM–George B. McClellan

GHT–George H. Thomas

GMG–George Mason Graham

GO–General Order

HBE–Hugh Boyle Ewing

HH–Henry Hitchcock

HST–Henry Smith Turner

Hunt–Huntington Library

HWH–Henry Wager Halleck

ISHL–Illinois State Historical Library

JAM–John Alexander McClernand

JAR–John A. Rawlins

JBM–James Birdseye McPherson

JHW–James Harrison Wilson

JS–John Sherman

LB–Letter Book

LC–Library of Congress

LSU–Louisiana and Lower Mississippi Valley Colections, LSU Libraries, Baton Rouge

M–When preceding a number, National Archives microfilm

MassHS–Massachusetts Historical Society

MDAH–Mississippi Department of Archives and History

MDM–Military Division of the Mississippi

MoHS–Missouri Historical Society

MOLLUS–Military Order of the Loyal Legion of the United States

MS, MSS–Manuscript(s), Papers

NA–National Archives

ND–University of Notre Dame

OHS–Ohio Historical Society

OOH–Oliver Otis Howard

OR–U.S. War Department. *The War of the Rebellion: A Compilation of the Official Records of the Union and Confederate Armies.* 128 vols. Washington: Government Printing Office, 1880–1901. (Series indicated by Roman numeral, volume by number, part by "Pt.," then page by number.)

PBE–Philemon Beecher Ewing

PTS–Philemon Tecumseh Sherman

RA–Robert Anderson

RG–Record Group

SCL–South Caroliniana Library

SF–Sherman Family

SFO–Special Field Order

SO–Special Order

Stan–Cecil H. Green Library, Stanford University

TE–Thomas Ewing

TEF–Thomas Ewing Family

TEJr–Thomas Ewing Jr.

TES–Thomas Ewing Sherman

UNC–Southern Historical Collection, University of North Carolina Library

USAMHI–U.S. Army Military History Institute

USG–Ulysses S. Grant

USMA–United States Military Academy

UVA–Alderman Library, University of Virginia

VHS–Virginia Historical Society

WLF–Walter L. Fleming Collection, MSS. 890, 893

WTS–William Tecumseh Sherman

## 1. LANCASTER

**Among the countless paths:** Hervey Scott, *A Complete History of Fairfield County, Ohio, 1795–1876* (Columbus, Ohio: Siebert and Lilley, 1877), 1–5; George Sanderson, *A Brief History of the Early Settlement of Fairfield County* (Lancaster, Ohio: Thomas Wetzler, 1851), 5–7.

**The land of the Wyandots:** Henry M. Wyn Koop, comp. and ed., *Picturesque Lancaster, Past and Present* (Lancaster, Ohio: Republic Printing, 1897), 2–5; Sanderson, *Brief History*, 5–7.

**The first log cabin:** C. M. L. Wiseman, *Centennial History of Lancaster, Ohio and Lancaster People* (Lancaster, Ohio: C. M. L. Wiseman, 1898), 10; Albert A. Graham, *History of Lancaster and Perry Counties* (Chicago: W. H. Beers, 1833), 34; Sanderson, *Brief History*, 14–16.

**What this burgeoning community:** Graham, *History*, 22.

**The new town, like:** Wiseman, *Centennial History*, 42; WTS, *Memoirs*, 2 vols. (New York: H. Appleton, 2nd ed., 1885), vol. 1, 14. Note: where the edition of Sherman's *Memoirs* is not given, reference is to the first edition, 1875.

**A class structure emerged:** Wiseman, *Centennial History*, 20, 55–56; Sanderson, *Brief History*, 19.

**In new communities such as Lancaster:** Sanderson, *Brief History*, 28; Wyn Koop, *Picturesque Lancaster*, 141.

**The population of Lancaster:** Graham, *History*, 22.

**As time passed:** Sanderson, *Brief History*, 101; Scott, *Complete History*, 32; Graham, *History*, 45.

**In the first decades:** Wiseman, *Centennial History*, 22–26; WTS, speech to the Ohio Society of New York, 1870, quoted in letter of TEJr to B. H. Liddell Hart, December 13, 1929, TEF MSS, LC.

**Hugh Boyle arrived in Lancaster:** TE, "Autobiography," Clement L. Martzolff, ed., *Ohio Archaeological and Historical Quarterly* 22 (1913): 136; clipping from *Cincinnati Commercial Gazette*, July 9, 1887, in clipping file, CE MSS, LC; Wiseman, *Centennial History*, 28.

**Thomas Ewing's entrance into:** TE, "Autobiography," 126–28.

**Ewing's physical development:** Charles C. Miller, *History of Fairfield County and Representative Citizens* (Chicago: Richmond-Arnold, 1912), 423; Lloyd Lewis, *Sherman: Fighting Prophet* (New York: Harcourt Brace, 1932), 14–16; TE, "Autobiography," 137–40, 154.

**Were this not occupation enough:** TE, "Autobiography," 153–62; Wiseman, *Centennial History*, 26–27; Graham, *History*, 81; Obituary of TE, *Cincinnati Commercial*, October 27, 1871.

**Among Ewing's close friends:** TE, "Autobiography," 163, 170; Lewis, *Sherman*, 19.

**About the Sherman family:** Address of General Sherman to the New England Society, December 22, 1870, printed in the *Army and Navy Journal*, January 7, 1871, 173; Jared W. Young, "General Sherman's Puritan Heritage," *Eugenical News*, 13 (August 1928), 106–09; Thomas Townsend Sherman, *Sherman Genealogy, Including Families of Essex, Suffolk, and Norfolk, England, Some Descendants of Captain John Sherman, Reverend John Sherman, Edward Sherman, and the Descendants of Honorable Roger Sherman and Honorable Charles Sherman* (New York: Tobias A. Wright, 1920), passim.

**That same year he:** Stanley P. Hirshson, *The White Tecumseh: A Biography of General William T. Sherman* (New York: John Wiley and Sons, 1997), ix–x, 2–3, 72, 106, 108, 342–43, 389.

**Charles Sherman soon left:** WTS, *Memoirs*, 2nd ed., vol. 1, 10; Taylor Sherman to Charles R. Sherman, August 24, 1810, MS VFM 478, OHS; William J. Reese, *Sketch of the Life of Charles R. Sherman* (n.p., n.d.), 4–5.

**Charles and Mary Sherman:** Reese, *Sketch*, 11; Graham, *History*, 97.

**Charles Sherman gave every:** JS, *John Sherman's Recollections of Forty Years in the House, Senate and Cabinet: An Autobiography*, 2 vols. (New York: D. Appleton, 1895) vol. 1, 15; WTS, "Autobiography," OHS, 2; Reese, *Sketch*, 2.

**One suspects that Judge Sherman:** W. McCrory, "Early Life and Personal Reminiscences of General W. T. Sherman," *Glimpses of the Nation's Struggle* (MOLLUS, Minnesota Commandery, 3, 1893), 314; WTS to William Stanley Hatch, November 22, 1872, WTS MSS, ISHL; WTS, *Memoirs*, 2nd ed., vol. 1, 11.

**With all his qualities:** Reese, *Sketch*, 11.

**Early in 1816:** James P. Boyd, *The Life of General William T. Sherman* (New York: Publishers' Union, 1891); JS, *Recollections*, vol. 1, 18–20; Wiseman, *Centennial History*, 336.

**Sometime after 1815 Charles's sister:** Betsy Sherman to Katherine Anne Rogers, January 7, —, Undated Letter File, TEF MSS, LC; WTS, *Memoirs*, 2nd ed., vol. 1, 10–11. WTS to DeWitt Talmadge, December 12, 1886, in Edward Bok, *The Americanization of Edward Bok: The Autobiography of a Dutch Boy Fifty Years After* (New York: Charles Scribner's

Sons), 217; JS, *Recollections*, vol. 1, 9; JS, "In Commemoration of General William T. Sherman," in *Personal Recollections of the War of the Rebellion* (MOLLUS, New York Commandery), (New York: G. P. Putnam's Sons, 1897), 49.

**For Charles Sherman's widow:** Young, "General Sherman's Puritan Heritage," 109.

**Nine-year-old Tecumseh:** McCrory, "Early Life," 17; untitled, undated newspaper clipping in TES scrapbook, OHS; WTS "Autobiography," 6.

**Two conflicting portraits:** WTS, *Memoirs*, 2nd ed., vol. 1, 12–13; Grenville M. Dodge, *Personal Recollections of Abraham Lincoln, General Ulysses S. Grant, and General William T. Sherman* (Council Bluffs, Iowa, 1914), 231; TE to Maria Ewing, December 13, 1831, TE MSS, LC; JS, *Recollections*, vol. 1, 32–33.

**In the first edition:** S. M. Bowman and R. B. Irwin, *Sherman and His Campaigns: A Military Biography* (New York: Charles B. Richardson, 1865), 12; WTS to EES, April 19, 1865, SF MSS, ND; Hiram Hitchcock in Thomas C. Fletcher, comp., *Life and Reminiscences of General William T. Sherman, by Distinguished Men of His Time* (Baltimore: B. H. Woodward, 1891), 298; WTS, *Memoirs*, 2nd ed., vol. 1, 14; McCrory, "Early Life," 13; Bok, *Americanization*, 216.

**Through all the years:** JS, *Recollections*, vol. 2, 1105; Bok, *Americanization*, 217; L. M. Dayton letter in New York *World*, February 22, 1891.

## 2. WEST POINT

**Charles and Mary Sherman:** WTS, *Memoirs*, 2nd ed., vol. 1, 16; TE to Lewis Cass, August 1, 1835, WTS file, folder 6143, 1835, U.S. Military Academy Cadet Application Papers, 1805–1866, M688, NA.

**A successful applicant:** *Regulations established for the Organization and Government of the Military Academy, at West Point, New-York, by Order of the President of the United States; to which is added the Regulations for the internal Police of the Institution; with an Appendix, containing the Rules and Articles of War, and Extracts from the General Regulations of the Army applicable to the Academy* (New York: Wiley and Putnam, 1839) 9.

**The letters were awarded:** James L. Morrison Jr., *"The Best School in the World:" West Point in the Pre-Civil War Years, 1833–1866* (Kent, Ohio: Kent State University Press, 1986), 4; WTS, *Memoirs*, 2nd ed., vol. 1, 16; WTS to JAR., Aug. 9, 1865, in WTS LB 23, Generals' Papers and Books, RG 94, E159, NA; hereafter cited as WTS LB 23 NA.

**The academy's Board of Visitors:** *Annual Report of the Board of Visitors of the United States Military Academy, West Point, New York, June, 1839* (Washington, D.C.: A. B. Claxton, 1839), 3; Morrison, *West Point*, 62; Cadet Application Papers, 1836, M688, NA, passim.

**Such things lent credence:** "Justitia," "To the Honorable Mr. Hawes," *Army-Navy Chronicle*, No. 25 (June 3, 1836), 386–89.

**When the examiners assembled:** Morrison, *West Point*, 65.

**For young William T. Sherman:** TE to Maria Ewing, June 28, 1836, TE MSS, LC; WTS, "Autobiography," OHS, 11; Bernarr Cresap, *Appomattox Commander: The Story of General E. O. C. Ord* (South Branch, N.J.: A. S. Barnes, 1981), 15–16.

**In September summer camp ended:** *Regulations*, 1839, 41; Joseph B. James, "Life at West Point One Hundred Years Ago," *Mississippi Valley Historical Review* 31 (June 1944): 27; WTS to EE, November 24, 1838, SF MSS, ND.

**A distinctive characteristic:** Roswell Park, *A Sketch of the History and Topography of West Point and the Military Academy* (Philadelphia: Henry Perkins, 1840), 1, 150.

**In some ways:** *Regulations*, 1839, 14; Park, *Sketch*, 106.

**Wherever feasible the learning:** Park, *Sketch*, 105; Edward S. Holden and W. E. Ostrander, comps., *Centennial of the United States Military Academy at West Point, New York, 1802–1902*, 2 vols. (Washington, D.C.: Government Printing Office, 1904), vol. 1, 379; WTS to PBE, May 18, 1839, PBE MSS, OHS.

**The third year:** WTS to JS, June 9, 1839, and January 14, 1840; WTS MSS, LC; Park, *Sketch,* 105.

**The cadets were presented:** Dennis Hart Mahan, *Advanced Out-Post, and Detachment Service of Troops, with the Essential Principles of Strategy and General Tactics, for the Use of Officers of the Militia and Volunteers* (New York: John Wiley and Sons, 1853), 30; Thomas E. Griess, "Dennis Hart Mahan: West Point Professor and Advocate of Military Professionalism, 1830–1871" (Ph.D. dissertation, Duke University, 1968), 217–18.

**The curriculum was overly ambitious:** *Regulations,* 1839, 6; Jasper Adams to Richard Delafield, February 3, 1840, Correspondence relating to the Military Academy, RG 94, E 212, NA; Holden and Ostrander, *Centennial,* vol. 1, 440; WTS to JS, August 31, 1839, WTS MSS, LC.

**Perhaps the most significant:** GBM to WTS, October 23, 1859, WLF Coll., LSU.

**While the academy was:** WTS, "Autobiography," OHS, 6; HBE, "Autobiography," HBE MSS, OHS, 44–45; Holden and Ostrander, *Centennial,* vol. 1, 36.

**In this era there:** Bowman and Irwin, *Sherman,* 12; John M. Schofield, *Forty-six Years in the Army* (New York: Century, 1897), 3.

**The corps of cadets:** WTS to George W. Childs, April 19, 1877, in George W. Childs, *Recollections* (Philadelphia: J. B. Lippincott, 1892), 153; WTS to Mahan, September 16, 1863, WTS LB 23, NA.

**In this little world:** Undated newspaper clipping, "About the Shermans," Clipping File, CE MSS, LC; "Thomas West Sherman", in Ezra J. Warner, *Generals in Blue: Lives of the Union Commanders* (Baton Rouge and London: Louisiana State University Press, 1995), 440–41.

**It appears from Sherman's letters:** WTS to PBE, April 29, 1838, WTS MSS, UHS; WTS, "Autobiography," 17; WTS to HBE, October 1, 1844, WTS MSS, OHS; McCrory, "Early Life," 314–15.

**The life of the cadet:** WTS to Maria Ewing, July 30, 1836, SF MSS, ND; William Whitman Bailey, *My Boyhood at West Point. Soldiers' and Sailors' Historical Society of Rhode Island Personal Narratives, Fourth Series, No. 12* (Providence: Published by the Society, 1891), 133; WTS to PBE, November 5, 1837, PBE MSS, OHS; *Forty-six Years,* 7.

**Within the cadet community:** Jeremy Adams to Richard Delafield, February 3, 1840, RG 94, E 212, NA; James W. Schureman to his sister, January 12, 1840, James W. Schureman MSS, LC.

**There were cadets who:** G. J. D. Kinsley to René Edward de Russy, June 3, 1836, RG 94, E 212, NA; entry for May 1836, Order Book, USMA, RG 94, E 614, NA.

**In September 1838 Major Richard Delafield:** James, "Life," 27; Richard Delafield to J. G. Totten, September 1, 1839, RG 94, E 212, NA; Erasmus D. Keyes, *Fifty Years Observations of Men and Events, Civil and Military* (New York: Charles Scribner's Sons, 1884), 193; W. S. Brown to Richard Delafield, June 21, 1837, RG 94, E 212, NA.

**Delafield believed:** James, "Life," 37.

**In the spring of 1840:** Richard Delafield to J. G. Totten, April 15, 1840, RG 94, E 212, NA.

**The name of William Tecumseh Sherman:** *St. Louis Republican,* March 11, 1866; Register of Cadet Hospital, USMA, West Point, January 1, 1838-July 3, 1840, Registers of Field Hospitals, RG 94, New York Register 604, NA, passim.

**Far more significant is:** WTS to JS, January 14, 1840, WTS MSS, LC.

**But the letters written:** J. Whitelaw Reid, *Ohio in the War,* 2 vols. (Columbus, Ohio: Eclectic, 1893), vol. 1, 419–20.

**The most voluminous correspondence:** WTS to EE, August 30, 1837, and November 1, 1839, SF MSS, ND.

**In everything he strives:** WTS to EE, May, 23, 1837, and July 10, 1837, SF MSS, ND.

**And as their correspondence:** WTS to EE, August 21, 1839, SF MSS, ND.

**Sherman also corresponded:** WTS to JS, January 4 and March 7, 1840, WTS MSS, LC.

**Sherman was already a voracious reader:** WTS to PBE, January 26, 1840, WTS MSS, OHS; WTS to Maria Ewing, October 15, 1836, TEF MSS, LC.

**The year 1840 would mark:** WTS to EE, June 22, 1839, SF MSS, ND.

**Then a boundary crisis:** WTS to EE, March 10, 1839, SF MSS, ND; Bowman and Irwin, *Sherman*, 13.

**In January 1840 Sherman:** WTS to JS, January 14, 1840, WTS MSS, LC; WTS to PBE, April 13, 1840, PBE MSS, OHS; WTS, "Autobiography," OHS, 15.

### 3. THE SOUTHERN YEARS

**In normal times:** Russell F. Weigley, *History of the United States Army*, enlarged edition (Bloomington: Indiana University Press, 1984), 163–64; Kevin Conley Ruffner, "History of the 2nd Battalion, 3rd Field Artillery Regiment" (unpublished MS, USAMHI Library), 1–4.

**Initially spokesmen for the Seminoles:** Maurice Matloff et al., *American Military History* (Washington, D.C.: Office of the Chief of Military History, United States Army, 1969), 159–61.

**The war Sherman found:** Edward M. Coffman, *The Old Army: A Portrait of the American Army in Peacetime, 1784–1861* (New York and Oxford: Oxford University Press, 1986), 50–51; WTS to JS, March 30, 1841, WTS MSS, LC.

**Sherman went out on a few patrols:** WTS to JS, March 30, 1841, WTS MSS, LC; WTS to Elisabeth Sherman Reese, January 16, 1841, WTS MSS, LC; Jane F. Lancaster, "William T. Sherman's Introduction to War, 1840–1842: Lesson for Action," *Florida Historical Quarterly* 72 (1993): 58–72; John David Waghelstein, "Preparing for the Wrong War: The United States Army and Low-Intensity Conflict, 1775–1890" (Ph.D diss., Temple University, 1990), 147–50.

**Sherman's solution, which he confided:** WTS to JS, March 30, 1841, WTS MSS, LC; Bowman and Irwin, *Sherman*, 14; Weigley, *Army*, 160–61; Waghelstein, "Wrong War," 159.

**Yet in Colonel Worth's command:** Coffman, *Old Army*, 51; Monthly Returns, Third Artillery, 1841–50, Regimental Returns, M727, NA, passim.

**Company G was soon transferred:** WTS to Willard Warner, February 5, 1879, Willard Warner Papers, ISHL.

**For this period:** WTS to AGO, March 16, 1842, M567, Letters Received by the Adjutant General's Office, NA.

**With the Second Seminole War:** WTS to JS, May 11, 1842, WTS MSS, LC.

**After the stir and action:** Weigley, *Army*, Appendix, 597.

**Since there was no retirement system:** Coffman, *Old Army*, 48.

**The disabled continued to serve:** Ibid., 49, 58–59; Marcus Cunliffe, *Soldiers and Sailors: The Martial Spirit in America* (Boston: Little, Brown, 1968), 47.

**Given these manifold disadvantages:** Morrison, *West Point*, 15; William B. Skelton, *An American Profession of Arms: The Army Officer Corps, 1784–1861* (Manhattan: University Press of Kansas, 1992), 158; WTS to EE, June 4, 1844, SF MSS, ND.

**But at bottom the army:** GO No. 65, 15 AC, August 15, 1863, in WTS, *Military Orders of General William T. Sherman, 1861–'65* (n.p., n.d.), 187.

**In the 1840s one problem:** Coffman, *Old Army*, 57; WTS to JS, May 23, 1843, WTS MSS, LC.

**Those officers who stayed:** WTS to EE, June 4, 1844, SF MSS, ND.

**Politics was a perennial subject:** Skelton, *American Profession*, 282–87; WTS to JS, October 11, 1841, WTS MSS, LC.

**Of Sherman's duties at Fort Moultrie:** Court Martial Trial Record of Private George Smith, Company G, Third Artillery, November, 1844, Records of the Judge Advocate General, Court Martial Case Files, RG 165, NA.

**Sherman had already started:** Ibid.

**If the enticements of Charleston:** Jacob Whitman Bailey to his brother, February 1833, Jacob Whitman Bailey MSS, Charleston Library Society; WTS to EE, February 10, 1845, SF MSS, ND.

**He seems to have accepted:** WTS to PBE, undated, in John Weatherford, ed., "Sherman Liked the South–Once," *Manuscripts* 8 (Winter 1956), 75; WTS to Annie Gilman Bowen, June 24, 1864, SCL.

**From the first Sherman:** WTS to Mrs. A. A. Draper, January 15, 1865, *Atlanta Journal-Constitution*, November 23, 1958.

**Sherman's stay at Fort Moultrie:** WTS to EE, February 8, 1844, WTS MSS, ND; WTS to TE, February 19, 1844, TE MSS, LC.

**Sherman's trip took him:** WTS, Diary, February 19 and 24, March 13 and April 3, 1844, SF MSS, ND.

**Sherman had already seen:** Ibid., April 11, 13 and 18, 1844, SF MSS, ND.

**Then there was Ellen:** WTS to EE, February 8, 1844, SF MSS, ND; WTS, Diary, March 5, 1844, SF MSS, ND.

**Few of Ellen's letters:** WTS to EE, February 8, 1844, SF MSS, ND; WTS, Diary, preliminary entry for November 1843, SF MSS, ND.

**But Ellen was apparently relentless:** WTS to EE, June 14, 1844, SF MSS, ND.

**Sherman then takes up:** Ibid.

**Ellen must have returned:** WTS to EE, September 17, 1844, SF MSS, ND.

**In November 1845 he wrote:** WTS to EE, November 19, 1845, SF MSS, ND.

## 4. CALIFORNIA

**Like most soldiers, Sherman:** WTS to EE, June 12, 1845, SF MSS, ND; WTS to AGO, December 19, 1845, WTS MSS, SCL.

**On January 15, 1846:** WTS to AGO, January 15, 1846, Thomas W. Morris Coll., Bancroft Library; WTS to EE, July 12, 1846, SF MSS, ND; WTS to Elisabeth Sherman, November 10, 1846, WTS MSS, LC.

**Sherman did not have:** WTS to Henry Benham, July 20, 1844, SCL; William Gates to AGO, April 5, 1846, M567, NA; WTS, "Autobiography," OHS, 53–54.

**Sherman read of these:** WTS to EE, June 11, 1846, SF MSS, ND.

**Then one day in June:** WTS, *Memoirs*, vol. 1, 11–12.

**The voyage lasted:** WTS to EE, September 12 and 16, 1846, SF MSS, ND; WTS to EE, October 27, 1846, SF MSS, ND; WTS to EE, January 25, 1847, SF MSS, ND.

**At Valparaiso the officers:** WTS to EE, January 27, 1847, SF MSS, ND.

**Today the California of 1847:** E. O. C. Ord to the Secretary of War John B. Floyd, June 2, 1857, E. O. C. Ord MSS, Cecil H. Green Library, Stanford University; Mary Floyd Williams, *History of the San Francisco Vigilante Committee of 1851: A Study of Social Control on the California Frontier in the Days of the Gold Rush* (New York: Da Capo, 1969), 54, footnote 1.

**The region's economy had:** Walter Colton, *Three Years in California* (New York: Arno, 1976), 22; James A. Hardie to WTS, January 12, 1848, WTS MSS, LC; WTS, "Autobiography," OHS, 71.

**As for viable political institutions:** Neal Harlow, *California Conquest: War and Peace in the*

*Pacific* (Berkeley: University of California Press, 1982), 280–82; Joseph Ellison, "The Struggle for Civil Government in California, 1846–1850," *California Historical Society Quarterly*, 10 (March 1931): 3–8, 23; Williams, *Vigilante Committee*, 43–44.

**With local legal practices:** Colton, *California*, 232.

**Sherman developed great admiration:** WTS to EE, October 8, 1847, SF MSS, ND.

**With not a thousand troops:** Mason to Roger Jones, September 28, 1847, Records of the Tenth Military District, M210, NA; WTS to Jas. Hardie, December 27, 1847, and January 18, 1848, Hardie MSS, LC.

**It is clear from:** WTS to EE, July 11, 1847, and February 3, 1848, SF MSS, ND; WTS, "Autobiography," OHS, 77–82; Mason to HWH, January 26, 1849, Generals' Papers and Books: HWH File, RG 94, E 159, NA.

**Because Sherman was known:** William Warner to WTS, March 1, 1849; J. C. Bonnycastle to WTS, February 15, 1849, WTS MSS, LC; W. E. Shannon to WTS, April 3, 1848, WTS MSS, LC; C. L. Kilborn to WTS, March 22, 1848, WTS MSS, LC; George Gibson to WTS, January 20, 1849, WTS MSS, LC.

**The function of assistant adjutant general:** WTS to EE, February 3, 1848, SF MSS, ND.

**Gradually Sherman extended:** WTS to EE, October 8, 1847, SF MSS, ND; Norman Bestor to WTS, March 9, 1849, WTS MSS, LC; N. V. Sanchez, "Grafting Romance on a Rose Tree: The True Story of Doña Maria Bonifacio and General Sherman at Monterey," *Sunset*, April 1916, 36–40.

**He and the other officers:** Cresap, *Ord*, 14; WTS to EE, February 3, 1848, SF MSS, ND.

**The ties of friendship:** Jas. Hardie to WTS, February 7, 1848, WTS MSS, LC; WTS to EE, March 12 and November 10, 1847, SF MSS, ND.

**He also sensed that:** WTS to EE, April 10, 1848, SF MSS, ND; WTS to JS, April 18, 1848, WTS MSS, LC.

**But Sherman saw something else:** WTS to EE, July 12, 1846, SF MSS, ND.

**He dashed off a letter:** WTS to EE, April 10, 1848, SF MSS, ND; WTS to EE, July 12, 1846, SF MSS, ND; EE to WTS, January 1 and February 5, 1849, SF MSS, ND.

**Ellen's health remained:** EE to WTS, April 25, 1847, SF MSS, ND; WTS to EE, November 10, 1847, and February 3, 1848, SF MSS, ND.

**Occasionally Ellen's endless preoccupation:** WTS to EE, October 8, 1847, SF MSS, ND.

**In Sherman's letters to Ellen:** Allen Nevins, *Frémont: The West's Greatest Adventurer*, 2 vols. (New York: Harper and Brothers, 1928), vol. 2, 430; WTS to EE, November 19, 1847, SF MSS, ND.

**By April 1848 Sherman's situation:** WTS to EE, April 10, 1848, SF MSS, ND; WTS to JS, June 18, 1848, WTS MSS, LC.

**Even as he wrote:** J. S. Holliday, *The World Rushed In: The California Gold Rush* (New York: Simon and Schuster, 1981), 34.

**The news seems to have passed:** Willard Brigham Farwell, "Recollections of Good Digging," *Society of California Pioneers Quarterly* 1 (1924): 19.

**The officials were Colonel Mason and Lieutenant Sherman:** Mason to AGO, December 27, 1848, M567, NA.

**It was not just men:** WTS, *A Letter of Lieutenant William T. Sherman Reporting on Conditions in California in 1848* (privately printed from the Collection of Thomas W. Norris, Carmel, Calif., 1947), 6; WTS to "Dear Friend," August 25, 1848, WTS MSS, LC.

**Nor was the army immune:** James L. Ord to WTS, July 31, 1848, WTS MSS, LC; WTS to "My Dear Friend," November 14, 1848, WTS MSS, LC.

**It was the profits:** WTS, *Letter of Lieutenant Sherman*, 7–8.

**California's governmental problems were:** WTS, "Autobiography," OHS, 108; Colton, *California*, 248.

Over the next few months: Cresap, *Ord*, 29.

In February 1849 Colonel Mason: Copies of Mason's endorsement of Sherman's application for leave, February 26, 1849, and General Smith's General Order No. 3, February 27, 1849, in WTS MSS, LC.

This action did not: WTS to TE, April 28, 1849, TEF MSS, LC; WTS to George Gibson, April 29, 1849, T. W. Norris Coll., Bancroft Library, University of California, Berkeley.

On May 15 General Smith: Cresap, *Ord*, 29; Johann Sutter to WTS, June 28, 1849, WTS MSS, LC.

Then word came that: WTS to Persifer Smith, June 12, 1849, WTS MSS, LC; WTS to Elisabeth Sherman Reese, October 31, 1849, WTS MSS, LC.

Curiously, there was no reaction: WTS, *Memoirs*, vol. 1, 73; WTS to AGO, March 15, 1850, M567, NA.

## 5. ENTR'ACTE

Traveling this time by steamship: WTS, *Memoirs*, vol. 1, 82–84.

The letters he addressed: WTS to EE, Mar. 21 and Mar. 27, 1850, SF MSS, ND.

Even as he was: James Hardie to WTS, June 14, 1850, WTS MSS, LC.

Now Sherman sent: WTS to George Gibson, March 26, 1850, WTS MSS, LC; WTS to AGO, March 26, 1850, M567, NA; Roy F. Nichols, ed., "William Tecumseh Sherman in 1850," *Pennsylvania Magazine of History* 75 (1951): 425–35.

The shift from line to staff duty: Michael R. Morgan, "Types and Traditions in the Old Army," *War Talks in Kansas* (MOLLUS, Kansas Commandery), 1 (1882), 386.

As the wedding date: WTS to EE, March 21, 27 and 29, 1850, SF MSS, ND.

After the wedding and honeymoon: WTS, *Memoirs*, vol. 1, 86–87.

St. Louis was a departmental headquarters: WTS, *Memoirs*, vol. 1, 88.

There was only one problem: WTS to AGO, October 2, 1850, M567; WTS to TE, October 27 and November 9, 1850, TEF MSS, LC.

For his immediate housing needs: WTS to HBE, Jan 5, 1851, WTS MSS, OHS; WTS to TEJr, July 16, 1851, TEF MSS, LC.

Without the company: WTS to TE, July 29 and August 18, 1852, TEF MSS, LC; WTS to TEJr, September 17, 1852, TEF MSS, LC.

He wrote steadily to Ellen: TE to EES, May 23, 1850, SF MSS, ND; WTS to EES, September 22 and 24, 1850, SF MSS, ND.

The couple had agreed: WTS to EES, November 1, 1850, and January 25, 1851, SF MSS, ND; WTS to TE, March 11, 1851, TEF MSS, LC.

The little family settled: WTS to EES, October 8 and November 1, 1850, SF MSS, ND; WTS to HBE, March 8, 1851, WTS MSS, OHS.

Ellen was less than charmed: EES to CE, April 27, 1851, CE MSS, LC.

If Ellen was afflicted: WTS to JS, May 19, July 16, and August 19, 1851, WTS MSS, LC; WTS to TE, December 4 and 19, 1851, and January 27, 1852, TEF MSS, LC.

The most frustrating aspect: WTS to JS, May 14 and November 26, 1851, February 2, April 19, and June 3, 1852, WTS MSS, LC; WTS to TE, August 12, 1851, TEF MSS, LC.

Meanwhile Major Lee had pursued: WTS to TE, October 11, 1851, TEF MSS, LC; WTS to HBE, February 15 and March 8, 1851, WTS MSS, OHS, LC; WTS to EES, May 30, 1852, SF MSS, ND; A. E. Shiras to WTS, May 17, 1851, WTS MSS, LC; WTS, *Memoirs*, vol. 1, 89.

Then Mary Sherman died suddenly: WTS to EES, September 30, 1852, SF MSS, ND.

Finally there was the separation: WTS to EES, August 14, 1852, SF MSS, ND.

The tone of the second letter: WTS to TE, September 2, 1852, TEF MSS, LC; WTS to TEJr, August 30, 1852, TEF MSS, LC.

For Sherman now knew: A. E. Shiras to WTS, September 4, 1852, WTS MSS, LC.

On his arrival in New Orleans: WTS to JS, November 17, 1852, WTS MSS, LC; WTS, *Memoirs*, vol. 1, 90–91.

In Washington commissary officials: Shiras to WTS, September 4, 1852, WTS MSS, LC.

When Lieutenant Shiras told Sherman: Shiras to WTS, September 4, 1852, WTS MSS, LC; WTS to George Gibson, November 19, 1852, M567, NA; WTS to JS, November 17, 1852, WTS MSS, LC; WTS to Isaac Bowen, February 7, 1853, Isaac Bowen MSS, USAMHI; WTS, *Memoirs*, vol. 1, 92.

Captain Sherman soon became: WTS to JS, November 17, 1852, WTS MSS, LC.

Sherman found time: WTS to EE, November 4, 1852, SF MSS, ND.

Sherman's letters to Ellen: WTS to EE, November 4 and 16, December 2 and 14, 1852, SF MSS, ND; WTS, *Memoirs*, vol. 1, 92–93.

Strangely enough, while he: WTS to JS, November 17, 1852, WTS MSS, LC; WTS to Isaac Bowen, February 7, 1853, Isaac Bowen MSS, USAMHI.

On November 21 Sherman: HST to WTS, December 7, 1852, WTS MSS, LC.

Then Turner announced that: Ibid.; WTS, *Memoirs,* vol. 1, 92–93.

Sherman had kept in touch: WTS to HBE, June 15, 1852, WTS MSS, OHS.

Not all of his dreaming: WTS to EE, December 2, 1852, SF MSS, ND; WTS to HST, December 17, 1852, WTS MSS, LC.

By the end of February: Lucas and Simonds to WTS, January 27 and 31, 1853, WTS MSS, LC; WTS to JS, March 4 and 24, 1853, WTS MSS, LC; WTS, *Memoirs*, vol. 1, 94.

## 6. RETURN TO CALIFORNIA

Sherman's return: WTS, *Memoirs*, vol. 1, 95–100; WTS to HST, May 16, 1855, WTS MSS, OHS.

The vast changes struck him: WTS to JS, June 3, 1853, WTS MSS, LC; Dwight L. Clarke, *William Tecumseh Sherman, Gold Rush Banker* (San Francisco: California Historical Society, 1969), 3; Ira B. Cross, *Financing an Empire: History of Banking in California,* 4 vols. (San Francisco and Los Angeles: S. J. Clarke, 1927), vol. 1, 30.

Sherman found his bank: Clarke, *Banker*, 23.

California's financial institutions: WTS to TE, December 8, 1854, TEF MSS, LC; John S. Littell, *The Commerce and Industries of the Pacific Coast of North America* (San Francisco: A. L. Bancroft, 1882), 125–26; Clarke, *Banker*, 7.

And this flood of gold: WTS, *Memoirs*, vol. 1, 132; Cross, *Financing,* vol. 1, 73; Clarke, *Banker*, 239.

In the freewheeling economy: Clarke, *Banker*, 5; Littell, *Commerce*, 125; Benjamin C. Wright, *Banking in California, 1849–1910* (San Francisco: H. S. Crocker, 1910), 8; Cross, *Financing*, vol. 1, 121–22.

Sherman saw very clearly: WTS to TE, April 17 and August 3, 1856, TEF MSS, LC.

The demand for credit: Clarke, *Banker*, 29; WTS to JS, June 3, 1853, WTS MSS, LC.

Then there was the dearth: WTS to TE, January 3, 1857, TEF MSS, LC; Littell, *Commerce*, 126; Cross, *Financing*, vol. 1, 91.

Banker Sherman explored: WTS to JS, June 3, 1853, WTS MSS, LC.

He must have come away: WTS to TE, August 12, TEF MSS, LC.

By the end of 1854: WTS to TE, December 8, 1854, TEF MSS, LC; WTS, *Memoirs*, vol. 1, 109–16; Cross, *Financing*, vol. 1, 184.

**Dwight L. Clarke, who:** Clarke, *Banker*, preface, xiii. WTS to HBE, WTS MSS, OHS, March 23, 1855; WTS, *Memoirs*, vol. 1, 104.

**Sherman made it a point:** Clarke, *Banker*, 29; WTS to HST, November 4, 1856, WTS MSS, OHS.

**He had a number of ideas:** Clarke, *Banker*, 82; WTS to JS, August 19, 1851, WTS MSS, LC; WTS to William M. Gardiner, July 27, 1855, William M. Gardiner MSS, UNC.

**Certain practices that other bankers:** Littell, *Commerce*, 127.

**Someone who knew him well:** *St. Louis Post-Dispatch*, February 22, 1891; WTS to S. Clay, February 6, 1858, letterpress copy, WTS MSS, LC.

**Some who dealt with him:** Isaac J. Wistar, *Autobiography of I. J. Wistar, 1827–1905: Half a Century in War and Peace* (Philadelphia: Wistar Institute of Anatomy and Biology, 1937), 324–25; James A. Garfield, *The Diary of James A. Garfield*, ed. with an introduction by Harry James Brown and Frederick D. Williams, 4 vols. (East Lansing: Michigan State University Press, 1967–81), May 25, 1875, vol. 3, 126.

**The town was not particularly well-run:** Doyce B. Nunis, ed., *The San Francisco Vigilance Committee, Three Views: William T. Coleman, William T. Sherman, and John O'Meara* (Los Angeles: Westerners U.S.A., 1971), 29.

**The two culprits:** Ibid.; WTS to HST, May 20, 1856, WTS MSS, OHS; WTS to TE, May 21, 1856, TEF MSS, LC.

**Sherman remained essentially:** WTS to HST, May 20, 1856, WTS MSS, OHS.

**The Committee of Vigilance:** Ibid.; Robert M. Denkewicz, *Vigilantes in Gold Rush California* (Stanford: Stanford University Press, 1985), 173.

**In this period he received:** WTS to HST, May 18, 1856, WTS MSS, OHS; Clarke, *Banker*, 216; *San Francisco Herald*, May 30, 1856.

**But as the chief military authority:** J. Neely Johnson to WTS, May 29 and June 12, 1856, Miscellaneous Communications Concerning the Vigilance Committee of San Francisco, 1856, in Governors' Papers: J. Neely Johnson, California State Archives; *Daily Evening Bulletin*, June 5, 1856; *Alta California*, June 4, 1856.

**Then there were the questions:** *Daily Herald*, June 6, 1856; *Daily Evening Bulletin*, June 4, 6, 7, and 9, 1856; WTS to HST, July 2, 1856, WTS MSS, OHS; J. Neely Johnson to WTS, June 12, 1856, Governors' Papers: J. Neely Johnson, California State Archives; Herbert G. Florken, ed. "The Law and Order View of the San Francisco Vigilance Committee of 1856," *California Historical Quarterly* 14 (1935): 372.

**In truth Sherman was:** *Daily Evening Bulletin*, May 20 and 29 and June 10, 1856; WTS, "Autobiography," OHS, 189–90; WTS to S. S. L'Hommedieu, July 7, 1862, in Joseph Ewing, ed., *Sherman at War* (Dayton, Ohio: Morningside Bookshop, 1992): 131.

**Sherman did not cut:** Wistar, *Autobiography*, 324–25.

**Then the act of resigning:** WTS to TE, June 16, 1856, TE MSS, LC; WTS to James Hardie, October 23, 1856, James Hardie MSS, LC.

**A decade earlier:** WTS to EE, July 12, 1846, SF MSS, ND.

**But there was another facet:** EES, Diary, 1854, passim.

**Sherman accepted that it:** EES, Diary, 1854, SF MSS, ND, passim.

**The greatest bone of contention:** WTS to HBE, March 23, 1855, HBE MSS, OHS; EES to TE, May 18 and 20, 1856, SF MSS, ND.

**Sherman for his part:** WTS to HBE, December 15, 1854, WTS MSS, OHS.

**Minnie was constantly:** TE to EES, October 15, 1853, SF MSS, ND; EES, Diary, January 1 and 2, 1854, SF MSS, ND.

**At the approach of each spring:** WTS to Hardie, February 10, 1854, and March 12, 1855, Hardie MSS, LC; EES to TE, February 25, 1855, SF MSS, ND.

While Ellen agonized: WTS to HST, December 23, 1854, WTS MSS, OHS; Clarke, *Banker*, 89.

Ellen's diary reveals: EES, Diary, 1854, passim, SF MSS, ND; WTS to EE, March 29, 1850, SF MSS, ND.

Sherman's own health was: WTS to HST, September 29, 1854, and August 24, 1857, WTS MSS, OHS.

Ellen's diary occasionally affords: EES, Diary, March 9 and May 14, 1854, SF MSS, ND.

In this period it is difficult: EES to TE, February 10 and 25, 1855, SF MSS, ND; WTS to HST, December 23, 1854, WTS MSS, OHS.

There was now another element: WTS to HBE, December 8, 1851, and July 15, 1854, WTS MSS, OHS; WTS to EES, January 30, 1851, SF MSS, ND.

But this detachment soon ended: WTS to HBE, January 1, 1853, and April 15, 1854, WTS MSS, OHS.

When William Tecumseh Sherman Jr.: WTS to TEJr, July 15, 1854, WTS MSS, OHS; Clarke, *Banker*, 185; WTS to TE, October 5, 1856, TEF MSS, LC.

A second son was born: WTS to HST, October 19, 1856, WTS MSS, OHS.

## 7. LOUISIANA

Closing the books on 1856: WTS to HST, December 18, 1856, WTS MSS, OHS; Clarke, *Banker*, 278–79.

Sherman was torn: WTS to TE, February 5, 1857, TEF MSS, LC.

Closing down a bank: WTS to JS, February 4, 1858, WTS MSS, LC; WTS to HST, January 18 and April 19, 1857, WTS MSS, OHS.

The Shermans sailed for New York: WTS to HST, January 18 and March 4, 1857, WTS MSS, OHS; WTS to TE, February 7, 1857, TEF MSS, LC.

Before June was out: WTS, *Memoirs*, vol. 1, 134; Kenneth Stampp, *America in 1857: A Nation on the Brink* (New York and Oxford: Oxford University Press, 1990), 217–20.

Among the first to succumb: Stampp, *1857*, 220–24; WTS to HST, August 27, September 17 and October 6, 9, and 13, 1857, WTS MSS, OHS; WTS, *Memoirs*, vol. 1, 137.

There was money: WTS to James Lucas, April 6, 1858, WTS MSS, MoHS; WTS to HST, February 10 and March 13, 1858, WTS MSS, OHS.

There was another matter: WTS to TE, April 15, 1858, TEF MSS, LC; WTS to TEJr, August 9, 1858, WTS MSS, OHS; WTS to J. G. Barnard, December 8, 1858, letterpress copy, WTS MSS, LC; WTS to Dan Garrison, December 11, 1888, WTS MSS, MoHS.

By July 1858 he had done: WTS to B. R. Alden, July 16, 1858, letterpress copy, WTS MSS, LC.

Thomas Ewing had renewed: WTS to TE April 2, 1858, TEF MSS, LC; WTS, *Memoirs*, vol. 1, 139–40; WTS to HST, March 4 and 21, 1858, WTS MSS, OHS.

In the end he chose: WTS to TEJr, August 9, 1858, TEF MSS, LC.

Sherman knew nothing of the law: WTS, *Memoirs*, vol. 1, 140.

He felt like an impostor: Samuel Adams Drake, "The Old Army in Kansas," *Civil War Papers* (MOLLUS, Kansas Commandery), vol. 1 (1900), 145; WTS, *Memoirs*, vol. 1, 140; WTS, "Memoirs," draft no. 3, Container 106, WTS MSS, LC; WTS "Autobiography," OHS, 201.

Tom had a house: WTS, *Memoirs*, vol. 1, 141; WTS, "Memoirs," draft no. 3, Container 106, 228, WTS MSS, LC; WTS to TE, October 3, 1858, TEF MSS, LC.

In March Ellen and the children: WTS to EES, September 18, 1858, SF MSS, ND.

From Ellen, who usually sustained him: EES to WTS, June 1, 1859, SF MSS, ND.

Sherman's diary in those spring months: WTS, Diary, April-July, 1859, WTS MSS, ND, passim.

**He had heard on the army grapevine:** WTS to Don Carlos Buell, June 11, 1859, letterpress copy, WTS MSS, LC.

**Buell replied that while:** Buell to WTS, June 17, 1859, WTS MSS, LC.

**Sherman sat down and addressed:** WTS to R. B. Wickliffe, July 1, 1859, letterpress copy, July 1, 1859, WTS MSS, LC.

**Here a curious chain:** WTS, *Memoirs*, vol. 1, 142; GMG to WTS, September 7, 1859, WTS MSS, LC.

**The selection of the teachers:** GMG to P. G. T. Beauregard, July 16, 1859, GMG MSS, VHS.

**Then Sherman's name came up:** GMG to P. G. T. Beauregard, January 26, 1866, GMG MSS, VHS; D. C. Buell to GMG, July 10, 1859, Buell MSS, Filson.

**Graham saw his chance:** Braxton Bragg to GMG, January 22, 1860, quoted in GMG to P. G. T. Beauregard, January 26, 1866, GMG MSS, VHS.

**The position as superintendent:** WTS to GMG, August 15 and 20 and September 7, 1859, GMG MSS, WLF Coll., LSU; Delafield to WTS, August 30, 1859, WTS MSS, LC; WTS, Diary, September 5 and 12, 1859, SF MSS, ND.

**As Sherman and Graham:** Buell to WTS, September 24, 1859, WTS MSS, LC; GBM to WTS, October 23, 1859, copy in GMG MSS, VHS.

**Sherman left for Louisiana:** WTS to EES, October 29, 1859, SF MSS, ND.

**The rising sectional antagonisms:** WTS to JS, September 1 and October 6, 1859, WTS MSS, LC; WTS to TEJr, February 17, 1860, TEF MSS, LC; WTS to HBE, August 13, 1859, WTS MSS, OHS.

**When he met George Mason Graham:** WTS to EES, November 12, 1859, SF MSS, ND; WLF, ed., *William T. Sherman as College President* (Cleveland: Arthur W. Clark, 1912), 24 and ff.

**The area surrounding the seminary:** WTS to EES, November 12, 19 and 25, 1859, SF MSS, ND; WTS, *Memoirs*, vol. 1, 145; WTS to HBE, April 15, 1860, TEF, LC.

**This was just as well:** WTS to TEJr, January 21, 1860, TEF MSS, LC; EES to TE, June 20, 1860, SF MSS, ND.

**Once he had taken his bearings:** WTS to EES, November 25 and December 2, 1859, SF MSS, ND; WLF, *Sherman*, 48; P. G. T. Beauregard to GMG, December 8, 1859, and January 26, 1860, GMG MSS, VHS.

**Though the seminary was:** WTS to EES, November 25, 1859, SF MSS, ND.

**No one had thought of advertising:** WTS to EES, November 25, 1859, SF MSS, ND; WTS, Diary, November 26, 1859, SF MSS, ND; WTS to GMG, December 25, 1859, GMG MSS, WLF Coll., LSU.

**The seminary could accommodate:** WTS, Diary, January 2, 1860, SF MSS, ND; WTS to EES, December 28, 1859, and January 4, 1860, SF MSS, ND.

**The faculty was spread:** GMG to WTS, August 3, 1859, WTS MSS, LC.

**Just as Sherman was beginning:** TE to WTS, LC, January 14, 1860, WTS MSS, LC; GMG to Thomas O. Moore, February 20, 1860, WTS MSS, LC.

**Sherman had now made his choice:** WTS to JS, December 12, 1859, WTS MSS, LC; JS to WTS, December 24, 1859, WTS MSS, LC.

**The *New Orleans Bulletin*:** WTS to EES, December 23, 1859, SF MSS, ND; GMG to WTS, January 15, 1860, WTS MSS, LC; EES to WTS, January 15, 1860, SF MSS, ND; WTS to TEJr, February 17, 1860, TEF MSS, LC.

**Within the seminary Sherman:** WTS to D. C. Buell, April 15, 1860, Filson; WTS to EES, January 27, 1860, SF MSS, ND; WTS to GMG, August 20, 1859, GMG MSS, WLF Coll., LSU.

**Sherman had at least a partial answer:** WTS to GMG, December 2, 1859, Records of the Office of the President, RG A0001, LSU Archives.

**Anyone who had been:** WTS to EES, December 25, 1859, SF MSS, ND.

**Sherman was not well-equipped:** WTS to GMG, January 20 and February 10, 1860, GMG MSS, WLF Coll., LSU; WTS, Diary, January 30, 1860, SF MSS, ND; WTS to EES, February 3, 1860, SF MSS, ND.

**In June a rash of misbehavior:** WTS to GMG, June 16 and 28, 1860, GMG MSS, WLF Coll., LSU.

**Here he ran into a stone wall:** WTS to St. John R. Liddell, June 26, 1860, Moses and St. John Richardson Liddell Family MSS, MS 531, WLF Coll., LSU.

**Ultimately this dispute:** Peter Tanner to GMG, May 8, 1860, GMG MSS, WLF Coll., LSU; GMG to S. A. Smith, February 13, 1860, GMG MSS, WLF Coll., LSU; GMG to James G. Campbell, June 16, 1860, GMG MSS, WLF Coll. LSU.

**Then too, the Board:** WTS to GMG, April 26, 1860, GMG MSS, WLF Coll., LSU.

**And thanks largely to Sherman's efforts:** Newspaper clipping, July 4, 1860, in DFB MSS, WLF Coll., LSU.

**Yet the subject was:** WTS to EES, November 23, 1860, SF MSS, ND; DFB, "Gen. W. T. Sherman. His Early life in the South and His Relations with Southern Men," *Confederate Veteran* 18 (1910): 412.

**One by one other Southern states:** Bragg to WTS, January 27, 1861, WTS MSS, LC.

**He sought to justify:** WTS to TE, February 3, 1861, TEF MSS, LC; WTS to EES, December 16, 1860, January 13 and 21, 1861, SF MSS, ND.

**He would have to find:** WTS to TEJr, February 3, 1861, TEF MSS, LC; WTS to EES, December 16, 1860, and January 13, 1861, SF MSS, ND; WTS to HBE, January 2, 1861, WTS MSS, OHS.

**Sherman's letters in this period:** WTS to EES, undated (1861) and January 13, 1861, SF MSS, ND; WTS to DFB, February 23, 1861, in DFB, "Sherman," 413.

**Sherman may have been remorseful:** WTS to EES, November 19, 1859, SF MSS, ND.

**Handling the day-to-day business:** WTS to EES, November 3 and 29, 1860, and January 8, 1861, SF MSS, ND; WTS to JS, December 9, 1860, WTS MSS, LC.

**The Board of Supervisors:** WTS to EES, January 20, 1861, SF MSS, ND.

**Finally, on February 19:** DFB, "Sherman," 413.

**He went to New Orleans:** WTS, Diary, February 22, 1861, SF MSS, ND; WTS to DFB, February 23, 1861, in DFB, "Sherman," 414.

## 8. WHAT MANNER OF MAN

**The other passengers:** DFB, "Sherman," 409; W. F. G. Shanks, "Recollections of Sherman," *Harper's New Monthly Magazine* 30 (1865): 646; Franc B. Wilkie, *Pen and Powder* (Boston: Ticknor, 1888), 160.

**By 1861 his skin:** *St. Louis Post Dispatch*, February 22, 1891.

**Putting the man's frame:** John Chipman Gray and John Codman Ropes, *War Letters, 1862–1865* (Boston and New York: Houghton-Mifflin, 1927), 427.

**People who met Sherman:** Theodore Lyman, *Meade's Headquarters, 1863–1865. Letters of Colonel Theodore Lyman from the Wilderness to Appomattox*, George Agassiz, ed. (Boston: Atlantic Monthly Press, 1922), 327.

**His clothing did little:** Shanks, "Recollections," 646.

**But even when he:** WTS, Diary, 24 August 1860, ND; WTS to Henry J. Reese, August 24, 1848, WTS MSS, LC; WTS to Brooks Brothers and Co., August 24, 1863, WTS LB 23, NA.

**Then anyone sitting near Sherman:** Gray and Ropes, *War Letters,* 427; Lyman, *Meade's Headquarters,* 327.

**Even when seated:** Shanks, "Recollections," 643; Childs, *Recollections*, 159.

**Sherman was in a boastful mood:** WTS to JS, April 19, 1852, WTS MSS, LC; WTS to EE, January 31, 1846, SF MSS, ND.

**This life of constant, restless activity:** Paul E. Steiner, *Medical-Military Portraits of Union and Confederate Generals* (Philadelphia: Whitmore Publishing Company, 1968), 62–63, 74, 75.

**Toward the end of that life:** WTS to EES, November 23, 1855, SF MSS, ND; WTS to PTS, January 19, 1888, WTS MSS, OHS.

**While the story:** Fletcher, *Life and Reminiscences*, 294; JS, *Recollections*, vol. 2, 407.

**Those who had occasion:** Wilkie, *Pen and Powder*, 160; Shanks, "Recollections," 642; Albert Castel, *Decision in the West: The Atlanta Campaign of 1864* (Lawrence: University Press of Kansas, 1992), 42; Joseph Hooker to Benjamin F. Wade, December 8, 1864, Joseph Hooker File, RG 94, E 159, NA.

**David French Boyd reached:** DFB, "Sherman," 411; Steiner, *Portraits*, 55; WTS to EES, August 14, 1852, SF MSS, ND; *Boston Daily Advertiser*, March 28, 1870; remarks of JS in "Commemoration," 33.

**Sherman himself acknowledged:** WTS to GMG, February 16, 1860, GMG MSS, VHS.

**In his writing:** Murat Halstead, "Recollections and Letters of General Sherman," *Independent* 5 (1899): 1611.

**Sherman told Boyd that:** DFB, "Sherman," 411.

**It is also true:** WTS to HST, August 10, 1857, WTS MSS, OHS.

**But generally there was:** WTS to GMG, December 2, 1859, SF MSS, ND.

**As with the man's words:** Clarke, *Banker*, 126.

**The man had what one observer:** Richard Wheeler, *We Knew Sherman* (New York: Thomas Crowell, 1977), 10; GO No. 4, October 28, 1863, *Military Orders*, 204.

**To admit error was harder:** Royal Cortissoz, *The Life of Whitelaw Reid*, 2 vols. (New York: Charles Scribner's Sons, 1921), vol. 2, 336.

**If Sherman were asked:** WTS to James B. Bidwell, April 24, 1864, WTS, LB 18, NA; WTS to JS, July 11, 1852, July 7, 1856, and February 1, 1861, WTS MSS, LC; Clarke, *Banker*, 273; WTS to J. W. Ripley, May 25, 1863, WTS, LB 22, NA.

**Curiously, though he was proud:** WTS to HBE, March 10, 1844, and January 5, 1851, WTS MSS, OHS.

**Some have argued that:** Eleanor Sherman Fitch, "Notes on the Sherman Papers," Preface to Reel 2, Microfilm Edition, SF MSS, ND; WTS to TEJr, August 9, 1858, WTS MSS, OHS.

**Like most conservatives:** Weatherford, "Sherman," 76.

**Sherman's idea of the well-ordered society:** WTS to William Everett, September 17, 1864, Edward Everett Papers, Massachusetts Historical Society.

**Since Andrew Jackson:** WTS to JS, March 4, 1859, WTS MSS, LC; WTS to GMG, January 5, 1861, DFB MSS, WLF Coll., LSU; WTS to David Stuart, August 1, 1863, WTS, LB 23, NA.

**In one respect Sherman:** WTS to Willard Warner, February 16, 1867, Willard Warner MSS, ISHL; WTS to Edward Bok, *Americanization*, 215; B. Anthony Gannon, "A Consistent Deist: Sherman and Religion," *Civil War History* 42 (1996): 307–21; WTS to EE, July 12, 1846. SF MSS, ND.

**Biographers have probably made:** WTS to Don Carlos Buell, June 11, 1859, letterpress copy, WTS MSS, LC; WTS to TE, December 14, 1850, TEF MSS, LC.

**Here and there one:** WTS, Diary, April 8, 1844, SF MSS, ND.

**Sherman's travel accounts:** WTS, Diary, November 20, 1843, March 29, 1850, SF MSS, ND.

**The reference to mulattos:** WTS to William McPherson, March 24, 1865, WTS, LB 15, NA.

**Such were his views:** WTS to HST, May 16, 1855, OHS.

**When he wrote that:** WTS to TEJr, January 8, 1860, TEF MSS, LC.

## 9. BULL RUN

**Among students of warfare:** Gamaliel Bradford, "Union Portraits III: William T. Sherman," *Atlantic Monthly*, 114 (1914): 319.

**Yet Sherman was not:** WTS to JS, December 9, 1860, WTS MSS, LC; WTS to CE, June 22, 1861, CE MSS, LC; WTS to EES, July 6, 1861, SF MSS, ND.

**While the popular view:** WTS to EE, April 10, 1848, SF MSS, ND; WTS to TEJr, May 1, 1861, TEF MSS, LC; WTS to PBE, July 12, 1861, in Ewing, *Sherman*, 59; WTS to EES, August 3, 1861, SF MSS, ND.

**These chilling estimates:** WTS to CE, undated, CE MSS, LC.

**The conquest and reincorporation:** WTS to TEJr, February 3, 1861, TEF MSS, LC; WTS to EES, December 10 and 16, 1860, SF MSS, ND; WTS to JS, January 18, 1861, and July 20, 1863, WTS MSS, LC.

**This change in attitude:** WTS to JS, December 9, 1860, WTS MSS, LC; WTS to HBE, December 18, 1860, WTS MSS, OHS.

**He may also have been:** WTS to TEJr, March 10, 1861, TEF MSS, LC.

**In January he wrote:** WTS to JS, January 18 and February 1, 1861, WTS MSS, LC.

**He reached the capital:** WTS, *Memoirs*, vol. 1, 167–68.

**There were probably some other words:** Ibid.; WTS to TEJr, April 26, 1861, TEF MSS, LC; JS, "Commemoration," 39.

**On March 8 he:** WTS, *Memoirs*, vol. 1, 168–69; WTS to JS, April 6, 1861, WTS MSS, LC.

**He threw himself into his work:** WTS to JS, May 22, 1861, WTS MSS. LC.

**He had learned that:** WTS to CE, March 9, 1861, CE MSS, LC; WTS to JS, March 22, 1861, WTS MSS, LC; Montgomery Blair to WTS, April 6, 1861, WTS MSS, LC; WTS to Montgomery Blair, April 8, 1861, WTS MSS, LC.

**He was in his office:** WTS to JS, May 20, 1861, WTS MSS, LC; WTS to TEJr, May 23 and June 3, 1861, TEF MSS, LC.

**Sherman was a spectator:** WTS, *Memoirs*, vol. 1, 172–74; WTS, Diary, April 14, 1861, SF MSS, ND; WTS to TEJr, May 11, 1861, TEF MSS, LC.

**Now, as the country:** TEJr to WTS, undated, WTS MSS, LC; James B. Fry to WTS, April 15, 1861, WTS MSS, LC; EES to JS, undated, WTS MSS, LC; WTS to TEJr, May 1 and 16, 1861, TEF MSS, LC.

**Sherman himself took up:** WTS to Simon Cameron, May 8, 1861, WTS MSS, LC.

**Congress was creating:** WTS to JS, May 22, 1861, WTS MSS, LC; WTS to TEJr, May 20 and June 3, 1861, TEF MSS, LC.

**When Sherman reached Washington:** WTS to JS, June 8, 1861, WTS MSS, LC; WTS to Simon Cameron, July 5, 1861, M567, NA; U. G. McAlexander, *History of the Thirteenth Regiment, United States Infantry, Compiled from Regimental Records and Other Sources*, (n.p.: Regimental Press, 1905), 4.

**Schuyler Hamilton, who had briefly worked:** DDP, Journal, DDP MSS, LC, 413; WTS to EES, June 18, 20 and 23, July 3, 1861, SF MSS, ND; WTS to JS, June 20, 1861, WTS MSS, LC.

**Scott well-remembered the young officer:** WTS to CE, June 22, 1861, CE MSS, LC; WTS, Diary, June 21, 1861, SF MSS, ND

**In the spring of:** DDP, Journal, DDP MSS, LC, vol. 1, 68.

**Fort Corcoran and its outworks:** Benjamin Franklin Cooling, *Symbol, Shield and Sword: Defending Washington During the Civil War*, 2nd rev. ed. (Shippensburg, Pa.: White Mane, 1997, 38; WTS to EES, July 3, 1861, SF MSS, ND; WTS, Diary, July 16, 1861, SF MSS, ND.

**Nor were the regiments:** William Todd, *The Seventy-ninth Highlanders New York Volunteers in the War of the Rebellion, 1861–1865* (Albany: Brandow, Burton, 1886), 13; Thomas Francis Meagher, *The Last Days of the Sixty-ninth in Virginia: A Narrative in Three Parts* (New York: Published at the office of the "Irish American," 1861), passim; W. Mark McKnight, *Blue Bonnets o'er the Border: The Seventy-ninth New York Highlanders* (Shippensburg, Pa.: White Mane, 1998, 36–40).

**Sherman's new assignment gave him:** WTS to James B. Fox, July 5, 1861, M567, NA; claim of John C. Brunaugh, File B 37, Claim Damages, RG 92, E 843, NA.

**But it was the 3,400 officers:** WTS to EES, August 13, 1861, SF MSS, ND.

**"Each of these regiments":** WTS to EES, July 6, 1861, SF MSS, ND; WTS, *Memoirs*, vol. 1, 180.

**Just before the army:** WTS to EES, July 16, 1861, SF MSS, ND.

**On the afternoon of July 16:** GO No. 17, July 16, 1861, *OR* I, 1, Pt. 2, 303.

**The men could not be hurried:** Irving McDowell to E. D. Townsend, July 18, 1861, *OR* I, 2, 306.

**Shepherding his regiments:** Todd, *Highlanders*, 120.

**The Third Brigade:** Ibid., 25, 31.

**He wrote Ellen:** WTS to EES, July 19, 1861, SF MSS, ND.

**On the twenty-first:** Ibid.

**Like all scenes of mass strife:** WTS, Report of July 25, 1861, *OR* I, 2, 368.

**Had the conviction spread:** Report of Captain Daniel P. Woodbury, July 30, 1861, *OR* I, 2, 339.

**Movement was imparted:** WTS to EES, July 28, 1861, SF MSS, ND; Irvin McDowell to E. D Townsend, July 22, 1861, *OR* I, 2, 316; Report of Irvin McDowell, August 4, 1861, *OR* I, 2, 321.

**But back in Washington:** William Thompson Lusk to his mother, August 5, 1861, in William Thompson Lusk, *War Letters of William Thompson Lusk, Captain, Assistant Adjutant General, United States Volunteers, 1861–1863* (New York: privately printed, 1911), 67; Todd, *Highlanders*, 52–53.

**By the twenty-third:** WTS, *Memoirs*, vol. 1, 188–91.

**The morale in the brigade:** Report of James Kelly, July 24, 1861, *OR* I, 2, 372; WTS to JS, December 24, 1861, WTS MSS, LC.

**And now, to fan the fires:** Todd, *Highlanders*, 53–54; Lusk, *Letters*, 68.

**On the twenty-fourth Sherman:** WTS to EES, June 24, 1861, SF MSS, ND; Winfield Scott to Irvin McDowell, July 24, 1861, Headquarters, U.S. Army, Letters Sent, RG 94, E 236, NA.

**Four days later, when he wrote:** WTS to EES, July 28, 1861, SF MSS, ND.

**In retrospect he could:** Ibid.

**Over the next few weeks:** WTS to EES, August 17, 1861, SF MSS, ND; McKnight, *Blue Bonnets*, 40–43.

**In the army traditionally:** WTS to EES, August 15, 1861, SF MSS, ND; WS to JS, August 19, 1861, WTS MSS, LC; Samuel Selden Partridge to Francis A. Macomber, August 13 and 14, 1861, in Blake McKelvey, *Rochester in the Civil War*, Rochester Historical Society *Publications* 22 (1944): 83–84.

**The customary usages of war:** WTS to S. Williams, August 1861, WTS to Stewart Van

Vliet, August 15, 1861, and WTS to GBM, August 1861, all in Telegrams Sent by the Field Office of the Military Telegraph and Collected by the Office of the Secretary of War, RG 37, E 36, NA.

**Journalists sounded no alarm:** WTS to EES, July 28, August 17, 1861, and undated (August 1861), SF MSS, ND.

**Even then he had his doubts:** WTS to EES, July 28, August 3 and 6, 1861, SF MSS, ND.

## 10. KENTUCKY

**Though Sherman was troubled:** WTS to EES, August 15, 1861, SF MSS, ND; TE to CE, August 21, 1861, TE MSS, LC: Report of Irwin McDowell, August 4, 1861, *OR* I, 4, 323; Report of Daniel Tyler, July 27, 1861, *OR* I, 4, 349.

**His promotion meant that:** WTS to EES, August 3, 1861, SF MSS, ND; Jared W. Young, "General Sherman on His Own Record: Some Unpublished Comments," *Atlantic Monthly*, 108 (1911): 293.

**Anderson had just been ordered:** WTS to EES, August 17, 1861, SF MSS, ND; WTS to JS, August 19, 1861, WTS, LC.

**Anderson and Sherman then met:** WTS, *Memoirs*, vol. 1, 192–93.

**Sherman seemed well-pleased:** Ibid., , 192; WTS to EES, letter fragment, August 1861, SF MSS, ND.

**The stakes were high:** Lowell H. Harrison and James C. Klotter, *A New History of Kentucky* (Lexington: University Press of Kentucky, 1997), 189; WTS to EES, September 15, 1861, SF MSS, ND.

**In May Kentucky had issued:** Lowell H. Harrison, "The Civil War in Kentucky: Some Persistent Questions," *Kentucky Historical Society Register*, 76 (1978): 3–4.

**By late summer what the diplomats call:** George Taylor *et al.* to Leonidas Polk, September 5, 1861, *OR* I, 4, 183.

**By September 1 Anderson:** WTS, Diary, September 1, 1861, SF MSS, ND.

**Sherman started out by rail:** WTS, Diary, September 17–25, 1861, SF MSS, ND; WTS to EES, September 26, 1861, SF MSS, ND; WTS to Oliver D. Greene, September 27, 1861, WTS MSS, LC.

**He wrote again:** WTS to EES, October 3, 1861, SF MSS, ND.

**He had other worries:** Ibid.; WTS to EES, October 6, 1861, SF MSS, ND.

**Only hours after he wrote:** WTS, *Memoirs*, vol. 1, 199; Winfield Scott to RA, October 6, 1861, *OR* I, 4, 296.

**Sherman had been concerned:** WTS to EES, October 12, 1861, SF MSS, ND.

**Whatever befell him:** WTS to Garrett Davis, October 8, 1861, *OR* I, 4, 299; WTS to Colonel Jackson, October 8, 1861, *OR* I, 4 299.

**His first orders to subordinates:** WTS to Colonel Jackson, October 8, 1861, *OR* I, 4, 298; WTS to William Ward, October 8, 1861, *OR* I, 4, 299; WTS to GHT, October 11, 1861, *OR* I, 4, 302–03.

**Sherman did not inherit:** E. G. Townsend to RA, August 24, 1861, RA MSS, LC; GHT to Oliver D. Greene, October 1, 1861, *OR* I, 4, 284–85; WTS to GHT, October 13, 1861, *OR* I, 4, 306; Albin Schoepf to George B. Flynt, November 2, 1861, *OR* I, 4, 329; WTS to EES, October 18, 1861, SF MSS, ND; Walter Q. Gresham to Matilda Gresham, October 24, 1861, Walter Q. Gresham MSS, LC.

**In this army of amateurs:** Circulars of October 10 and 15, 1861, *Military Orders*, 2, 5; GO No. 14, October, 21, 1861, Orders and Special Orders, Department of the Ohio (Cumberland), 1861, RG 94, vol. 219, NA.

**Yet the chief deficiency:** RA to Solomon Chase, September 1, 1861, *OR* I, 4, 255–56; Oliver

D. Greene to I. B. Turchin, October 2, 1861, Letters Sent, Department of the Cumberland, RG 393, Pt. 1, E 866, NA.

**For that other basic element:** Thomas L. Crittenden to WTS, October 30, 1861, Thomas L. Crittenden Miscellaneous Papers, Filson; Jeremiah T. Boyle to WTS, November 4, 1861, Jeremiah T. Boyle Papers, Filson.

**The state had put off:** WTS to EES, September 18, 1861, SF MSS, ND; Oliver P. Morton to Thomas A. Scott, September 2, 1861, *OR* I, 4, 256.

**Of raw manpower:** Report of Stephen G. Burbridge, October 31, 1861, *OR* I, 4, 219-20; George G. Curtiss to RA, August 30, 1861, RA MSS, LC; WTS to TE, September 30, 1861, Ewing, *Sherman,* 32; R. M. Kelly, "Holding Kentucky for the Union," in Robert Underwood Johnson and Clarence Clough Buel, eds., *Battles and Leaders of the Civil War,* 4 vols. (New York: 1887-88), vol. 1, 380.

**There were endless delays:** Oliver P. Morton to RA, October 17, 1861, and Oliver P. Morton to Thomas A. Scott, October 10, 1861, Governor Oliver P. Morton Telegraph Books, Indiana State Library; WTS to Alexander McCook, October 15, 1861, RG 393, Pt. 1, E 866, NA.

**Sufficient arms:** WTS to TE, September 15, 1861, TEF MSS, LC. WTS to GBM, October 8, 1861, Unbound Telegrams Collected by the Office of the Secretary of War, M504, NA; WTS to AL, October 10, 1861, *OR* I, 4, 300.

**The whole party went:** WTS, *Memoirs,* vol. 1, 200-03; Statement of General Thomas J. Wood, August 24, 1866, reproduced in ibid., 210-14; Report of Lorenzo Thomas, October 21, 1861, *OR* I, 4, 313-14; Erwin Stanley Bradley, *Simon Cameron, Lincoln's Secretary of War: A Political Biography* (Philadelphia: University of Pennsylvania Press, 1966), 191-93.

**In the conversation that followed:** WTS, *Memoirs,* vol. 1, 204-08.

**Cameron had indeed seemed:** Simon Cameron to AL, October 16, 1861, *OR* I, 4, 308; Simon Cameron to Thomas A. Scott, October 16 1861, *OR* I, 4, 308; Alexander K. McClure, *Recollections of Half a Century* (Salem, Mass.: Salem Press, 1902), 332.

**During his stay in Washington:** Meagher, *Sixty-ninth in Virginia,* 6; Warner, *Generals in Blue,* 317-18; WTS to TE, September 15, 1861, TEF MSS, LC.

**One of Sherman's first acts:** John Jefferson Diary, September 19, 1861, Kentucky Historical Society; *Cincinnati Commercial,* November 9, 1861; Emmet Crozier, *Yankee Reporters, 1861-1865* (New York: 1956), 173-74.

**At least one journalist:** Henry Villard, *Memoirs of Henry Villard, Journalist and Financier,* 2 vols. (Boston and New York: Houghton Mifflin, 1904), vol. 1, 209.

**The *Cincinnati Daily Commercial*:** Donald W. Curl, *Murat Halstead and the Cincinnati Commercial* (Boca Raton: University Presses of Florida, 1980), 19; *Cincinnati Commercial,* September 24 and November 12, 1861.

**The *Commercial*'s Louisville correspondent:** *Cincinnati Commercial,* October 25, 1861.

**The most punishing attack:** Report of Lorenzo Thomas, October 21, 1861, *OR* I, 4, 313-14; *New York Tribune,* October 30, 1861; *Cincinnati Commercial,* November 2, 1861.

**Sherman's reaction might have been:** WTS to EES, October 23 and November 1, 1861, SF MSS, ND.

**The situation seemed to grow:** WTS to GBM, November 3, 1861, RG 107, E 36, NA.

**The next day Sherman:** WTS to GBM, November 4, 1861, Stephen W. Sears, ed., *The Civil War Papers of George B. McClellan. Selected Correspondence, 1860-1865* (New York: Ticknor and Fields, 1998), 127, note 1; WTS to GBM, November 6, 1861, RG 107, E 36, NA.

**Sherman sent the adjutant:** WTS to Lorenzo Thomas, November 6, 1861, M567, NA; WTS to GBM, November 6, 1861, *OR* I, 4, 340-41; WTS to RA, November 5, 1861, RA MSS, LC.

On the morning of November 8: EES to JS, November 10, 1861, WTS MSS, LC.

She decided to go: TEJr to B. H. Liddell Hart, June 3, 1929, TEF MSS, LC.

The day after her arrival: EES to JS, November 10, 1861, WTS MSS, LC.

McClellan had by this time: GBM to WTS, November 8, 1861, Sears, ed., *McClellan*, 127.

Sherman in the meantime: WTS to GHT, November 12, 1861, *OR* I, 4, 353-54.

There was no threat: Harrison, "Civil War in Kentucky," 7-8.

The damage done was not: WTS to GHT, November 12, 1861, *OR* I, 4, 353; GHT to WTS, November 12, 1861, *OR* I, 4, 354.

It was at this juncture: WTS to JS, November 14, 1861, WTS MSS, LC.

It should be pointed out that: Gideon Pillow to William Mackall, November 13, 1861, *OR* I, 4, 550; Leonidas Polk to Samuel Cooper, November 14, 1861, *OR* I, 4, 553; Gideon Pillow to John S. Bowen, November 16, 1861, *OR* I, 4, 557.

Sherman stayed on a week: WTS, *Memoirs*, vol. 1, 214-16; EES, Diary, November 26-29, 1861, SF MSS, ND; EES to CE, December 2, 1862, CE MSS, LC; EES to JS, December 10, 1862, SF MSS, ND.

Despite his soothing words: HWH to GBM, December 2, 1861, *OR* I, 52, Pt. 1, 198.

Sherman's spirits seemed to improve: EES to JS, December 10, 1861, SF MSS, ND.

There was worse to come: *Cincinnati Commercial*, December 11, 1861; *Cincinnati Gazette*, December 12, 1861; Halstead, "Recollections," 1611; Whitelaw Reid, *The "Agate" Dispatches of Whitelaw Reid 1861-1865*, ed. with an introduction and notes by James G. Smart, 2 vols. (Memphis: Memphis State University Press, 1988), vol. 1, 80.

Philemon Ewing was in Lancaster: PBE to TE, December 13, 1861, TE MSS, LC.

On the precise question: Ibid.; WTS to TE, undated, TE MSS, LC.

It is clear from Sherman's letters: PBE to TE, December 13, 1862, TE MSS, LC; WTS to Colonel Hazzard, November 8, 1861, Franklin D. Roosevelt Library; JS, "Commemoration," 41; Steiner, *Portraits*, 66-73.

This "breakdown" in Kentucky: WTS to HST, January 5 and February 4, 1856, WTS MSS, OHS.

The episode in Kentucky: Stephen Ambrose, "Sherman: A Reappraisal," *American History Illustrated* 1, no. 9 (January, 1967): 6.

The craft of history: Janann Sherman, "The Jesuit and the General: Sherman's Private War," *Psychohistory Review* 21 (1993): 266.

According to current theory: Stephen M. Johnson, *Humanizing the Narcissistic Style* (New York: W. W. Norton, 1987), 39; James F. Masterson, *The Search for the Real Self: Unmasking the Personality Disorders of Our Age* (New York: Free Press, 1988), 90-92.

There is no simple litmus test: Masterson, *Real Self*, 94; American Psychiatric Association, *Diagnostic and Statistical Manual of Mental Disorders, IV*, 4th ed. (Washington, D.C.: American Psychiatric Association, 1994), 78-79; Norman Cameron and Joseph F. Rychlak, *Personality Development and Psychopathology: A Dynamic Approach*, 2nd. ed. (Boston: Houghton Mifflin, 1985,) 460; Theodore Millon, *Disorders of Personality, DSM-III: Axis II* (New York: John Wiley and Sons, 1981), 98.

One frequently cited hallmark: American Psychiatric Association, *Diagnostic and Statistical Manual of Mental Disorders,* 3rd ed., rev. (Washington, D.C.: American Psychiatric Association, 1987), 351; Dodge, *Recollections*, 198.

This coin has another side: WTS to Mrs. Lloyd Tilghman, August 26, 1863, WTS LB 23, NA.

"The narcissist is unequivocal": Andrew Morrison, *Shame: The Underside of Narcissism* (Hillsdale, N.J., Analytic Press, 1989), 124-25.

In the Kentucky episode: WTS to EES, November 1, 1861, SF MSS, ND.

## 11. SHILOH

**Though he had any number:** WTS to JS, January 8, 1862, WTS MSS, LC; WTS to EES, January 5, 1862, SF MSS, ND.

**For once he could not lose himself:** WTS to EES, January 1, 1862, SF MSS, ND; Lucien B. Crooker, "Episodes and Characters in an Illinois Regiment," *Military Essays and Recollections* (MOLLUS, Illinois Commandery, I), Chicago: McClure, 1891, 35.

**Sherman could not take his mind away:** WTS to EES, January 11, 1862, SF MSS, ND; WTS to JS, Dec. 24, 1861, WTS MSS, LC.

**Were this not enough:** WTS to JS, December 24, 1861, WTS MSS, LC.

**What were his options:** WTS to JS, January 4, 1862, WTS MSS, LC; JS to WTS, January 18, 1862, LC; WTS to EES, January 11, 1862, SF MSS, ND.

**Lizzie was in this period:** WTS to EES, January 19 and 29, 1862, SF MSS, ND.

**Disturbing as these passages:** WTS to EES, January 1 and March 12, 1862, SF MSS, LC.

**And as he was frank:** WTS to EES, January 1, 1862, SF MSS, ND; WTS to PBE, January 20, 1862, Ewing, *Sherman*, 131; WTS to TE, December 24, 1861, TEF MSS, LC; WTS to JS, January 2, 1862, WTS MSS, LC.

**Yet even as he wrote:** TE to WTS, undated, TE MSS, LC; WTS to TE, December 24, 1862, TEF, LC; WTS to JS, January 4, 1862, WTS MSS, LC.

**His hopes of returning:** WTS to EES, January 1, 1864, SF MSS, ND.

**In the end it was Ellen:** EES to AL, January 10, 1862, copy in EES to WTS, January 29, 1862, SF MSS, ND; EES to CE, April 13, 1862, CE MSS, LC.

**But in her mind:** EES to CE, April 13, 1862, CE MSS, LC.

**As early as the second half of January:** HWH to WTS, January 22, 1862, WTS MSS, LC.

**Sherman was quick to appreciate:** WTS to EES, February 17, 1862, SF MSS, ND.

**New riverborne expeditions were:** HWH to GBM, February 8, 1862, *OR* I, 7, 595; EMS to AL, February 14, 1862, EMS MSS, LC.

**In the end the expedition:** HWH to WTS, February 9, 1862, *OR* I, 52, Pt. 1, 209; HWH to WTS, February 17, 1862, *OR* I, 7, 629; WTS to HWH, March 6, 1862, *OR* I, 10, Pt. 2, 12.

**Sherman did get to lead:** WTS, *Memoirs*, vol. 1, 226–27; GO No. 7, March 12, 1862, *Military Orders*, 14–15; Circular to Commanders of Brigades, March 12, 1862, WTS LB 5, NA.

**The mission was a washout:** WTS, Report, March 15, 1862, *OR* I, 10, 22–24; WTS, Report, March 17, 1862, *OR* I, 10, 24–25.

**There was periodic gunfire:** GO No. 18, March 23, 1862, *Military Orders*, 20.

**Sherman's conduct during both days:** James Lee McDonough, *Shiloh: In Hell Before Night* (Knoxville: University of Tennessee Press, 1977), 116.

**Such was the judgment of his superiors:** HWH to Ethan Allen Hitchcock, April 18, 1862, copy in TEF MSS, LC; HWH to EMS, April 13, 1862, *OR* I, 10, Pt. 1, 98; Report of John A. McClernand, April 24, 1862, *OR* I, 10, Pt. 1, 117.

**Though Halleck did not mention it:** John T. Taylor, "Reminiscences of Service as an Aide-de-Camp with General William Tecumseh Sherman," *War Talks in Kansas* (MOLLUS, Kansas Commandery), 1892, 132.

**News of the great battle:** *Daily Missouri Democrat*, April 9, 10 and 14, 1862; *Cincinnati Commercial*, April 9, 1862; J. Cutler Andrews, *The North Reports the Civil War* (Pittsburgh: University of Pittsburgh Press, 1955), 177–81.

**The more disturbing aspects:** *Cincinnati Commercial*, April 21, 1862; *Ironton* (Ohio) *Register*, reprinted in *Cincinnati Commercial*, April 28, 1862.

**There had to be some other explanation:** *Cincinnati Commercial*, April 12 and 23, 1862; *Daily Missouri Democrat*, April 15, 1862; *Louisville Journal*, undated column reproduced in the *Cincinnati Commercial*, April 15, 1862.

**Some of the journalists:** *St. Louis News* article, no date, reprinted in the *Cincinnati Commercial*, April 21, 1862; *Daily Missouri Democrat*, April 11, 1862; *Cincinnati Commercial*, April 18, 1862.

**A Confederate cavalry probe:** Orders, No. 19, April 4, 1862, *Military Orders*, 21; Peter John Sullivan to O. M. Poe, June 18, 1878, Peter John Sullivan Letters, MS 3228l, Louisiana and Lower Mississippi Valley Coll., LSU.

**Yet he vehemently denied:** WTS to CE, April 22, 1862, CE MSS, LC.

**The most telling evidence:** Order No. 20, April 12, 1862, *Military Orders*, 22–24; WTS to C. F. Smith, April 17, 1862, M504, NA; WTS to Colonel Dickey, May 9, 1862, WTS LB 1, NA.

**The other important measure:** Michael S. Green, "Picks, Spades and Shiloh: The Entrenchment Question," *Southern Studies* 3 (Spring 1992), 13–54; Court Martial of Colonel Thomas Worthington, Forty-sixth Ohio Infantry, Court Martial Case Files, RG 153, NA, 7; Orders No. 20, April 12, 1862, *Military Orders*, 23; Circular from Camp No. 3, May 5, 1862, WTS LB 5, NA.

**Why Sherman did not show:** McDonough, *Shiloh*, 53.

**But the newspapers had:** WTS to TE, April 27, 1862, TEF MSS, LC; WTS to Henry Coppee, January 5, 1864, in *Army and Navy Journal*, December 31, 1864, 299.

**This assertion is somewhere short:** WTS, *Memoirs*, vol. 1, 227.

**If Smith approved the spot:** WTS to USG, March 17, 1862, *OR* I, 10, Pt. 1, 27.

**It was no doubt:** WTS to Jacob Lauman, March 20, 1862, *OR* I, 10, Pt. 2, 54.

**In the hue and cry:** WTS to CE, April 25, 1862, CE MSS, LC.

**His anger was particularly acute:** WTS to TE, April 27, 1862, TEF MSS, LC.

**Colonel Rodney Mason:** WTS to PBE, May 16, 1862, TEF MSS, LC.

**When the *Cincinnati Gazette*:** *Louisville Journal*, May 10, 1862; *Cincinnati Commercial*, May 5, 1862.

**Eventually Sherman himself:** WTS, *Memoirs*, vol. 1, 246; WTS to Benjamin Stanton, June 20, 1862; Ewing, *Sherman*, 56 and 62; WTS to Benjamin Stanton, July 12, 1862, WTS MSS, LC; EES to CE, May 28, 1862, CE MSS, LC; WTS to PBE, July 14, 1862, TEF MSS, LC.

**Hardly had the repercussions:** *Cincinnati Commercial*, August 15, 1862; WTS to USG, *OR* I, 17, Pt. 2, 188; WTS to S. S. L'Hommedieu, August 20, 1862, WTS LB 3, NA; WTS to W. H. H. Taylor, August 25, 1862, WTS LB 3, NA; WTS to EES, August 20, 1862, SF MSS, ND.

**Sherman received letters:** WTS to Miss Elbit, August 30, 1862, WTS LB 3, NA.

**If his official acts:** Order No. 20, April 12, 1862, *Military Orders*, 22–24.

**Regiments that had given way:** WTS to PBE, July 13, 1862, Ewing, *Sherman*, 58; James McPherson, *For Cause and Comrades: Why Men Fought in the Civil War* (New York and Oxford: Oxford University Press, 1997), 49.

**The officers of these regiments:** WTS to JAR, April 23, 1862, WTS LB 1, NA; WTS to ___ Tod, April 26, 1862, WTS LB 1, NA; Colonel Jesse J. Appler, Fifty-third Ohio Volunteers, File 10860 V.S. 1886, Letters Received by the Volunteer Service Branch, RG 94, NA.

**Sherman prepared extensive lists:** Robert W. McCormick, "Challenge of Command: Worthington vs. Sherman," *Timeline: A Publication of the Ohio Historical Society* 8, no. 3 (June-July 1991): 26–39; Thomas Worthington Court Martial File, RG 153, NA, passim.

**In preferring charges:** Sixty-fifth Article of War and addenda, in *Revised Regulations for the Army of the United States, 1861* (By Authority of the War Department. Philadelphia: J. G. L. Brown, 1861), 509; WTS to James Worthington, September 18, 1862, WTS LB 3, NA; "A National Disgrace," *Capital*, March 24, 1878; WTS to JS, March 1, 1862, WTS

MSS, LC; T. Worthington, *Shiloh: The Only Correct Military History of U.S. Grant and the Missing Army Records, for Which He Is Alone Responsible to Conceal His Defeat of the Union Army at Shiloh, April 6, 1862* (Washington, D.C.: n.p., 1872), 31.

## 12. MEMPHIS

**Though Sherman looked forward:** WTS to HWH, July 16, 1862, *OR* I, 17, Pt. 2, 100–01.

**As for Sherman:** WTS, *Memoirs*, vol. 1, 255; EES to CE, May 27, 1862, CE MSS, LC; WTS to EES, October 1, 1862, SF MSS, ND.

**After the fall of Corinth:** WTS to PBE, July 13, 1862, Ewing, *Sherman*, 138; WTS to EES, June 27, 1862, SF MSS, ND; WTS, *Memoirs*, vol. 1, 257; Ernest Walter Hooper, "Memphis, Tennessee: Federal Occupation and Reconstruction, 1862–1870" (Ph.D. diss., University of North Carolina, 1957), 1–4; J. H. Parks, "A Confederate Trade Center Under Federal Occupation: Memphis, 1862 to 1865," *Journal of Southern History* 7 (1941): 296.

**The people of Memphis:** *Memphis Daily Argus*, June 10, 1862; *Memphis Union Appeal*, July 3, 6, 13, and 20, 1862; *Memphis Daily Appeal*, August 8, 1862; Hooper, "Memphis," 14.

**On his arrival:** Orders, No. 54, July 19, 1862, *OR* I, 17, Pt. 2, 106; *Cincinnati Commercial*, August 12, 1862.

**Sometime that first afternoon:** *Cincinnati Commercial*, July 28, 1862.

**Jarring as this introduction was:** WTS to John Park, July 27, 1862, *OR* I, 17, Pt. 2, 127.

**Sherman's Order No. 61:** Orders, No. 61, July 24, 1862, *OR* I, 17, Pt. 2, 117; WTS to William S. Hillyer, July 2, 1862, William S. Hillyer MSS, UVA.

**The general also contributed:** WTS to Memphis Board of Trade, August 24, 1862, WTS LB 3, NA; *Cincinnati Commercial*, August 23 and 26, 1862; WTS to EES, August 20, 1862, SF MSS, ND.

**The new commandant of Memphis:** Minutes of the Meeting of the Mayor and Board of Aldermen, July 29, 1862, Memphis-Shelby County Library and Information Center; *Cincinnati Commercial*, August 5, 1862.

**Porter had himself announced:** DDP, Journal, vol. 1, 435–37, DDP MSS, LC.

**Still, Porter had waited:** Ibid.

**In his first days:** WTS to EES, July 31, 1862, WTS MSS, NA; WTS to John T. Swayne, November 12, 1862, WTS LB 3, NA.

**He needed laborers at Fort Pickering:** Order No. 67, August 8, 1862, *Military Orders*, 158–61; WTS to Frederick Steele, September 13, 1862, WTS LB 3, NA; WTS to EES, August 5, 1862, SF MSS, ND; WTS to JAR, August 14, 1862, *OR* I, 17, Pt. 2, 169–70.

**The most important:** *Cincinnati Commercial*, August 1, 1862.

**The military in general:** William S. McFeely, *Grant: A Biography* (New York and London: W. W. Norton, 1981), 122–23; *Cincinnati Commercial*, August 5, 1862; WTS to Leslie Couch, December 17, 62, WTS LB 4, NA; WTS to Lorenzo Thomas, August 11, 1862, WTS LB 3, NA.

**He protested to the adjutant general:** WTS to Lorenzo Thomas, August 11, 1862, WTS LB 3, WTS MSS, LC; *Cincinnati Commercial*, August 5, 1862.

**The *Commercial*'s Memphis correspondent:** *Cincinnati Commercial*, August 5, 1862; HWH to WTS, August 25, 1862, *OR* I, 17, Pt. 2, 186; HWH to USG, August 2, 1862, *OR* I, 17, Pt. 2, 150; WTS to J. C. Kelton, August 29, 1862, WTS LB 2, NA.

**Sherman had in the meantime:** WTS to HWH, August 18, 1862, WTS LB 3, NA; WTS to J. C. Kelton, September 4, 1862, *OR* I, 17, Pt. 2, 200; WTS to USG, August 11, 1862, WTS LB 3, NA; WTS to I. F. Quinby, August 18, 1862, WTS LB 3, NA.

**The cotton trade became:** A. A. Van _____ to William S. Hillyer, October 22, 1862, William S. Hillyer MSS, UVA.

**His efforts to prevent:** WTS to W. H. H. Taylor, August 25, 1862, WTS LB 3, NA.

What angered him most: Ibid.; WTS to EES, August 20, 1862, SF MSS, ND.

Smuggling was rampant: WTS to Lorenzo Thomas, August 31, 1862, WTS LB 3, NA; WTS to HWH, August 18, 1862, WTS LB 3, NA.

That summer the most critical: *Cincinnati Commercial*, August 22, 1862; WTS to C. D. Anthony, August 17, 1862, WTS LB 3, NA.

Sherman acknowledged: WTS to Lorenzo Thomas, August 13, 1862, WTS LB 3, NA.

But on occasion the commission: WTS to C. D. Anthony, August 17, 1862, WTS LB 3, NA.

The most systematic press critique: JS to WTS, September 23, 1862, WTS MSS, LC.

Sherman had no choice: WTS to Frederick Steele, September 13, 1862, WTS LB 3, NA.

Running Memphis took a great deal: WTS to EES, August 5 and 20 and December 14, 1862, SF MSS, ND; WTS to Minnie, August 16, 1865, in Minnie Sherman, "My Father's Letters," *Cosmopolitan* 12 (1891): 67.

Sherman's life in Memphis: WTS to EES, September 22 and October 1 and 4, 1862, SF MSS, ND; James H. Otey to WTS, December 1, 1862, and Otey's Diary, October–November, 1862, passim, James H. Otey MSS, UNC.

Then there was Felicia Shover: WTS to Felicia Shover, various dates, 1862, in Thornton Family MSS, VHS; WTS to EES, August 10, 1862, SF MSS, ND.

In his impromptu speech: WTS to John Park, July 27, 1862, *OR* I, 17, Pt. 2, 127.

Though a profound admirer: WTS to J. J. Gant, September 23, 1862, Supplementary Correspondence, Container 105, WTS MSS, LC; Bowman and Irwin, *Sherman*, 74; WTS to EES, August 20, 1862, SF MSS, ND; WTS to Frederick Steele, September 13, 1862, WTS LB 3, NA; WTS to EES, August 20, 1862, SF MSS, ND.

The structure of justice: WTS to James Wickersham, August 10, 1862, WTS LB 3, NA; GO No. 63, July 26, 1862, WTS LB 5, NA; GO, No. 90, October 25, 1862, *Military Orders*, 68–70; L. M. Dayton to D. C. Anthony, September 18, 1862, LB 3, NA.

The military commission heard cases: GO No. 76, August 25, 1862, WTS LB 5, NA; *Cincinnati Commercial*, August 5, 1862.

The military commission was: WTS to John T. Swayne, November 17, 1862, *OR* I, 17, Pt. 2, 863–65.

Perhaps the most bizarre instance: C. D. Anthony, Report of October 2, 1862, in William S. Hillyer MSS, UVA.

Predictably, Sherman's relations with: WTS to Reverend Samuel Sawyer, July 24, 1862, *OR* I, 17, Pt. 2, 116–17.

The newspapers were expected: WTS to Editors of the *Bulletin* and the *Appeal*, August 21, 1862, WTS LB 3, NA.

A month later Sherman had: WTS to C.D. Anthony, September 6, 1862, WTS LB 3, NA; WTS to Frederick Steele, September 6, 1862, LB 3, NA.

The turn of the *Argus*: WTS to D. C. Anthony, November 5, 1862, WTS, LB 4, NA.

It would be wrong: SO No. 166, August 20, 1862, LB 6, NA.

Memphis had more than its share: Hooper, "Memphis," 26; To All Commanders of Regiments and Companies, October 30, 1862, WTS LB 3, NA.

For Memphis prosperity lay: WTS to EMS, January 25, 1863, WTS LB 22, NA.

He boasted that whatever else: WTS to EES, September 22, 1862, SF MSS, ND; WTS to C. D. Anthony, November 5, 1862, WTS LB 3, NA.

Then when several local men: WTS to W. H. H. Taylor, August 19, 1862, LB 3, NA; WTS to John W. Leftwich, August 20, 1862, LB 3, NA.

In justice to the man: WTS to Salmon P. Chase, Oct. 25, 1863, WTS LB 8, NA; WTS to JS, December 6, 1862, WTS MSS, LC.

Such was his belief: WTS to JS, December 6, 1862, WTS LC; WTS to EES, December 11, 1862, SF MSS, ND.

## 13. VICKSBURG

**In the spring of 1863:** Charles Dana Gibson with E. Kay Gibson, *Assault and Logistics: Union Army Coastal and River Operations, 1861–1866,* Army's Navy Series, vol. 2 (Camden, Me.: Ensign Press, n.d.), 52.

**On the morning of December 8:** HWH to USG, December 8, 1862, and USG to WTS, December 8, 1862, in WTS, *Memoirs,* vol. 1, 280–81.

**Sherman hurried out of College Hill:** Ibid., 281–83; WTS to HWH, December 8, 1862, WTS MSS, LC; WTS to Benjamin Grierson, December 9, 1862, MDAH; "John Alexander McClernand," Warner, *Generals in Blue,* 293–94; WTS, *Memoirs,* vol. 1, 283; Gibson and Gibson, *Assault,* 151–52.

**The expedition had sailed:** GO No. 8, December 18, 1862, Military Orders, 87–88.

**If civilian reporters were found:** Ibid.; Ewing, *Sherman,* 82; Wilkie, *Pen and Powder,* 237.

**On the twenty-third:** WTS to Commanders of Divisions, December 23, 1862, *OR* I, 17, Pt. 1, 616.

**A complication of a different sort:** JAM to AL, December 17, 1862, *OR* I, 17, Pt. 2, 420; HWH to USG, December 18, 1862, *OR* I, 24, 476.

**McClernand thus had a clear mandate:** JAM to AL, December 29, 1862, JAM MSS, ISHL.

**On the day after Christmas:** L. B. Parsons to HWH, December 15, 1862, *OR* I, 17, Pt. 2, 496–97.

**This show of equanimity:** WTS to TE, January 16, 1863, WTS MSS, USAMHI.

**It would not end:** WTS to JAR, January 5, 1863, and WTS, Report of January 5, 1863, *OR* I, 17, 613.

**Admiral Porter had tried:** DDP, Journal, DDP MSS, LC, vol. 1, 457–58.

**After two days:** SO No. 37, December 28, 1862, *OR* I, 17, Pt. 1, 622.

**The attack was made:** WTS, *Memoirs,* vol. 1, 291–92; George W. Morgan, "The Assault on Chickasaw Bluffs," *Battles and Leaders,* vol. 3, 467–68.

**It was a profoundly dejected Sherman:** A. A. Stuart, *Iowa Colonels and Regiments: Being a History of Iowa Regiments in the War of the Rebellion, and Containing a Description of the Battles in Which They Have Fought* (Des Moines: Mills, 1865), 120; DDP, Journal, DDP MSS, LC, vol. 1, 464–68.

**But Sherman's cup was:** JAM to EMS, January 3, 1863, JAM MSS, ISHL.

**There was the possibility:** DDP, Journal, DDP MSS, LC, vol. 1, 485.

**The next day McClernand:** Edwin C. Bearss, "The Battle of Arkansas Post," *Arkansas Historical Quarterly* 18 (1959): 237–45.

**The taking of the Post of Arkansas:** WTS, *Memoirs,* vol. 1, 296–303; WTS to JS, January 17, 1863, WTS MSS, LC; DDP, Journal, DDP MSS, LC, vol. 1, 486.

**McClernand was not surprised:** JAM to AL, January 16, 1863, *OR* I, 17, Pt. 2, 566–67.

**McClernand pleaded with Lincoln:** WTS to JAM, January 10, 1863, JAM MSS, ISHL.

**Infighting of this sort:** WTS, *Memoirs,* vol. 1, 296; DDP, Journal, vol. 1, 485.

**Nor is it possible:** Bearss, "Arkansas Post," 245; Richard L. Kiper, "John Alexander McClernand and the Arkansas Post Campaign," *Arkansas Historical Review* 56 (1997): 56–79.

**This time calamity:** WTS to EES, April 12 and 29, 1863, SF MSS, ND.

**Sherman also gave vent:** WTS to JS, January 17, 29, and 31, 1863, WTS MSS, LC; WTS to EES, February 6, 1863, SF MSS, ND.

**At the same time:** WTS to EES, January 24 and 28, 1863, SF MSS, ND; WTS to JS, May 29, 1863, WTS MSS, LC.

**In a long letter:** WTS to Ethan Allen Hitchcock, January 25, 1863, WTS LB 22, NA; WTS to EES, February 4, 1862, SF MSS, ND.

His relations with his chief critic: Andrews, *North Reports*, 378; *Indianapolis Daily Journal*, February 4, 1863.

Here too Sherman went: WTS to EES, February 4, 1863, SF MSS, ND; TE to EMS, March 9, 1863, WTS MSS, LC; Thomas Gantt to WTS, February 6, 1863, WTS MSS, LC.

Among the correspondents: *New York Herald*, January 18, 1863; Taylor, "Reminiscences," 138.

Knox reported that after the failed assault: Morgan, "Chickasaw Bluffs," 469; Knox to WTS, February 1, 1863, LB 22, NA.

Knox's article was soon: WTS to Benjamin Grierson, February 9, 1863, WTS MSS, MDAH; Andrews, *North Reports*, 379–83.

But Sherman learned something: WTS to Murat Halstead, April 8, 1863, *OR* I, 17, Pt. 2, 896.

And to this idea: Ibid.; WTS to EES, February 26, 1863, SF MSS, ND; WTS to TE, January 16, 1863, WTS MSS, USAMHI.

So he explained it: WTS to EES, April 10, 1863, SF MSS, ND.

But others at Grant's headquarters: C. A. Dana to EMS, April 1, 1863, *OR* I, 24, Pt. 1, 69–70; Dana to EMS, April 12, 1863, *OR* I, 24, Pt. 1, 73–74; WTS to EES, April 6, 1863, SF MSS. ND.

Grant tended to be: USG, *Personal Memoirs*, 2 vols. (New York: Charles L. Webster, 1885), vol. 1, 49–50.

Between the end of January: Ibid., 445–55; WTS to EES, March 30, 1863, SF MSS, ND; WTS to Harry McDougall, July 19, 1863, WTS LB 23, NA.

As early as January 1863: USG, *Memoirs*, vol. 1, 456–64; WTS to L. B. Parsons, July 19, 1863, WTS LB 23, NA.

Early in April Sherman: WTS to USG, April 8, 1863, *OR* I, 24, Pt. 3, 179–80; Bowman and Irwin, *Sherman*, 471; WTS to JS, April 26, 1863, WTS MSS, LC.

Eventually Union engineers opened: WTS to EES, April 17, 1863, SF MSS, ND; Dana to EMS, April 17, 1863, *OR* I, 24, Pt. 1, 76.

By then Grant's army: WTS to EES, April 23 and 29, 1863, SF MSS, ND; WTS, *Memoirs*, vol. 1, 319.

Grant's complicated movement: WTS to James Madison Tuttle, May 5, 1863, *OR* I, 24, Pt. 3, 274; WTS to EES, April 29, 1863, SF MSS, ND.

Grant had Sherman bring: USG to WTS, May 3, 1863, *OR* I, 24, Pt. 3, 268–69; WTS to EES, May 9, 1863, SF MSS, ND.

Now mobile and with no supply line: USG to HWH, June 23, 1863, *OR* I, 24, Pt. 1, 43; USG, Report, July 6, 1863, *OR* I, 24, Pt. 1, 55.

Sherman was not present: WTS to EES, July 5 and July 15, 1863, SF MSS, ND; WTS to USG, July 4, 1863, *OR* I, 24, Pt. 3, 472; USG to WTS, July 4, 1863, *OR* I, Pt. 1, 24, Pt. 3, 473.

Sherman advanced toward Jackson: WTS, *Memoirs*, vol. 1, 331.

On the night of July 16-17: WTS to JAR, July 11, 1863, WTS LB 23, NA.

Halleck was following the situation: HWH to USG, July 22, 1863, *OR* I, 24, Pt. 3 , 542; USG to HWH, July 4, 1863, *OR* I, 24, Pt. 1, 44; USG to WTS, July 17, 1863, *OR* I, 24, Pt. 3, 528.

Was the criticism justified: WTS to USG, July 17, 1863, *OR* I, 24, Pt. 2, 528; WTS to DDP, July 17, 1863, *OR* I, 24, Pt. 3, 531; WTS to Ord, August 3, 1863, *OR* I, 24, Pt. 3, 528.

## 14. CHATTANOOGA

Moving back toward Vicksburg: HWH to WTS, August 4, 1863, WTS, LB 23, NA; Mahan to WTS, August 28, 1863, WTS MSS, LC.

**In his camp Sherman:** WTS to JS, August 3, 1863, WTS MSS, LC; WTS, *Memoirs*, vol. 1, 344–45.

**Sherman got first news:** WTS, *Memoirs*, vol. 1, 346–47.

**Ellen and the children:** Ibid., 348.

**In death Willie Sherman:** Ibid., 349; WTS to DDP, October 13, 1862, WTS LB 3, NA.

**Sherman spent a week:** USG, *Memoirs*, vol. 2, 18–19.

**Grant was under great pressure:** USG to WTS, October 24, 1863, *OR* I, 31, Pt. 1, 713; same to same, November 13, *OR* I, 31, Pt. 3, 140; WTS, *Memoirs*, vol. 1, 359–61.

**He reached Chattanooga:** WTS, *Memoirs*, vol. 1, 361.

**Grant outlined the plan:** Ibid., 362–63; W. F. Smith to WTS, February 20, 1886, John C. Ropes Coll., Dept. of Special Collections, BU.

**He would be given a sizable force:** Smith to WTS, February 20, 1886, Ropes Coll., BU; JHW, *Under the Old Flag: Recollections of Military Operations in the War for the Union, the Spanish War, the Boxer Rebellion,* 2 vols. (New York and London: D. Appleton, 1912), vol. 1, 292.

**What happened to Sherman:** WTS, *Memoirs*, vol. 1, 364.

**Given the state of the roads:** WTS to JBM, November 18, 1863, *OR* I, 31, Pt. 3, 188–89; JHW, *Old Flag*, vol. 1, 290–91.

**Nor did they reach:** JHW, Diary, November 21, 1863, DHS; JHW, *Old Flag*, vol. 1, 291; Dana to EMS, No. 20, 1863, *OR* I, 1, Pt. 2, 39; same to same, November 23, 1863, *OR* I , 31, Pt. 2, 63; USG to Woods, November 23, 1863, *OR* I, 31, Pt. 2, 42.

**Sherman's failure to keep:** HBE, Diary, November 12, 1863, HBE MSS, OHS; WTS to HBE, November 18, 1863, HBE MSS, OHS; Dana to EMS, November 23, 1863, *OR* I, 31, Pt. 2, 64.

**On November 23 Sherman:** WTS to USG, November 23, 1864, *OR* I, 31, Pt. 2, 41; Dana to EMS, November 23, 1863, *OR* 31, Pt. 2, 64.

**He was almost:** Dana to EMS, November 24, 1863, *OR* I, 31, Pt. 2, 66–67; JHW, *Old Flag*, vol. 1, 483.

**The day was overcast:** James Lee McDonough, *Chattanooga—A Death Grip on the Confederacy* (Knoxville: University of Tennessee Press, 1984), 120–22.

**Here he made a serious error:** Peter Cozzens, *The Shipwreck of Their Hopes: The Battles for Chattanooga* (Urbana and Chicago: University of Illinois Press, 1994), 149–51, 208.

**In Baldy Smith's plan:** JHW, Diary, November 24, 1863, Delaware Historical Society; JHW, *Old Flag*, vol. 1, 295; Oliver Otis Howard, *Autobiography of Oliver Otis Howard, Major General United States Army,* 2 vols. (New York: Baker and Taylor, 1907) vol. 1, 483; Cozzens, *Shipwreck*, 18–50.

**When Sherman's troops renewed:** Schurz, *Reminiscences*, vol. 3, 75.

**Sherman's problems were visible:** Dana to EMS, November 26, 1863, *OR* I, 31, Pt. 2, 69.

**Grant was happy to have won:** USG to WTS, November 25, 1863, *OR* I, 31 ; USG, Report, December 23, 1863, *OR* I, 31, Pt. 2, 34; WTS, Report, December 19, 1863, *OR* I, 31, Pt. 2, 575.

**This view of the battle:** USG, *Memoirs*, vol. 2, 78.

**In his *Memoirs* Sherman:** WTS, *Memoirs*, vol. 1, 364.

**If this was the plan:** WTS, Report, December 19, 1863, *OR* I, 31, Pt. 2, 575.

**The days following the battle:** JHW, Diary, Delaware Historical Society, November 30 and December 4, 1863; Burnside to WTS, December 5, 1863, in S. H. M. Byers, ed., "Some War Letters," *North American Review* 143 (1886): 375.

**In this instance Sherman:** WTS to EES, December 8, 1863, SF MSS, ND.

**If in the end:** WTS to EES, November 8, 1863, SF MSS, ND.

The explanation may likely lie: WTS to EES, July 5, 1863, SF MSS, ND.

Sherman's letter to Grant: WTS to USG, July 4, 1863, WTS LB 23, NA.

"Even at West Point": Mahan to WTS, Aug. 28, 1863, WTS MSS, LC.

Nor is his performance: HBE, Diary, November 25, 1863, HBE MSS, OHS; JHW, *The Life of John A. Rawlins, Lawyer, Major General of Volunteers, and Secretary of War* (New York: Neale, 1916), 190.

Through Dana, who was: HBE, Diary, November 25, 1863, HBE MSS, OHS; JHW, *Rawlins*, 148; Dana to USG, January 21, 1864, Dana MSS, LC.

## 15. GOOD WAR, BAD WAR

In some ways: WTS to Charles W. Noble, August 20, 1864, MDM, Letters Sent, RG 393, Pt. 1, E 2499, NA; WTS to Willard Warner, January 16, 1866, Willard Warner MSS, ISHL.

In keeping with his legal interest: Emmerich de Vattel, *The Law of Nations, or Principles of the Laws of Nature as Applied to the Conduct and Affairs of Nations and Sovereigns*, 6th American ed. (Philadelphia: T. and J. W. Johnson, 1844); WTS to USG, July 20, 1862, WTS, LB 3, NA; WTS to JAR, August 28, 1862, *OR* I, 30, Pt. 3, 197.

Sherman's concern for doing: WTS to J. C. Pemberton, November 18, 1862, *OR* I, 17, Pt. 2, 872–73; WTS to G. Robertson and Richard A. Buckner, *Cincinnati Commercial*, October 17, 1861; WTS to Jesse Reed and W. B. Anderson, August 3, 1863, *OR* I, 24, Pt. 3, 571–72; WTS to Benjamin Grierson, November 8, 1862, WTS LB 4, NA; GO No. 50, September 16, 1863, *OR* I, 24, Pt. 3, 570–71.

This man whom several generations: WTS to James B. Bidwell, April 24, 1864, WTS LB 8, NA; WTS to TE, May 6, 1865, TEF MSS, LC; WTS to JS, April 5, 1865, WTS MSS, LC; WTS to EES, Mar. 23, 1865, SF MSS, ND.

If Sherman shunned "trophies": WTS to EES, July 28, 1861, SF MSS, ND; WTS to DDP, November 16, 1862, WTS LB 4, NA; WTS to HWH, November 17, 1862, *OR* I, 17, Pt. 2, 351.

Some argued that these: WTS to EMS, January 25, 1863, WTS LB 22, NA.

It was the great mass: Christopher Thomas Losson, "Jacob Dolson Cox: A Military Biography" (Ph.D. diss., University of Mississippi, 1993), 129; Milo Hascall to John M. Schofield, May 23, 1864, *OR* I, 38, Pt. 4, 297.

As for Sherman, his order books: *Military Orders*, 23–81, passim; WTS to HWH, September 22, 1862, WTS LB 3, NA; WTS to Frederick Steele, October 13, 1862, WTS, LB 3, NA.

He continued his efforts: WTS to EES, September 25, 1862, SF MSS, ND.

On July 24, 1863: General Court Martial Case File of Captain William B. Keeler, Lieutenant Henry Blanck and Private Ora Tebow, Company A, Thirty-fifth Iowa Infantry, RG 163, NA, passim.

Sherman had him arrested: Ibid.

The court deliberated and found: Ibid.; WTS to JAR, August 4, 1863, *OR* I, 24, Pt. 3, 540–42.

While his regiments bivouacked: GO No. 65, August 9, 1863, July 24, 1863, *Military Orders*, 182–88.

The army that settled: WTS to JS, August 3, 1863, WTS MSS, LC; ; WTS to PBE, July 28, 1863, Ewing, *Sherman*, 115.

It may well be: GO No. 2, October 23, 1863, *Military Orders*, 200–01.

There probably was a conscious decision: SFO NO. 17, June 4, 1864, *Military Orders*, 151–53.

Five months later: WTS, *Memoirs*, vol. 2, 174–76.

**By 1863 heavy-handed and relentless foraging:** Orders, No. 44, June 18, 1862, *Military Orders*, 43; WTS to Leslie Coombs, August 1, 1864, WTS LB 11, NA.

**From the legal standpoint:** HWH, *Elements of Military Art and Science, or Course of Instruction in Strategy, Fortification, Tactics of Battles, &c.; Embracing the Duties of Staff, Infantry, Cavalry, Artillery, and Engineers. Adapted to the Use of Volunteers and Militia* (Westport, Conn.: Greenwood, n.d.), 90.

**By then Sherman was:** WTS to Daniel Butterfield, May 30, 1863, WTS LB 22, NA; WTS to George Crook, November 6, 1863, *OR* I, 31, Pt. 3, 69–70.

**If the "eating out":** WTS to G. M. Dodge, November 9, 1863, *OR* I, 31, Pt. 3, 100.

**These partisans were a Southern contribution:** WTS to JBM, August 29, 1862, WTS, LB 3, NA; WTS to A. B. Wells, August 15, 1862, WTS LB 3, NA.

**Among the first units:** Andrew Brown, "The First Mississippi Partisan Rangers, C.S.A.," *Civil War History* 1 (1955): 375.

**Once launched, the partisan movement:** *A Digest of the Military and Naval Laws of the Confederate States from the Commencement of the Provisional Congress to the End of the First Congress Under the Permanent Constitution*, W. W. Lester and W. J. Brown, Eds. (Columbia, S.C.: Evans and Cogswell, 1864), 84.

**Ever sensitive about the Mississippi:** WTS to Charles Walcutt, September 24, 1862, WTS LB 3, NA; WTS to USG, September 25, 1862, *OR* I, 17 Pt. 1, 145.

**Grant approved:** JAR to WTS, October 29, 1862, *OR* I, 17 Pt. 2, 307; WTS to Valeria Hurlbut, September 25, 1862, WTS LB 4, NA.

**He assigned Colonel William Bissell:** WTS to Bissell, August 30, 1862, August 30, 1862, WTS LB 3, NA.

**While Bissell was in the neighborhood:** Ibid.

**The general didn't seem:** WTS to J. McArthur, August 13, 1864, *OR* I, 32, Pt. 5, 486; WTS to Salmon Chase, August 10, 1862, WTS LB 3, NA; WTS to H. H. Taylor, Aug. 25, 1862, WTS LB 22, NA; WTS to Horatio Wright, Sept. 20, 1862, WTS LB 3, NA.

**Their legal status was:** WTS to James Guthrie, April 6, 1864, WTS LB 3, NA.

**For a time he seems:** GO No. 50, August 1, 1863, *OR* I, 24, Pt. 3, 570; WTS to USG, August 30, 1863, *OR* I, 30, Pt. 3, 226; WTS to Joseph Holt, April 6, 1864, *OR* II, 7, 118–19.

**Though the general was:** WTS to USG, April 16, 1864, *OR* I, 32, Pt. 3, 382; S. A. Hurlbut to USG, July 25, 1863, *OR* I, 24, Pt. 3, 552; WTS to USG, Apr. 16, 1864, RG 393, Pt. 1, E2400.

**But Sherman's goal now:** WTS to Bissell, August 30, 1862, WTS LB 3, NA; WTS to PBE, July 13, 1862, in Ewing, *Sherman*, 59: WTS to H. E. Wright, September 6, 1862, WTS LB 3, NA; WTS, *Memoirs*, vol. 1, 359–60.

**If partisans pursued by Federal cavalry:** WTS to Editors of the *Memphis Bulletin and Appeal*, August 20, 1862, WTS LB 3, NA.

**Since in the general's view:** S. A. Hurlbut to USG, July 25, 1863, *OR* I, 24, Pt. 3, 552; GO No. 3, October 25, 1863, *Military Orders*, 201.

**In that summer and fall:** USG to WTS, August 6, 1863, *OR* I, 24, Pt. 3, 578; WTS to Ord, August 3, 1863, WTS MSS, USAMHI.

**Halleck was intrigued:** GO No. 50, August 1, 1863, *OR* I, 24, Pt. 3, 570–71; WTS to HWH, September 17, 1863, *OR* I, 30, Pt. 3, 694–700.

**The fourth class was:** WTS to HWH, September 17, 1863, *OR* I, 30, Pt. 3, 696.

**Halleck was apparently impressed:** HWH to WTS, October 1, 1863, in S. H. M. Byers, "Some War Letters," *North American Review* 143 (1886): 499.

**In the period between:** WTS to H. W. Hill, September 7, 1863, *OR* I, 30, Pt. 3, 402.

**While steadfastly refusing:** Ibid., 402–03.

**Here lay the chief reason:** WTS to J. B. Bingham, January 26, 1863, WTS LB 3, NA.

**Conquest, by its very nature:** WTS to JBM, November 18, 1863, *OR* I, 31, Pt. 3, 188; WTS to R. B. Buckland, January 27, 1864, WTS LB 7, NA; Benjamin P. Thomas and Harold Hyman, *Stanton: Life and Times of Lincoln's Secretary of War* (New York: Alfred A. Knopf, 1962), 410; William A. Russ Jr., "Administrative Activities of the Union Army during and After the Civil War," *Mississippi Law Journal* 17 (1945): 71–89.

**When Confederate cavalry:** WTS to A. G. Locklear, August 25, 1863, WTS LB 23, NA.

**In September he was:** WTS to Alexander Asboth, September 9, 1863, *OR* I, 30, Pt. 3, 475.

**As the year 1863:** WTS to Grenville Dodge, November 9, 1863, *OR* I, 31, Pt. 3, 100.

**He began to speak:** Ibid.

**At the end of 1863:** WTS to Daniel Butterfield, May 30, 1863, WTS LB 22, NA; WTS to USG, December 1, 1863, *OR* I, 31, Pt. 3, 297; WTS to George Crook, November 6, 1863, *OR* I, 31, Pt. 3, 69.

**In that summer of 1863:** WTS to TE, June 14, 1863, WTS MSS, OHS.

**January 1864 saw another:** WTS to R. M. Sawyer, January 31, 1864, *OR* I, 32, Pt. 2, 278–81; WTS to JS, April 11, 1864, WTS MSS, LC; JS to WTS, April 17, 1864, WTS MSS, LC; "Revolutionary Document: To the Soldiers of the Army of Tennessee," Rare Books and Manuscripts Department LC.

## 16. ATLANTA

**Though the year 1864:** WTS, *Memoirs*, vol. 1, 387–402.

**Though his campaign was marred:** Ibid., 394.

**Sherman considered the Valley:** WTS to JS, April 2, 1863, WTS MSS, LC.

**When Sherman and Grant:** USG to WTS, March 4, 1864, *OR* I, 32, Pt. 3, 49.

**Soon Sherman was summoned:** WTS, *Memoirs*, vol. 2, 15–16.

**Their plan contained no timetable:** USG, Report, July 22, 1865, *OR* I, 38, Pt. 1, 1.

**That April logistics was:** James A. Huston, "Logistical Support for the Federal Armies," *Civil War History* 7 (1961): 41–43.

**Those accidents could easily happen:** WTS to DDP, October 25, 1863, WTS LB 7, NA; R. M. Sawyer to J. S. Donaldson, March 28, 1864, RG 393, Pt. 1, E 2480, vol. 1; Montgomery Meigs to WTS, April 20, 1864, M745, NA; Langdon Easton to Amos Beckwith, November 5, 1864, RG 93, Pt. 1, E 2487, NA.

**Throughout the campaign Sherman:** OOH in Fletcher, *Life and Reminiscences*, 294; WTS to PBE, October 12, 1863, in Ewing, *Sherman*, 120.

**Through April and into May:** Meigs to WTS, April 20 and 26, 1864, M745, NA; Huston, "Logistical Support," 40; OOH to E. P. Smith, RG 393, Pt. 1, E 521, NA.

**He began issuing orders:** EMS to WTS, October 4, 1864, *OR* I, 39, Pt. 3, 63; WTS to Thomas Bramlette, May 6, 1864, WTS LB 13, NA; HH, *Marching with Sherman: Passages from the Letters and Diary of Henry Hitchcock, Major and Assistant Adjutant General of Volunteers*, ed. with an introduction by M. A. DeWolfe Howe (New Haven: Yale University Press, 1927), 35; WTS to James Yeatman, April 10, 1864, WTS LB 7, NA; WTS to Adams and Co. October 1, 1864, WTS LB 13, NA; J. D. Webster to WTS, September 24, 1864, RG 393, Pt. 1, E 2480, Pt. 1, NA; J. L. Donaldson to L. B. Brown, July 17, 1864, RG 393, Pt. 1, E 2521.

**It was May before the Federal Army:** WTS, *Memoirs*, vol. 2, 33.

**But McPherson never made it:** Ibid., 34; Andrew Hickenlooper, "Reminiscences," Hickenlooper MSS, Cincinnati Historical Society, 210; WTS, *Memoirs*, vol. 2, 34.

**Ten days after McPherson's failure:** Castel, *Decision*, 198–202.

The campaign was to be: *OR* I, 38, Pt. 1, 54–55; WTS to E. R. S. Canby, July 7, 1864 *OR* I, 38, Pt. 5, 84–85.

That spring and summer: HWH to WTS, July 16, 1864, *OR* 38, Pt. 5, 150–51.

In past campaigns: DDP, Journal, DDP MSS, LC, 438; WTS to EES, February 22, 1863, SF MSS, ND.

The terrain his army encountered: DDP, Journal, DDP MSS, 439–40.

Sherman had no intelligence officer: Correspondence, Reports, Appointments, and Other Records Relating to Individual Scouts, Guides, Spies, and Detectives, Records of the Provost Marshal's Bureau, RG 110, E 36, passim; Correspondence, Reports, Accounts, and Related Records of Two or More Scouts, Guides, Spies, and Detectives, RG 110, E 31, passim; Thomas G. Dyer, *Secret Yankees: The Union Circle in Confederate Atlanta* (Baltimore: Johns Hopkins University Press, 1999), 43–46.

Those who worked most closely: SFO No. 1, May 3, 1864, *Military Orders*, 239; Charles A. Dana to EMS, July 13, 1863, Dana MSS, LC.

Life in the general's official family: HH, *Marching*, 182.

It was apparently his custom: WTS to EES May 26 and July 5, 1863, SF MSS, ND.

At Sherman's headquarters: WTS to EES, April 10, 1863, SF MSS, ND. Matilda Gresham, *Life of Walter Quinton Gresham, 1832–1895,* 2 vols. (Chicago: Rand McNally, 1919), vol. 1, 315.

Beyond his staff Sherman's: WTS to Schofield, August 2 and 4, 1864, M504, NA; WTS to Schofield August 21, 1864, *OR* I, 38, Pt. 5, 623; WTS to GHT, *OR* I, 38, Pt. 4, 75; Henry Slocum in Fletcher, *Life and Reminiscences*, 346–47.

As Sherman's legions pushed: WTS to USG, July 12, 1864, *OR* I, 38, Pt. 5, 123; WTS to HWH, June 27, 1864, *OR* I, 38, Pt. 4, 607.

As the mountain loomed: WTS to HWH, June 13, 1864, *OR* I, 38, Pt. 4, 446; WTS to HWH, June 16, 1864, *OR* I, 38, Pt. 4, 492.

By June 24: SFO No. 28, June 24, 1864, *OR* I, 38, Pt. 4, 588; GHT to WTS, June 24, 1864, *OR* I, 38, Pt. 4, 503.

The Union attack came: GHT to WTS, June 27, 1864, *OR* I, 38, Pt. 4, 581.

There was another consideration: WTS to USG, July 12, 1864, *OR* I, 38, Pt. 5, 123; WTS to EES, June 30, 1864, SF MSS, ND.

In Richmond there was increasing dissatisfaction: Craig L. Symonds, *Joseph E. Johnston: A Civil War Biography* (New York: W. W. Norton, 1992), 319.

That was certainly not the case: Richard McMurry, *John Bell Hood and the Southern War for Independence* (Lexington: University Press of Kentucky, 1982), 23, 84.

In the Battle of Atlanta: WTS, *Memoirs*, vol. 2, 82.

Sherman's own role: Castel, *Decision*, 413.

Sherman's lack of involvement: WTS to GHT, July 28, 1864, *OR* I, 38, Pt. 5, 279 and note.

For Sherman the Battle of Atlanta: WTS to ____ Angus, June 8, 1882, WTS MSS, Boston Public Library; testimorny of Willard Warner in article in the *Daily Express*, Jan. 30, 1879, Scrapbook, WTS MSS, LC.

Sherman's choice was the commander: Warner, *Generals in Blue*, 237; Hooker to Henry Wilson, December 8, 1864; Hooker File, RG 94, E 159, NA; WTS to C. C. Washburn, Aug. 2, 1864, RG 393, Pt. 1, E 2902, NA .

After the Battle of Ezra Church: WTS to JS, July 31, 1864, WTS MSS, LC.

There was a great outcry: USG to HWH, July 15, 1864, *OR* I, 38, Pt. 5, 143–44; Michael Fellman, *Inside War: The Guerrilla Conflict in Missouri During the American Civil War* (New York and Oxford: Oxford University Press, 1989), 95; SO No. 198, October 20, 1863, *Military Orders*, 198–99.

**Whatever questions were raised:** WTS to JS, August 12, 1864, WTS MSS, LC.

**Before the year was over:** GBM to WTS, September 26, 1864, Sears, ed., *McClellan*, 604; SFO No. 22, June 10, 1864, *Military Orders*, 256.

**The taking of Atlanta:** Lee Kennett, *Marching Through Georgia* (New York: HarperCollins, 1994), 100.

**Sherman had already extended:** WTS to JBM, June 16, 1864, *OR* I, 39, Pt. 2, 123; WTS to C. C. Washburn, August 7, 1864, *OR* I, 39, Pt. 2, 233.

**He had exhausted his patience:** WTS to Joseph Holt, April 6, 1864, *OR* II, 7, 18–19; WTS to Commanding Officer, Marietta, Georgia, July 14, 1864, *OR* I, 39, Pt. 2, 142.

**"The use of torpedoes":** WTS to James B. Steedman, June 23, 1864, *OR* I, 38, Pt. 4, 579.

**That June Sherman also:** WTS to JS, April 11, 1864, WTS MSS, LC.

**The dispossession and deportation:** WTS to Stephen Burbridge, June 21, 1864, *OR* I, 39, Pt. 2, 135–36.

**This decree was to be:** WTS to Burbridge, June 21, 1964, *OR* I, 39, Pt. 2, 135.

**Sherman addressed a copy:** EMS to WTS, July 12, 1864 *OR* I, 39, Pt. 2, 157.

**In the end it appears:** Michael D. Hitt, *Charged with Treason: Ordeal of 400 Mill Workers During Military Operations in Roswell, Georgia in 1864–1865* (Monroe, N.Y.: Library Research Associates, 1992), passim.

**Strictly speaking Sherman had:** Ibid. WTS to AL, September 17, 1864, *OR* I, 39, Pt. 3, 395–96.

**It was not to be:** J. H. Parks, *Joseph E. Brown of Georgia* (Baton Rouge: Louisiana State University Press, 1977), 295–99. *Philadelphia Weekly Times*, December 6 and 13, 1879.

## 17. THE GREAT MARCH

**Initially the conqueror of Atlanta:** Horace Porter, *Campaigning with Grant* (New York: Century, 1897), 289–90.

**Now conversation was replaced:** Ibid.

**Grant and Sherman had given:** WTS to S. R. Curtis, April 9, 1864, WTS LB 7, NA; USG to HWH, July 15, 1864, *OR* I, 38, Pt. 5, 143; USG to WTS, September 12, 1864, *OR* I, 39, Pt. 2, 364–65; WTS to E. R. S. Canby, September 10, 1864, RG 393 Pt. 1, E 2502, NA; WTS to USG, September 10, 1864, *OR* I, 39, Pt. 2, 355.

**First, something had to be done:** WTS to EMS, September 20, 1864, RG 393, Pt. 1, E 2502.

**On its way back through Georgia:** WTS to EES, October 24, 1864, SF MSS, ND; WTS to Slocum, October 23, 1864, *OR* I, 39, Pt. 3, 406.

**During October heavy traffic:** GHT to USG, October 4, 1864, *OR* I, 39, Pt. 3, 78–79; Edward G. Longacre, *From Union Stars to Top Hat: A Biography of the Extraordinary James Harrison Wilson* (New Harrisburg, Pa.: Stackpole, 1972), 165.

**As for Lincoln:** USG to WTS, October 11, 1864, *OR* I, 39, Pt. 3, 202; EMS to USG, October 12, 1864, *OR* I, 39, Pt. 3, 39, 222; USG to WTS, October 13, 1864, *OR* I, 39, Pt. 3, 240.

**By October 28 Sherman:** WTS to EMS, October 18, 1864, RG 393, Pt. 1, 2502, NA.

**The withdrawal from North Georgia:** WTS to O. M. Poe, November 1, 1864, *OR* I, 39, Pt. 3, 577; WTS to Jacob Cox, October 7, 1864, *OR* I, 39, Pt. 3, 136–37; WTS to Slocum, November 7, 1864, *OR* I, 39, Pt. 3, 681; S. M. Budlong to T. T. Heath, October 30, 1864, *OR* I, 39, Pt. 3, 513.

**On Monday, November 14:** HH, *Marching*, 53.

**In Atlanta later that same day:** Orlando Poe, Diary, November 15, 1864, Orlando Poe MSS, LC; Joseph Wheeler to Hood, November 13, 1864, *OR* I, 38, Pt 3, 918.

**The expedition had been prepared:** WTS to _____ Tyler, November 8, 1864, *OR* I, 39, Pt. 3, 700.

The army's ordnance specialists: HH, *Marching*, 76.

Sherman had promised Governor Brown: *Guide to Civil War Materials in the National Archives* (Washington, D.C.: Government Printing Office, 1986), 564–65.

For both tactical and logistical reasons: WTS, *Memoirs*, vol. 2, 174–76.

Special Field Order, No. 120: Ibid.

Considering the patterns of foraging: Ibid., 174; HH, *Marching*, 69.

That Sherman was preparing: *Chicago Times*, November 9, 1864; Wheeler to Hood, October 28 and November 2, 1864, *OR* I, 39, Pt. 3, 859 and 878.

There was one imponderable: WTS to HWH, November 8, 1864, *OR* I, 39, Pt. 3, 697; HH to his wife, November 10, 1864, HH, *Marching*, 42; WTS to commanding officers, all posts, November 8, 1864, *OR* I, 39, Pt. 3, 700.

By contrast the march: Samuel Mahon, "The Foragers in Sherman's Last Campaigns," *War Sketches and Incidents* (MOLLUS, Iowa Commandery), vol. 2 (Des Moines: n.p. 1898), 193; Julian Hinkley, *A Narrative of Service with the Third Wisconsin Infantry* (Madison: Wisconsin Historical Commission, 1912), 193.

But Sherman had his Boswell: HH, *Marching*, 1–11.

After only forty-eight hours: Ibid., 30.

Hitchcock also offers: Ibid., 101, 112–14.

Somewhat to Hitchcock's disappointment: Ibid., 149.

Columbia, South Carolina, was not: Ibid., 269; Marion Brunson Lucas, *Sherman and the Burning of Columbia* (College Station and London: Texas A. and M. University, 1976), 129–62; Charles Royster, *The Destructive War: William Tecumseh Sherman, Stonewall Jackson and the Americans* (New York: Alfred A. Knopf, 1991), 3–33.

When the flames finally: WTS, Diary, February 18, 1865, SF MSS, ND.

Hitchcock was a shocked witness: HH, *Marching*, 23, 137, 140.

Hitchcock noticed other things: Ibid., 64.

Then much of the foraging: Samuel Merrill to Emily, December 15, 1864, Samuel Merrill MSS, INSL.

Much of what had been brought: Schurz, *Reminiscences*, vol. 3, 248

Here Sherman rarely intervened: WTS to Wheeler, February 8, 1865, *OR* I, 47, Pt. 2, 342; WTS to Kilpatrick, February 8, 1865, *OR* I, 47, Pt. 2, 351.

Here and there, then: HH, *Marching*, 62, 134

But the more Hitchcock: Ibid., 86, 130–31.

And Hitchcock himself was struck: Ibid., 60.

Sherman's passivity, his laissez-faire attitude: Ibid., 77, 83.

But this attitude of seeming indifference: Ibid., 124.

At first Hitchcock believed: Ibid., 35, 168.

But here again the general: HH, *Marching*, 83; *Sunny South* (Atlanta), November 30, 1901; WTS to Mrs. Caroline Carson, January 20, 1865, SCL; WTS, *Memoirs*, vol. 2, 284–86.

To do otherwise would be: WTS to Ethan Allen Hitchcock, January 25, 1863, WTS LB 22, NA; OOH to F. Blair, February 20, 1865, *OR* I, 47, Pt. 2, 505–06.

He left Atlanta with the plaudits: Edward D. Killoe to WTS, December 26, 1864, WTS MSS LC; Ethan Allen Hitchcock to WTS, December 26, 1864, WTS MSS, LC.

And his reaction was: WTS to HWH, January 12, 1865, *OR* I, Pt. 1, 36; WTS to USG, January 21, 1865, *OR* I, 47, Pt. 1, 103.

With his family: WTS to TE, December 31, 1864, M. A. DeWolfe Howe, ed., *Home Letters of General Sherman* (New York: Charles Scribner's Sons, 1909), 103; WTS to JS, December 31, 1864, and January 2, 1865, WTS MSS, LC.

**Since Ellen understood little:** WTS to EES, December 31, 1864, and January 9 and 21, 1865, SF MSS, ND.

**Sherman's tendency to enlarge:** SFO No. 25, Mar. 7, 1864, Special Field Orders, Army and Department of the Tennessee, 1864, RG 92, vol. 444.

**These actions provoked:** WTS to John Brough, September 15, 1864, RG 393, Pt. 1, E 2502.

**Governor Morton wrote a letter:** EMS to WTS, October 4, 1864, *OR* I, 39, Pt. 3, 63; WTS to EMS, October 20, 1864, *OR* I, 39, Pt. 3. 369.

**Then the Southern newspapers:** WTS to Mrs. A. A. Draper, January 15, 1865, *Atlanta Journal-Constitution*, November 23, 1958.

**Sherman briefly described:** WTS, *Memoirs*, vol. 2, 324–31; DDP, Journal, LC, vol 2, 420–21.

**Thereafter events came:** EMS to USG, March 3, 1865, *OR* I, 47, Pt. 3, 285.

**The news was some time:** Zebulon Vance to WTS, April 10, 1865, *OR* I, 47, Pt. 3, 178; HH, *Marching*, 302; Joseph E. Johnston to WTS, April 14, *OR* I, 47, Pt. 3, 206–7; WTS to HWH or USG, April 15, 1865, *OR* I, 47, Pt. 3, 221.

**Sherman rode out to meet:** WTS, *Memoirs*, vol. 2, 346–47; Joseph E. Johnston, *Narrative of Military Operations Directed During the Late War Between the States* (New York: D. Appleton, New York, 1874), 400–404.

**Once they were alone:** Sherman, *Memoirs*, vol. 2, 347–50; Johnston, *Narrative*, 347–50.

**What Sherman saw:** WTS to EES, April 13, 1865, SF MSS, ND.

**The Confederates were to agree:** Memorandum, or Basis of Agreement, April 18, 1865, in WTS, *Memoirs*, vol. 2, 356–57.

**At Appomattox Lee's army:** WTS to John W. Draper, November 6, 1868, John W. Draper MSS, LC.

**Copies of the agreement:** WTS, *Memoirs*, vol. 2, 353.

**His immediate subordinates:** Slocum, "Final Operations of Sherman's Army," *Battles and Leaders*, vol. 4, 755; HH, *Marching*, 304; Schurz, *Reminiscences*, vol. 3, 114.

**Sherman, now in a flurry:** WTS to HWH, April 18, 1865, *OR* I, 47, Pt. 3, 245; WTS to Easton, April 18, 1865, *OR* I, 47, Pt. 3, 246; SFO No. 58, April 19, 1865, *OR* I, 47, Pt. 3, 250.

**It was Sherman's letters:** WTS to EES, April 18 and 22, 1865, SF MSS, ND.

**He was anxiously awaiting:** HH, *Marching*, 308.

## 18. THE FIRST YEARS OF PEACE

**Sherman was understandably concerned:** WTS to EMS, April 25, 1865, *OR* I, 47, Pt. 3, 301; Sam Milligan to Andrew Johnson, Apr. 29, 1865, in Leroy P. Graf, ed., *The Papers of Andrew Johnson*, 16 vols. (Nashville: University of Tennessee Press, 1967–2000), vol. 7, 664–65.

**But Sherman was isolated:** Sam Milligan to A. Johnson, April 29, 1865, *A.J. Papers, vol.7, 664–65;* George W. Childs to WTS, May 27, 1865, WTS MSS, LC, 1865, Johnson MSS, LC, *OR* I, 47, Pt. 3, 301.

**Not until April 28:** *New York Times*, April 24, 1865; EMS to John Dix, April 22, 1865, *OR* I, 47, Pt. 3, 285–86; HWH to EMS, April 26, 1865, *OR* I, 47, Pt. 3, 311.

**The general reacted with rage:** Schurz, *Reminiscences*, vol. 3, 116–17.

**It was a still-seething Sherman:** HWH to WTS, May 8 and 9, 1865, *OR* I, 47, Pt. 3, 435, 454; WTS to HWH, May 10, 1865, *OR* I, 47, Pt. 3, 454–55.

**Nor was that all:** WTS, Report, May 9, 1865, *OR* I, 47, Pt. 1, 30–40 passim.

**The report was a gesture:** WTS to USG, May 10, 1865, WTS MSS, LC.

**But in the next few days:** USG to WTS, May 6, 1865, WTS MSS, LC; JS to WTS, May 16, 1865, WTS MSS, LC; Henry Stanbery to WTS, May 25, 1865, WTS MSS, LC;

Thomas and Hyman, *Stanton*, 412; WTS to JAR, May 22, 1865, MS HM 23285, Huntington Library.

**By the time he reached Washington:** "Testimonial to Maj. Gen. William Tecumseh Sherman. Meeting of the Citizens of Ohio, in the Hall of the House of Representatives at Columbus, January 11, 1865," broadside in Special Colls., OHS Library; WTS to JS, July 23, 1863, WTS MSS, LC.

**On May 22 Sherman:** "Testimony of Major General William T. Sherman, Washington, May 22, 1865," *Report of the Joint Committee on the Conduct of the War*, 38th Congress, 2nd Session (Washington, D.C.: GPO, 1865), vol. 3, 532; WTS to John W. Draper, November 6, 1868, John W. Draper MSS, LC.

**He could not yet:** WTS to T. S. Bowers, May 24, 1865, WTS LB 18, NA; WTS to USG, May 28, 1865, WTS MSS, LC.

**He now felt sufficiently emboldened:** WTS to EES, May 23, 1865, SF MSS, ND.

**By May 16 he:** WTS to EES, May 16, 1865, M504, NA; WTS to JAR, May 19, 1865, *OR* I, 47, Pt. 3, 531.

**He had known that:** Untitled, undated newspaper clipping, WTS Scrapbook, OHS.

**Sherman's headquarters would be:** "Resolution of the New York City Board of Aldermen, May 29, 1865," WTS MSS, LC; *New York Times*, June 5, 1865.

**The general slipped away:** *New York Times*, June 5, 1865.

**The Shermans' trip ended:** WTS, Diary, June 24, 1865, SF MSS, ND; WTS to EES, March 31, 1865, SF MSS, ND; "Solicitations for the Sherman Fund," undated, WTS MSS, LC; WTS to PBE, January 29, 1865, Ewing, *Sherman*, 155; Charles Anderson to WTS, November 16, 1865, WTS MSS, LC.

**But St. Louis:** HST John I. Rowes et al. to WTS, August 19, 1865, WTS MSS, LC; EES to TE, September 7, 1865, SF MSS, ND.

**Sherman's home life:** EES to TE, February 7 and 23, 1867, and October 17, 1868, SF MSS, ND.

**Ellen carefully watched over:** EES to TE, December 14, 1868, SF MSS, ND.

**But Sherman had little patience:** EES to TE, March 5, 1868, SF MSS, ND.

**Sherman's most frequent complaint:** WTS to Montgomery Meigs, March 12, 1866, RG 92, E 225, NA; EES to TE, January 11, 1869, SF MSS, ND.

**Sherman doted on the children:** EES to TE, February 23, 1867, December 5, 1868, and January 16, 1869, SF MSS, ND.

**In the house on Garrison Avenue:** Martin Hardwick Hall, "The Campbell-Sherman Diplomatic Mission to Mexico," *Bulletin of the Historical and Philosophical Society of Ohio* 13 (1995): 258–59 and passim; WTS to JS, October 31, 1865, WTS MSS, LC.

**This Western command:** Robert G. Athern, *William Tecumseh Sherman and the Settlement of the West* (Norman: University of Oklahoma Press, 1956), xvi.

**To begin with, the Military Division:** USG to WTS, March 14, 1866, WTS MSS, LC.

**When he was at home:** EES to TE, September 25, 1868, SF MSS, ND.

**Though he came to know:** Athern, *Sherman*, 36, 93; WTS to JS, June 17, 1868, WTS MSS, LC; PTS, "Reminiscences," 10, SF MSS, ND.

**Sherman's command embraced:** Weigley, *Army*, 267; Athern, *Sherman*, 104, 149.

**The fighting was only part:** Athern, *Sherman*, xv–xvi.

**Frustrating as Sherman's contacts:** Sherry L. Smith, *The View from Officers' Row: Army Perceptions of Western Indians* (Tucson: University of Northern Arizona Press, 1990), 2–10; Thomas C. Leonard, "Red, White and Army Blue: Empathy and Anger in the American West," in Peter Karsten, ed., *The Military in America: From the Colonial Period to the Present*, new revised ed. (New York: Free Press, 1986), 226–38.

**Some of the generals:** Dudley Acker Jr., "Nantan Lupan: George Crook on America's Frontiers" (Ph.D. diss., Northern Arizona University, 1995), 458–62; Athern, *Sherman*, 459.

**Lieutenant Sherman did not know:** Smith, *Officers' Row*, 47–48; Paul Andrew Hutton II, "Philip H. Sheridan and the Army in the West" (Ph.D. diss., Indiana University, 1980), 201.

**As for Sherman, his notions:** Clipping from *Montreal Daily Witness*, September 20, 1882, WTS Scrapbook, LC.

**In the case of the Western Indians:** WTS to JS, June, 1868, WTS MSS, LC; WTS to EES, July 10, 1866, SF MSS, ND.

**But what if racial distinctions:** Athern, *Sherman*, 205; WTS to JS, June 11, 1868, WTS MSS, LC.

**As for solving the Indian problem:** *New York Times*, May 31, 1867; WTS to JS, August 3, 1867, WTS LB 21, NA.

**The railroad also aided:** WTS to JAR, October 23, 1865, MDM, Letters Sent, 1865–66, NA; WTS to Sheridan, April 9, 1869, Sheridan MSS, LC; *Athern*, Sherman, 196.

**Sherman saw for the Indians:** WTS to JS, July 30 and September 23, 1868, WTS MSS, LC; Hutton, "Sheridan," 208.

**It should be said:** Athern, *Sherman*, 82.

**In the summer of 1868:** Ibid., 203–04; Samuel F. Tappan to WTS, July 8, 1868, WTS MSS, LC; Daniel Haas Moore, "Chiefs, Agents, and Soldiers: Conflict on the Navajo Frontier, 1866–1880" (Master's thesis, Northern Arizona University, 1988), 40–65.

**In the 1870s, as commanding general:** *Montreal Daily Witness*, July 20, 1882.

**Sherman sent clear signals:** WTS to Andrew Johnson, February 11, 1866, WTS MSS, LC; WTS to William Church, July 18, 1874, LB, WTS MSS, LC.

**Sherman was clearly not angling:** WTS to Andrew Johnson, February 1 and 11, 1866, LB, WTS MSS, LC.

**President Johnson kept him:** WTS to Andrew Johnson, January 27, 1868, LB, WTS MSS, LC.

**He briefly considered turning down:** Hamilton Fish et al. to WTS, March 3, 1869 (facsimile), WTS MSS, MoHS; WTS to JS, LB, WTS MSS, LC.

## 19. COMMANDING GENERAL

**Then too, he found himself:** Robert F. Stohlman Jr., *The Powerless Position: The Commanding General of the Army of the United States, 1864–1903* (Manhattan, Kansas: Military Affairs, 1975), 1–9; Schofield to WTS, December 20, 1878, WTS MSS, LC.

**There was a long-standing line-staff controversy:** Matloff et al., *American Military History*, 291–92; Robert M. Utley, *Frontier Regulars: The United States Army and the Indian, 1866–1890* (New York: Macmillan, 1973), 30–32.

**Staff officers were better paid:** Richard Andrew Allen, "Years of Frustration: William T. Sherman, the Army and Reform, 1869–1883" (Ph.D., Northwestern University, 1968) 31.

**At high command level:** Weigley, *Army*, 246–50.

**As line officers:** USG to EMS, January 29, 1866, *Grant Papers*, vol. 16, 37; USG to WTS, September 18, 1867, Ibid., 216.

**Now, with Grant moving:** Stohlman, *Powerless Position*, 52–57; WTS to Ord, May 22, 1874, WTS MSS, MoHS; JHW, *Rawlins*, 355–56.

**With the new system:** JHW, *Rawlins*, 352.

**Sherman knew Rawlins exercised:** WTS to USG, March 26, 1869, LB, WTS MSS, LC; WTS to Henry Benham, November 17, 1873, Benham-McNeil MSS, LC; WTS to John Pope, April 24, 1876, LB, WTS MSS, LC.

It may be that Sherman: Schofield, *Forty-six Years*, 421; JHW, *Rawlins*, 336; Allen, "Years of Frustration," 13.

Rawlins's tenure at the War Department: Hutton, "Sheridan," 469; W. W. Belknap to WTS, August 8, 1865, WTS MSS, LC.

Relations between Belknap and Sherman: Stohlman, *Powerless Position*, 57–64; WTS to C. C. Augur, C. C. Augur MSS, ISHL; L. D. Ingersoll, *A History of the War Department of the United States with Biographical Sketches of the Secretaries* (Washington, D.C.: Francis Mohun, 1879), 569.

Now not only was Sherman: WTS to JS, March 21, 1870, LB, WTS MSS, LC.

Logan's bill passed the House: Allen, "Years of Frustration," 78–80.

Sherman was feeling: WTS to Belknap, August 17, 1870, LB, WTS MSS, LC; WTS to C. C. Augur, March 18, 1871, Augur MSS, ISHL; WTS to JS, July 27, 1870, WTS MSS, LC; WTS to TE, July 8, 1871, TE MSS, LC.

Their itinerary took them: Joseph Audenreid, "Notes of Travel in Europe: General Sherman in Europe and the East," WTS MSS, OHS, passim; J. C. Audenreid, "General Sherman in Europe and the East," *Harper's* 47 (1873): 225–42, 481–95, 652–71; Sheridan to WTS, August __, 1870, Sheridan MSS, LC.

Little was changed in Washington: Stohlman, *Powerless Position*, 64; WTS to S. A. Hurlbut, May 29, 1874, LB, WTS MSS, LC; WTS to Henry A. DuPont, November 20, 1873, Henry A. DuPont MSS, Hagley-Eleutherian Mills Foundation Library; WTS to William E. Church, July 18, 1874, LB, WTS MSS, LC.

In 1874 Sherman decided: WTS to C. W. Moulton, March 9, 1876, LB, WTS MSS, LC; Stohlman, *Powerless Position*, 66.

Actually he did not: WTS to John Pope, April 24, 1876., LB, WTS MSS, LC; Allen, "Years of Frustration," 120; WTS to Sheridan, LB, WTS MSS, LC; WTS to HST, January 11, 1877, WTS MSS, OHS; WTS to Hon. D. H. Armstrong, March 31, 1879, LB, WTS MSS, LC.

His correspondence for the seventies: WTS to C. C. Augur, March 18, 1871, C. C. Augur Papers, ISHL; Allen, "Years of Frustration," 221.

Sherman did have some allies: Donald Nevius Bigelow, *William Conant Church and the Army and Navy Journal* (New York: Columbia University Press, 1952), 203.

Sherman's removal to St. Louis: Allen, "Years of Frustration," 111–12, WTS to C. W. Moulton, March 9, 1876, LB, WTS MSS, LC.

But the problem of Belknap: Stohlman, *Powerless Position*, 67; WTS to C. C. Augur, April 21, 1876, ISHL; WTS to JS, November 17, 1875, WTS MSS, LC.

Sherman gave up his: WTS to Ord, June 25, 1876, LB, WTS MSSS, LC.

Belknap was succeeded by: Stohlman, *Powerless Position*, 68–69, Weigley, *Army*, 287; Coffman, *Old Army*, 254.

The years of Sherman's command: WTS to EES, April 22, 1865; Peter Karsten, "Armed Progressives: The Military Reorganizes for the American Century," in Karsten, ed., *The Military in America*, 266.

Though the "revolution" Sherman: WTS to C. C. Augur, March 12, 1871 Augur MSS, ISHL ; Jerry M. Cooper, "The Army and Civil Disorder, 1877–1900," *Military Affairs* 33 (1969): 255.

To Sherman it seemed: WTS to T. T. Gantt, May 6, 1878, LB, WTS MSS, LC.

The academy was still: WTS to Hon. Heister Clymer, February 15, 1878, LB, WTS MSS, LC; WTS to Ord, May 14, 1882, WTS MSS, MoHS.

Sherman's first reaction: Allen, "Years of Frustration," 248–49; WTS to Ord, May 14, 1882, WTS MSS, MoHS; "Proposal of Captain E. S. Godfrey to Introduce New Exercises and Gymnastics in the Teaching of Cadets Horsemanship, December 8, 1881," with notation of Sherman, WTS MSS, USMA.

**Sherman opposed racial integration:** John A. Carpenter, *Sword and Olive Branch: A Biography of O. O. Howard* (Pittsburgh: University of Pittsburgh Press, 1964) 272; Schofield, *Forty-six Years*, 445-46; WTS to John T. Doyle, May 7, 1880, WTS MSS, CalHS; Halstead, *Recollections*, 684.

**He was nonetheless the agent:** WTS to ___ Thomas, October 12, 1869, WTS MSS, USMA; Robert Allen Wooster, "The Military and United States Indian Policy" (Ph.D. diss., University of Texas, 1985), 72; WTS to Alexander A. Rice, November 22, 1878, LB, WTS MSS, LC.

**To Sherman's way of thinking:** WTS to George Hoyt, 1887, in *Pointer View* (USMA, West Point), April 8, 1977; Timothy Nenninger, *The Leavenworth Schools and the Old Army: Education, Professionalism, and the Officer Corps of the United States Army, 1881-1918* (Westport, Conn.: Greenwood, 1978), passim; Weigley, *Army*, 273-74.

**In such practical matters:** WTS to Sheridan, February 20, 1873, LB, WTS MSS, LC; WTS to C. T. Christiansen, n.d., WTS MSS, USMA; William Ralph Crites, "The Development of Infantry Tactical Doctrine" (M.A. thesis, Duke University, 1968), 52-53; Sidney B. Brinkerhoff and Pierce Chamberlain, "The Army's Search for a Repeating Rifle," *Military Affairs*, 32 (1968): 15-23; Sheridan to WTS, April 5, 1871, WTS MSS, LC.

**Then for a decade:** Peter S. Michie, *The Life and Letters of Emory Upton, Colonel of the Fourth Artillery and Brevet Major-General, U.S. Army*, with an introduction by JHW (New York: D. Appleton, 1885), 197-204; Mark R. Grandstaff, "'Preserving the Habits and Usages of War': William Tecumseh Sherman, Professional Reform, and the U.S. Army Officer Corps, 1865-1881, Revisited," *Journal of Military Affairs*, 62 (1998): 521-45.

**As early as 1873:** WTS to Upton, June 20, 1873, M857, NA; Hutton, "Sheridan," 249; WTS to Sheridan, November 22, 1881, Sheridan MSS, LC.

**Sherman did not believe:** WTS to George Bancroft, July 11, 1872, George Bancroft MSS, MassHS.

**Sherman's solution:** WTS to John Gibbon, April 3, 1877, John Gibbon MSS, Maryland Historical Society.

**In 1863 he had reestablished:** Bowman to WTS, March 30, 1865, WTS MSS, LC; Charles B. Richardson to WTS, November 21, 1865, WTS MSS, LC; Bowman and Irwin, *Sherman and His Campaigns*, 479-80; Young, "General Sherman on His Own Record," 493; TES, Diary, January 24 and February 18, 1872, SF MSS, ND.

**Though Sherman often misjudged:** Weigley, *Army*, 273.

**Sherman awaited the reaction:** *New York Times*, August 24, 1902; USG to WTS, January 29, 1876, LB, WTS MSS.

**The nation's newspapers had:** WTS to George Bancroft, January 23, 1875, Bancroft MSS, MassHS.

**At the outset:** WTS, *Memoirs*, vol. 1, 394; WTS to William Sooey Smith, July 11, 1875, M857, NA; WTS, *Memoirs*, vol. 1, 292; Garfield, *Diary*, August 25, 1875, vol. 3, 134.

**Friends warned Sherman:** James Hardie to WTS, January 14, 1876, WTS MSS, LC; *Philadelphia Evening Star*, Apr. 9, 1878.

**In his description of the fighting:** WTS, *Memoirs*, vol. 1, 362.

**While Sherman's proper task:** *St. Louis Globe-Democrat*, Oct. 11, 1875.

**Moreover Boynton was well-connected:** *Washington Post*, February 19, 1880; WTS to JS, December 6, 1875, WTS MSS, LC.

**Then, to compound the mischief:** *Sherman's Historical Raid. The Memoirs in Light of the Record* (Cincinnati: Wilstach, Baldwin, 1875), 8, passim.

**The impact of Boynton's charges:** WTS to JS, December 29, 1875, WTS MSS, LC.

**Bowman wanted to stand firm:** Bowman to WTS, January 3, 1876, WTS MSS, LC; WTS to D. Appleton and Co., November 30, 1875, WTS MSS, LC.

**Early in 1876 Moulton:** *The Review of General Sherman's Memoirs Examined, Chiefly in Light of Its Own Evidence* (Cincinnati: Robert Clarke, 1875); *Baltimore Gazette*, January 4, 1876.

## 20. THE BANQUET YEARS

**In his last years:** Murat Halstead to WTS, March 8, 1879, Miscellaneous Letters File, WTS MSS, OHS.

**Though still possessing remarkable energy:** PTS, "Reminiscences of Early Days," SF MSS, ND, 2.

**In this setting:** WTS to James C. Welling, November 7, 1878, M857, NA.

**As Sherman was mellowing:** WTS to John T. Doyle, January 2, 1890, CalHS; newspaper clipping dated May 18, 1871 in TES Scrapbook, WTS MSS, OHS.

**But Ellen gradually dropped:** PTS, "Reminiscences," SF MSS, ND, 6, 13, and 25; EES to TE, January 1, 1867, SF MSS, ND.

**The family group gathered:** PTS, "Reminiscences," SF MSS, ND, 3.

**In the 1870s Elly:** TES, Diary, 1872, SF MSS, ND, passim.

**Tom's delight was to work:** WTS to Henry Benham, February 4, 1874, Benham-McNeil Family MSS, LC; WTS to Felicia Shover, June 30, 1875, Thornton Family MSS, VHS.

**Things went as Sherman planned:** TES to WTS, May 20, 1878, SF MSS, ND.

**Coming out of the blue:** WTS to JS, May 29, 1878, WTS MSS, LC; WTS to HST, May 27, 1878, WTS MSS, OHS.

**Then the general had been offered:** WTS to J. C. Lynes, June 15, 1878, WTS MSS, LC; WTS to W. S. Hancock, August 8, 1885, WTS MSS, LC.

**Though Sherman lived long enough:** WTS to John T. Doyle, June 15, 1885, CalHS; Chauncy Depew, *My Memories of Eighty Years* (New York: Charles Scribner's Sons, 1899), 380–81.

**Nowhere in the surviving texts:** Undated newspaper clipping, WTS Scrapbook, WTS MSS, OHS.

**The general would on occasion:** WTS to John T. Doyle, June 15, 1885, CalHS.

**Of all the fraternal groups:** *Report of the Proceedings of the Society of the Army of the Tennessee, Fourth Meeting*, 1869, 344–45.

**The society's meetings took place:** *Report of the Proceedings of the Society of the Army of the Tennessee, Twelfth Meeting*, 1878, 113; *Report of the Proceedings of the Society of the Army of the Tennessee, Fifteenth Meeting*, 1882, 329.

**In 1879 Sherman communed:** WTS to George W. McCrary, February 5, 1879. LB, WTS MSS, LC.

**Despite the hollowness:** PTS, "Reminiscences," SF MSS, ND, 14, 25, 28–29; Garfield, *Diary*, December 16, 1873, vol. 2, 238.

**There was now less contact:** WTS to CE, January 4 and February 14, 1866, CE MSS, LC.

**Perhaps one night:** Kathryn Allamong Jacob, *Capital Elites: High Society in Washington, D.C. After the Civil War* (Washington, D.C., and London: Smithsonian Institution Press, 1995), 37–42, 66–78.

**As the nation's second:** DDP to WTS, August 11, 1867, WTS MSS, LC.

**Clark Carr, who encountered:** Clarke E. Carr, *My Day and Generation* (Chicago: A. C. McClurg, 1908), 125.

Others who moved: Garfield, *Diary*, November 9, 1872, vol. 2, 113; Jacob, *Capital Elites*, 126.

The rich tapestry: Sotheby-Parke-Benet, Inc., Catalog 3982.

In the case of sculptress: Vinnie Ream Hoxie MSS, LC, passim; Michael Fellman, *Citizen Sherman: A Life of William Tecumseh Sherman* (New York: Random House, 1995), 355–70.

The general's interest: Marszalek, *Sherman*, 416–17; "William Tecumseh Sherman," *Illustrated American* (March 7, 1891), 119; New York *World*, February 22, 1891; WTS to —— Angus, June 8, 1882, Boston Public Library, Rare Books and Manuscripts Department, courtesy of the Trustees.

By the mid-eighties: St. Louis newspaper clipping, September 20, 1884, WTS Scrapbook, WTS MSS, LC.

The year 1884 brought: TES, "Notes for Mrs. Fitch on General Sherman," November 17, 1892, SF MSS, ND.

He thus remained firm: WTS to David Stuart, August 1, 1863, LB 23, WTS MSS, LC; WTS to JS, January 15, 1869, WTS MSS, LC.

For most people retirement: WTS to John T. Doyle, August 8, 1886, CalHS.

The blow was telling: New York *World*, February 8, 1890.

But the ranks were thinning: Remarks of Mrs. John A. Logan, *Report of the Proceedings of the Society of the Army of the Tennessee, Thirty-fourth Meeting*, 1903, 112.

"Sherman was never in a happier mood": Edward S. Ellis, "Reminiscences of General Sherman," *Chautauquan* 27 (1898): 474–75.

Such lapses must have been: WTS to JS, February 3, 1891, in JS, *Recollections*, vol. 2, 1102.

The following evening he: PTS, Diary, 1891; Lizzie Sherman, Diary, February 5–11, 1891, both in SF MSS, ND; Account of Dr. C. T. Alexander in Fletcher, *Life and Reminiscences*, 452–53; Interview with Walter Pharr, MD, October 26, 1999.

## 21. TO POSTERITY

The funeral service: *National Tribune*, February 26, 1891.

While his progeny gradually disappeared: James L. Ehrenberger and Francis G. Gschwind, *Sherman Hill* (Callaway, Neb.: E and G, 1973), 11–24; John C. and Winona C. James, *The Prairie Pioneers of Western Kansas and Eastern Colorado* (Boulder, Col.: Johnson, n.d.), 13.

The general was further immortalized: Kathryn Allamong Jacob, *Testament to Union: Civil War Monuments in Washington, D.C.* (Baltimore and London: Johns Hopkins University Press, 1998), 91–99.

But the general's death: *National Tribune*, April 16, 1891; John C. Ropes, "General Sherman," *Atlantic Monthly* 68 (1891): 191–204.

In thirteen pages Ropes: Ropes, "Sherman," 192, 194.

To Ropes George H. Thomas: Ibid., 192, 194, 202.

With this article Ropes: *Buffalo Enquirer*, August 14, 1891; *Dubuque Telegraph*, August 3, 1891; *New York Advertiser*, August 7, 1891; *Portland Oregonian*, August 3, 1891.

Ropes's article was particularly well-received: *Richmond Times-Dispatch*, July 29, 1891; *Charleston News and Courier*, August 13, 1891; *Atlanta Constitution*, August 29, 1891.

The other place where Ropes's article: Henry M. Cist to John C. Ropes, July 29, 1891. Ropes Coll., BU.

Then Cist turned his attention: Ibid.

In the popular mind: *New York Times*, June 3, 1934.

A stronger memory: *Confederate Veteran* 8 (1906): 300; *New York Times*, July 15, 1911.

**"But the Civil War aficionados":** Joel Chandler Harris, *Life of Henry W. Grady, Including the Writings and Speeches* (New York: Cassell Publishing Company, 1890), 87.

**In 1937 the Postal Service:** "The Stamp Issues," *Bulletin of the Society for Correct Civil War Information*, No. 19 (April 1937), 49–53.

**At the same time:** John F. Marszalek, "Philatelic Pugilists," *Cump and Company: A Newsletter for Friends and Fanciers of General William T. Sherman* 6, no. 2 (October 1999): 28–35.

**The thirties also saw:** Lewis, *Sherman.*

**But even more important:** Drew Gilpin Foust, "Clutching the Chains That Bind: Margaret Mitchell and *Gone with the Wind*," *Southern Cultures* 5, no. 5 (1999): 6.

**Were all these enthusiasts:** Doris A. Walker, "Message from Headquarters," *Cump and Company* 4, no. 2 (September/October 1997), 1–2.

**In the past few years:** Telephone interview with David Hatton, Springfield, Ill., February 5, 2000.

**This was the case:** Carol Reardon, *Soldiers and Scholars: The United States Army and the Uses of Military History, 1865–1890* (Lawrence: University Press of Kansas, 1990), 63.

**Sherman had frequently used:** Ibid., 96–97.

**Just what was his peculiar genius:** Dennis Hart Mahan, "Cadet Life of Grant and Sherman," newspaper clipping, March 1, 1866, WTS Scrapbook, OHS.

**Had caution alone governed:** L. B. Parsons to HWH, December 20, 1862, *OR* I, 17, Pt. 2, 441.

**Had the general been called on:** Mahan, "Cadet Life."

**"The U.S. Army's Current":** Russell F. Weigley, *The American Way of War: A History of United States Military Strategy and Policy* (New York: Macmillan Publishing Co., 1973), 149

**In the aftermath:** *New York Times,* January 6, 1925.

**In that same year:** *Paris, or the Future of War* (New York: E. P. Dutton, 1925).

**This new path to victory:** B. H. Liddell Hart, *Sherman: Soldier, Realist, American,* republication of the New York edition of 1929, with a new introduction by Jay Luvaas (New York: Da Capo, 1993), preface, viii.

**Writing to Philemon Ewing:** WTS to PBE, January 20, 1865, PBE MSS, OHS; O. O. Howard in Fletcher, *Life and Reminiscences,* 307.

**There is no doubt:** WTS to EMS, January 19, 1865, *OR* I, 47, Pt. 2, 87.

**It's too bad that:** Gary Gallagher, *The Confederate War* (Cambridge and London: Harvard University Press, 1998), passim.

**If one may thus debate:** E. Merton Coulter, "Sherman and the South," *North Carolina Historical Review* 8 (January 1931), 42.

**Then when the *Dictionary*:** Oliver Lyman Spaulding to Dumas Malone, October 16 and 27, 1931, Dumas Malone to Oliver Lyman Spaulding, October 24 and 29, 1931, *Dictionary of American Biography* File, Learned Societies MSS, LC.

**Of course it was:** *New York Times,* June 8, 1915; James Reston Jr., *Sherman's March and Vietnam* (New York: Macmillan, 1984).

**In the specific case:** Instructions on the government of Armies of the United States, GO No. 100, April 24, 1863, Section 1, Article 22, *OR* III, 3, 151.

**Sherman's record in the Civil War:** J. M. Spaight, *War Rights on Land* (Macmillan, 1911), 110.

**There is, moreover, evidence:** Correspondence of the Record and Pension Office, File R&P 688464, RG 94, E 501, NA.

**Finally, America's most:** *New York Times,* January 6, 1925; B. H. Liddell Hart, *The Liddell Hart Memoirs,* 2 vols. (New York: G. P. Putnam's Sons, 1965), vol. 1, 169–70; Norman H. Schwarzkopf with Peter Petrie, *It Doesn't Take a Hero: General Norman H. Schwarzkopf, the Autobiography* (New York: Bantam, 1992), 430.

# Bibliography

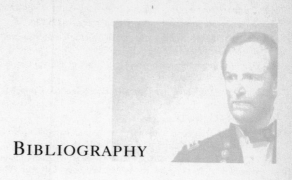

## Manuscript Sources

### DOCUMENTS
#### State and Local

Minutes of the Meetings of the Mayor and Board of Aldermen, 1862, Memphis–Shelby County Library and Information Center.

Minutes of Meetings of the Atlanta City Council, 1864, Atlanta History Center, Atlanta.

Governors' Papers: J. Neely Johnson, California State Archives, Sacramento.

Governor Oliver P. Morton Telegraph Books, Indiana State Library. Indianapolis.

Letter Books of Governor Joseph E. Brown, Georgia Department of Archives and History, Atlanta.

Minutes of Board of Governor's Meetings, 1865–1902, Louisiana State University.

#### Federal
*National Archives Microfilm Series:*

M 182 Letter Books of the Military Governor of California

M 210 Records of the Tenth Military District

M 345 Union Provost Marshal's File of Papers Relating to Individual Civilians

M 416 Union Provost Marshal's Files of Papers Relating to Two or More Civilians

M 504 Unbound Telegrams Collected by the Office of the Secretary of War

M 565 Adjutant General's Office, Letters Sent, 1800–1890

M 567 Adjutant General's Office, Letters Received, 1822–1860

M 619 Adjutant General's Office, Letters Received, 1861–1890

M 621 Reports and Decisions of the Provost Marshal General, 1863–1866

M 668 Cadet Application Papers, USMA, 1805–1866

M 682 Letters Received by the Secretary of War

M 707 Monthly Regimental Returns, 1821–1901

M 857 Letters Sent by the Headquarters of the Army, 1849–1866 and 1873–1903

M 997 Annual Reports of the War Department, 1822–1907

M 1395 Letters Received by the Appointment, Commission and Personnel Board of the Adjutant General's Office, 1871–1904

RG 37, E 36 Telegrams Sent by the Field Office of the Military Telegraph and Collected and Collected by the Secretary of War

RG 92, E 225 Office of the Quartermaster General, Consolidated Correspondence File, 1794–1915

RG 92, E 843 Claims Damages

RG 92, E 1671 Telegrams Sent by A. Anderson, General Superintendent of the U.S. Military Railroads, MDM, February–November 1864

RG 92, E 1672 Letters Received and Forwarded by A. Anderson, General Superintendent of the U.S. Military Railroads, MDM, February–October 1864

RG 92, E 1673 Letters and Orders Received and Forwarded by A. Anderson, General Superintendent of the U.S. Military Railroads, MDM, February–October 1864

RG 92, E 1674 Letters Received by A. Anderson, General Superintendent of the U.S. Military Railroads, MDM, February-October 1864

RG 92, E 1675 Letters Received from the Quartermaster General by A. Anderson, General Superintendent of the U.S. Military Railroads, MDM, June–October 1864

RG 94, E 159 Generals' Papers and Books:
Manning F. Force
Henry Halleck
Joseph Hooker
William T. Sherman

RG 94, E 206 Letters Sent Relating to the Military Academy, 1818–1867

RG 94, E212 Correspondence Relating to the Military Academy, 1819–1866

RG 94, E 219 Order Book, United States Military Academy

RG 94, E 225 Military Regulations, United States Military Academy

RG 94, E 227 Semi-annual Muster Rolls of Cadets, 1818–1849

RG 94, E 230 Merit Rolls, 1818–1866

RG 94, E 232 Monthly Class Reports and Conduct Rolls, 1831–1866

RG 94, E 258 Correspondence Relative to a Court of Inquiry at West Point to Examine into the Moral Condition of the Military Academy, and also into the Conduct on a Certain Occasion of Lieutenant Shiras (1839)

RG 94, E 501 Correspondence of the Record and Pension Office

RG 94, E 544 Field Records of Hospitals: New York, Register 604 (Cadet Hospital, USMA, 1838–1840)

RG 94, E 1525 Annual Reports of General D. C. McCallum, 1863–65

RG 94, Vol. 219, Orders and Special Orders, Department of the Ohio (Cumberland), 1861

RG 94, Vol. 444, Special Orders, Department and Army of the Tennessee

RG 107, E 36 Unbound Telegrams Collected by the Office of the Secretary of War

RG 110, E 31 Correspondence, Reports, Account and Related Records of Two or More Scouts, Guides, Spies and Detectives, 1862–1866

RG 110, E 36 Correspondence, Reports, Appointments, and Other Records Relating to Individual Scouts, Guides, Spies and Detectives, 1862–1866

RG 153, E 15 Court Martial Case Files, 1809–1894:
Private George Smith, 3rd Artillery, 1844

Colonel Thomas Worthington, 42nd Ohio Infantry, 1862

Private Ora Tebow, Lieutenant Henry Blanck and Captain William B. Keeler, 35th Iowa Infantry, 1863

RG 391, E155 Descriptive Book for Battery F, Third Artillery, 1833–1848

RG 393–1, E 866 Department of the Cumberland, Letters Sent, September 13–November 16, 1861

RG 393–1, E 2481 MDM Telegrams Received and Sent, March 1864–May 1865

RG 393–1, E 2484 MDM Letters sent in the Field, 1863–64

RG 393–1, E 2498 MDM Letters sent in the Field, April– May 1864

RG 393–1 E 2499 MDM Letters sent in the Field, May 1864–January 1865

RG 393–1 E 2504 MDM Special Field Orders, May 1864–May 1865

RG 393–1 E 2517 MDM Letters sent by the Provost Marshal General, March 1864–January 1865

GR 393–1 E 2520 MDM Report of Property Seized from Citizens, July-December 1864

RG 393–1, E 2521 MDM Records of the Provost Marshal General, MDM, RG 393–1 E 2902 MDM, Telegrams Sent in the Field, 1863–64

RG 393–1, E 2538 Division of the Missouri, Letters Sent, 1868–69

RG 393–2, E 2651 District of the Etowah, Letters Sent June 1864–June 1865

RG 393–2 E 2658 District of the Etowah, Reports 1865

RG 393–2 E 2671 List of Civilians to Whom Permits Have Been Issued to Live Within Three Miles of the Railroad

RG 393–2, E 5862 Letters and Telegrams Sent, 15th AC Dec. 1863–May 1865

RG 393–2, E 5864, Endorsements, 15th AC

### Personal Papers

*William T. Sherman Papers*

Over the last century General Sherman's letters have been so widely spread that libraries and archives of any size can generally produce at least one Sherman item. The sheer volume of his correspondence is in part responsible; then his fame gave what he wrote a certain monetary value–a good Sherman letter would bring $10 in 1900. It will suffice here to note the major collections, citing the holdings of other repositories in the notes.

The Library of Congress collection of Sherman papers, acquired through various bequests and purchases, contains some 18,000 items, measuring twenty-six linear feet, which have been reduced to fifty-one reels of microfilm; half of the materials concern the post-1865 period, and a full quarter of the holdings relate to the last seven years of Sherman's life. There are several closely related collections in the same repository, particularly the large Thomas Ewing Family Collection. From 1857 on Sherman began to preserve his correspondence systematically, first by use of the letterpress and then in registers or letter books; a number of these are found in the Library of Congress collection. In theory he only retained copies of "official" letters, but in fact many of essentially private nature were copied as well.

The University of Notre Dame collection, the William Tecumseh Sherman Family Papers, has as its nucleus items given to it by the general's granddaughter, Eleanor Sherman Fitch. In addition to this collection available on microfilm (fifteen reels), six additional cartons of Sherman family materials have recently been inventoried and made available to researchers.

The Ohio Historical Society has a sizable quantity of Sherman-related materials in several collections, including the general's extensive correspondence with Henry S. Turner and with various members of the Ewing family. The same repository has a considerable collection of Sherman memorabilia.

The National Archives has a collection of Sherman material that was particularly valuable for this study: twenty-three letter books containing the general's wartime correspondence–official and unofficial–from March 1862 to May 1865. Sherman had lent the volumes to the compilers of the *Official Records*; he never reclaimed them, so they were eventually placed in the National Archives under the Generals' Papers and Books rubric cited above.

While much of the research was done in Sherman materials, a number of other manuscript collections were useful. In alphabetical order, these and their locations are as follows:

Robert Anderson, LC
C. C. Augur, ISHL
Jacob Whitman Bailey, Charleston
  Library Society
George Bancroft, MassHS
John Milton Bancroft, Auburn
  University Library
Jesse S. Bean, UNC
Benham-McNeil Family, LC
William K. Bixby, MoHS Coll.
Isaac Bowen, USAMHI
David French Boyd, WLF Coll., LSU
Jeremiah T. Boyle, Filson and UNC
John Bragg, UNC
Orville H. Browning, ISHL
Don Carlos Buell, Filson
Horace Capron, LC
Caleb Carlton, LC
Salmon P. Chase, Historical Society of
  Pennsylvania
George S. Childs, UVA
William C. Church, LC
William Coffee, OHS
Philip St. George Cooke Family, Library
  of Virginia
Thomas L. Crittenden, Filson
William Dennison, OHS
John T. Doyle, CalHS
John W. Draper, LC
Henry A. DuPont, Hagley Museum and
  Library
Robert P. Effinger, CalHS
Edward Everett, MassHS
Charles Ewing, LC
Hugh Boyle Ewing, OHS
Philemon Beecher Ewing, OHS
Thomas Ewing, LC
Thomas Ewing Family, LC

Thomas Ewing Jr., Kansas Historical
  Society (microfilm)
Edward Everett, MassHS
Walter L. Fleming, New York Public
  Library
Manning F. Force, LC
William M. Gardiner, UNC
John Gibbon, Maryland Historical
  Society
George Mason Graham, VHS and WLF
  Coll., LSU
Walter Q. Gresham, LC
Benjamin H. Grierson , ISHL
James A. Hardie, LC
John Marshall Harlan, LC
Heth-Selden Collection, UVA
Andrew Hickenlooper MSS, Cincinnati
  Historical Society
William S. Hillyer, UVA
Josiah Hinds, Memphis Public Library
  and Information Center
Hitchcock Family, MoHS
Henry Hitchcock, LC
Vinnie Ream Hoxie, LC
John Jefferson Diary, Kentucky
  Historical Society
J. Neely Johnson, CalHS
Joseph E. Johnston, Library of Virginia
Henry C. Lay, UNC
R. E. Lee Headquarters, VHS
Louis Alexander Leonard, Emory
  University Library
Moses and St. John Richardson Liddell
  Family, LSU
John A. Logan, LC
James H. Lucas, MoHS (microfilm)
William T. Lusk, Huntington
Thaddeus A. Marshall, OHS

Lafayette McLaws, UNC
John Alexander McClernand, ISHL
McLane-Fisher, Maryland Historical
  Society
Absolom H. Markland, OHS
Montgomery Meigs, LC
Samuel Merrill, Indiana State Library
E. O. C. Ord, Stan
James H. Otey, UNC
John A. Palmer, ISHL
Lewis B. Parsons, ISHL
James Louis Pettigru, LC
Orlando Poe, LC
David Dixon Porter, LC
Richard Henry Pratt, Beineke Rare
  Books and Manuscripts,
Library, Yale University
James Robinson, OHS
John C. Ropes, BU
John M. Schofield, LC
James W. Schureman, LC
Philip Sheridan, LC
Charles R. Sherman, OHS
John Sherman, LC and OHS
Taylor Sherman, OHS

William Henry Smith, OHS
Edwin M. Stanton, LC
Stephen Z. Starr, Cincinnati Historical
  Society
Frederick Steele, Stan
John Peter Sullivan, LSU
James M. Swank, Historical Society of
  Pennsylvania
Thornton Family, VHS
Tomkins Family, VHS
Henry S. Turner, MoHS
William C.Tuttle, Kentucky Historical
  Society
Union Miscellany, Emory University
  Library
John Van Duzer, Duke University
Zebulon Vance, UNC
Willard Warner, ISHL
Elihu B. Washburne, LC
Jacob Perry Welch, Meridian Public
  Library
William Ashbury Whitaker, UNC
Bell Wiley, Emory University Library
Maria McKinne Winter, LSU

# BOOKS AND ARTICLES

Adams, John Q. "Hold the Fort!" *War Sketches and Incidents.* MOLLUS, Iowa Commandery 2(1893): 164–72.

Allen, Thomas. "The Second Wisconsin at the First Battle of Manassas or Bull Run." *War Papers.* MOLLUS, Wisconsin Commandery 1(1891): 374–93.

Ambrose, Stephen E. *Halleck: Lincoln's Chief of Staff.* Baton Rouge: Louisiana State University Press, 1962.

———. "Sherman: A Reappraisal." *American History Illustrated* 1, no. 9 (Jan. 1967): 5–11, 54–57.

American Psychiatric Association, *Diagnostic and Statistical Manual of Mental Disorders.* 4th ed. Washington, D.C.: American Psychiatric Association, 1994.

Andrews, J. Cutler. *The North Reports the Civil War.* Pittsburgh: University of Pittsburgh Press, 1955.

———. "In Commemoration of General William Tecumseh Sherman, April 6, 1892." *Personal Recollections of the War of the Rebellion.* MOLLUS, New York Commandery 2(1897): 30–56.

———. "The Stamp Issues." *Bulletin of the Society for Correct Civil War Information* 19 (April 1937): 49–53.

———. "William Tecumseh Sherman." *Illustrated American,* Mar. 7, 1891: 119.

Athern, Robert G. *William Tecumseh Sherman and the Settlement of the West.* Norman: University of Oklahoma Press, 1956.

Audenreid, J. C. "General Sherman in Europe and the East." *Harper's Monthly Magazine* 47(1873): 225–42, 481–95, 652–71.

Badeau, Adam. *Military History of U.S. Grant from April, 1861 to April, 1865.* 3 vols. New York: D. Appleton & Company, 1868–1881.

Bailey, William Whiting. "My Boyhood at West Point." *Soldiers and Sailors' Historical Society of Rhode Island Personal Narratives.* Fourth series, no. 12. Providence: Soldiers and Sailors' Historical Society of Rhode Island, 1891.

Baldwin, Alice Blackwood. *Memoirs of the Late Frank D. Baldwin, Major General, U.S.A.* Los Angeles: Wetzel Publishing Company, Inc., 1929.

Ballard, Michael B. *Pemberton: A Biography.* Jackson & London: University Press of Mississippi, 1991.

Barnard, John A. *Portrait of a Hero: The Story of Absalom Baird, His Family, and the American Military Tradition.* Philadelphia: Dorrance & Co., 1972.

Barr, Ronald J. *The Progressive Army: U.S. Army Command and Administration.* New York: St. Martin's Press, Inc., 1998.

Barrett, John G. *Sherman's March through the Carolinas.* Chapel Hill: University of North Carolina Press, 1956.

Bearss, Edwin C. "The Battle of Post of Arkansas." *Arkansas Historical Review* 18(1959): 237–79.

———. *Fort Moultrie, No.3, Fort Sumter National Monument, Sullivan's Island, South Carolina.* Washington, D.C.: National Park Service, Office of Architecture and Historic Preservation, December 30, 1968.

Bigelow, Donald Nevius. *William Conant Church and the Army and Navy Journal.* New York: Columbia University Press, 1952.

Bigelow, John. *The Principles of Strategy.* New York: G. P. Putnam's Sons, 1891.

Bok, Edward. *The Americanization of Edward Bok: The Autobiography of a Dutch Boy Fifty Years Later.* New York: Charles Scribner's Sons, 1923.

Bonner, James C. "Sherman at Milledgeville in 1864." *Journal of Southern History* 22(1956): 273–91.

Bowman, S. M. "Major-General William T.Sherman." *United States Service Magazine* 2(1864): 113–24, 240–55.

———. "Sherman's Atlanta Campaign." *United States Service Magazine* 3(1865): 304–323.

———. "Sherman's Georgia Campaign–from Atlanta to the Sea." *United States Service Magazine* 3(1865): 426–446.

———, and R. Irwin. *Sherman and his Campaigns.* New York: C. B. Richardson, 1865.

Boyd, David F. "General W. T. Sherman: His Early Life in the Southand his Relations with Southern Men." *Confederate Veteran* 18(1910): 409–14.

Boyd, James P. *The Life of General William T. Sherman.* Philadelphia: Publishers Union,1891.

Boynton, Henry Van Ness. *Sherman's Historical Raid. The Memoirs in Light of the Record.* Cincinnati: Wilstach, Baldwin & Co., 1875.

Bradford, Gamaliel. "Union Portraits III: William T. Sherman." *Atlantic Monthly* 114(1914): 318–29.

Bradley, Erwin Stanley. *Simon Cameron, Lincoln's Secretary of War: A Political Biography.* Philadelphia: University of Pennsylvania Press, 1966.

Brinkerhoff, Sidney B., and Pierce Chamberlin. "The Army's Search for a Repeating Rifle: 1873–1903." *Military Affairs* 32(1968): 20–30.

Brown, Andrew. "The First Mississippi Rangers, C.S.A." *Civil War History* 1(1955): 371–99.

Browning, Orville Hickman. *The Diary of Orville Hickman Browning*. Edited by Theodore Coleman Pease and James G. Randall. Collections of the Illinois State Historical Library, vols. 20 and 21. Springfield: Illinois State Historical Library, 1925.

Burt, Jesse C., Jr. "Sherman: Railroad General." *Civil War History* 2(1956): 45–54.

Burton, Katherine. *Three Generations: Maria Boyle Ewing (1801–1864), Ellen Ewing Sherman (1824–1888), Minnie Sherman Fitch (1851–1913)*. New York: Longmans, Green, 1947.

Byers, S. H. M. "Some Personal Recollections of General Sherman." *McClure's Magazine* 18(1901, 270–75.

———. "Some War Letters." *North American Review* 145(1887): 553–54.

Bynum, Hartwell T. "Sherman's Expulsion of the Roswell Women in 1864." *Georgia Historical Quarterly* 54(1970): 169–82.

"Cadet Life Before the Civil War." *Bulletin* 1. West Point, N.Y.: United States Military Academy Press, n.d.

Cadle, Cornelius. "An Adjutant's Recollections." *Sketches of War History, 1861–1865*. MOLLUS, Ohio Commandery 5(1903): 384–401.

Cameron, Norman, and Joseph F. Rylchak. *Personality Development and Psychopathology: A Dynamic Approach*. 2nd ed. Boston: Houghton Mifflin Company, 1985.

Capers, Gerald Mortimer, Jr. *The Biography of a River Town: Memphis in its Heroic Age*. Chapel Hill: University of North Carolina Press, 1939.

Carpenter, John A. *Sword and Olive Branch: Oliver Otis Howard*. Pittsburgh: University of Pittsburgh Press, 1964.

Carr, Clark E. *My Day and Generation*. Chicago: A.C. McClurg & Co., 1908.

Carter, John Denton, "Thomas Sim King, Vigilante Editor," *California Historical Society Quarterly*, 21(1942): 23–38.

Carter, Samuel III. *The Final Fortress: The Struggle for Vicksburg, 1862–1863*. New York: St. Martin's Press, 1980.

Castel, Albert. *Decision in the West: The Atlanta Campaign of 1864*. Lawrence: University Press of Kansas, 1992.

———. "Prevaricating through Georgia: Sherman's Memoirs as a Source for the Atlanta Campaign." *Civil War History* 40(1994): 48–71.

Childs, George William. *Recollections*. Philadelphia: J.B. Lippincott, 1892.

Church, Albert E. *Personal Reminiscences of the Military  Academy from 1824 to 1831*. West Point, N.Y.: United States Military Academy Press, 1879.

Clarke, Dwight L. *William Tecumseh Sherman: Gold Rush Banker*. San Francisco: California Historical Society, 1969.

Clauss, Errol MacGregor. "Sherman's Rail Support in the Atlanta Campaign." *Georgia Historical Quarterly* 50(1966): 413–20.

Coffman, Edward M. *The Old Army: Portrait of the American Army in Peacetime, 1784–1898*. New York and Oxford: Oxford University Press, 1986.

Colton, Walter. *Three Years in California*. 1850; reprint, New York: Arno Press, 1976.

Conyngham, David P. *Sherman's March through the South*. New York: Sheldon & Co., 1865.

Cooling, Benjamin Franklin. *Symbol, Sword and Shield: Defending Washington during the Civil War*. 2nd rev. ed. Shippensburg, Pa.: White Mane Publishing Company, Inc., 1997.

Cooper, Jerry M. "The Army and Civil Disorder, 1877–1900." *Military Affairs* 33(1969): 390–98.

[Coppee, Henry]. "Sherman's Truce." *United States Service Magazine* 3 (1865): 497–99.

Cortissoz, Royal. *The Life of Whitelaw Reid*. 2 vols. New York: Charles Scribner's Sons, 1921.

Coulter, E. Merton. "Sherman and the South." *North Carolina Historical Review* 8 (1931): 41–54.

Cox, Jacob D. *Atlanta.* New York: Charles Scribner's Sons, 1882.

Cozzens, Peter. *The Shipwreck of Their Hopes: The Battle of Chattanooga.* Urbana & Chicago: University of Illinois Press, 1994.

Cresap, Bernarr. *Appomattox Commander: The Story of General E. O. C. Ord.* South Branch, N.J.: A.S. Barnes, 1981.

Crooker, Lucian B., "Episodes and Characters in an Illinois Regiment." *Military Essays and Recollections.* MOLLUS, Illinois Commandery 1 (1891): 33–49.

Cross, Ira B. *Financing an Empire: History of Banking in California.* 2 vols. Chicago, San Francisco, and Los Angeles: S. J. Clarke Publishing Co., 1912.

Cullum, George W. *Biographical Register of the Officers and Graduates of the Military Academy at West Point, New York, since its Establishment in 1802.* Cambridge, Mass.: Riverside Press, 1901.

Cunliffe, Marcus. *Soldiers and Civilians: The Martial Spirit in America.* Boston: Little, Brown and Company, 1968.

Curl, Donald W. *Murat Halstead and the Cincinnati Commercial.* Boca Raton: University Presses of Florida, 1980.

Daniel, John W. "The Great March." *DeBow's Review.* Apr. 1868: 337–46.

Davis, Theodore R. "With Sherman in his Army Home." *Cosmopolitan* 12 (1891): 195–205.

De Laubenfels, D. J. "Where Sherman Passed By." *Geographical Review* 47(1957): 381–95.

——. "With Sherman through Georgia." *Georgia Historical Quarterly* 41(1957): 288–300.

Denkewicz, Robert M. *Vigilantes in Gold Rush California.* Palo Alto, Calif.: Stanford University Press, 1985.

Depew, Chauncy. *My Memories of Eighty Years.* New York: Charles Scribner's Sons, 1922.

Dodge, Grenville M. *The Battle of Atlanta and Other Campaigns: Addresses, etc.* Council Bluffs, Iowa: Monarch Printing Company, 1910.

——. *Personal Recollections of President Abraham Lincoln, General Ulysses S. Grant and General William T. Sherman.* Council Bluffs, Iowa: Monarch Printing Company, 1914.

Drake, Samuel Adams. "The Old Army in Kansas." *War Talks in Kansas.* MOLLUS, Kansas Commandery 1 (1900): 141–52.

Dufour, Charles L. *The Mexican War, a Compact History, 1846–1848.* New York: Hawthorn Books, 1968.

Durkin, Joseph T. *General Sherman's Son: The Life of Thomas Ewing Sherman.* New York: Farrar, Strauss and Cudahy, 1959.

Dyer, Thomas G. *Secret Yankees: The Union Circle in Confederate Atlanta.* Baltimore and London: Johns Hopkins University Press, 1999.

Ehrenberger, James K. and Francis G. Gschwind. *Sherman Hill.* Callaway, Neb.: E & G. Publications, 1973.

Ellis, Edward S. "Reminiscences of General Sherman." *Chatauquan* 27 (1898): 474–75.

Ellison, Joseph. "The Struggle for Civil Government in California, 1846–50." *California Historical Society Quarterly* 10 (1930): 3–26.

Ewing, Joseph H. *Sherman at War.* Dayton, Ohio: Morningside Press, 1992.

Ewing, Thomas. "The Autobiography of Thomas Ewing." Edited by Clement L. Martzloff. *Ohio Archeological and Historical Society Publications* 22 (1913): 126–204.

Farwell, Willard Brigham. "Recollections of Good Digging." *Society of California Pioneers Quarterly* I (1924): 17–27.

Fellman, Michael. *Citizen Sherman: A Life of William Tecumseh Sherman.* New York: Random House, 1995.

——. *Inside War: The Guerrilla Conflict in Missouri during the American Civil War.* New York and Oxford: Oxford University Press, 1989.

Field, Stephen L. *Personal Experiences of Early Days in California.* 1893; reprint, New York: Da Capo Press.

Fisher, Noel C. "Prepare Them for My Coming: General William T. Sherman, Total War, and Pacification in West Tennessee." *Tennessee Historical Quarterly* 51(1992): 74–86.

Fitzgerald, David. "Annotations by General Sherman." *Journal of the Military Service Institution of the United States.* Sept. 1893: 978–79.

Fleming, Robert H. "The Battle of Shiloh as a Private Saw It." *Sketches of War History.* MOLLUS, Ohio Commandery 6(1908): 13–19.

Fleming, Walter L. *William T. Sherman as College President.* Cleveland: Arthur W. Clark Company, 1912.

[Fletcher, Thomas Clement]. *Life and Reminiscences of General Wm. T. Sherman, by Distinguished Men of his Time.* Baltimore: R.H. Woodward Company, 1891.

Florken, Herbert G., ed. "The Law and Order View of the San Francisco Vigilance Committee of 1856." *California Historical Society Quarterly* 14(1935): 350–74.

Force, Manning F. "Marching across Carolina." *Sketches of War History, 1861–1865.* MOLLUS, Ohio Commandery 1(1886): 1–18.

Foreman, Sidney, ed. *Cadet Life before the Mexican War.* West Point, N.Y.: United States Military Academy Printing Office, 1945.

Foust, Drew Gilpin. "Clutching the Chains that Bind." *Southern Cultures* 5, no. 1(1999), 6–20.

French, Giles. *The Golden Land. A History of Sherman County, Oregon.* Portland: Oregon Historical Society, 1958.

Friedel, Frank. "General Orders 100 and Military Government." *Mississippi Valley Historical Review* 32(1946): 541–56.

Gallagher, Gary W. *The Confederate War.* Cambridge, Mass., and London: Harvard University Press, 1997.

Gannon, B. Anthony. "A Consistent Deist: Sherman and Religion." *Civil War History* 42(1996), 307–21.

Garesché, Louis. *Biography of Lieutenant Colonel Julius Garesché, Assistant Adjutant General, United States Army.* Philadelphia: J.B. Lippincott Company, 1887.

Garfield, James A. *The Diary of James A. Garfield.* Edited by Harry James Brown and Frederick D. Williams. 4 vols. East Lansing: Michigan State University Press, 1967–1981.

Gibson, Charles Dana, with E. Kay. *Assault and Logistics: Union Army Coastal and River Operations, 1861–1866.* The Army's Navy Series, volume 2) Camden, Me: Ensign Press, n.d.

Glatthaar, Joseph T. *The March to the Sea and Beyond: Sherman's Troops in the Georgia and Carolina Campaigns.* New York: New York University Press, 1985.

Goslin, Charles R. *Crossroads and Fence Corners: Historical Lore of Fairfield County.* Lancaster, Ohio: Fairfield Historical Association, 1976.

Graham, Albert A. *History of Fairfield and Perry Counties, Ohio.* Chicago: W.H. Beers & Co., 1883.

Graham, Stephen. "Marching through Georgia. Following Sherman's Footsteps Today." *Harper's Magazine* 140(1920): 612–20, 813–23.

Grandstaff, Mark R. "'Preserving the Habits and Usages of War:' William Tecumseh Sherman, Professional Reform, and the U.S. Army Officer Corps, 1865–1881, Revisited." *Journal of Military History* 62(1998): 521–45.

Grant, Ulysses S. *The Papers of Ulysses S. Grant.* Edited by John Y. Simon. 23 vols. to date. Carbondale and Edwardsville: Southern Illinois University Press, 1967–.

——. *Personal Memoirs of U.S. Grant.* 2 vols. New York: Charles L. Webster & Company, 1885.

Graves, Ralph A. "Marching through Georgia Sixty Years After." *National Geographic* 50 (1926): 259–311.

Gray, John Chipman, and John Codman Ropes. *War Letters, 1862–1865.* Boston and New York: Houghton-Mifflin Company, 1927.

Green, Michael S. "Picks, Spades and Shiloh: The Entrenchment Question." *Southern Studies* 3, Spring, 1992): 13–54.

Gresham, Matilda. *Life of Walter Quinton Gresham, 1832–1895.* 2 vols. Chicago: Rand McNally, 1919.

Grimsley, Mark. *The Hard Hand of War: Union Military Policy toward Southern Civilians, 1861–1865.* Cambridge, New York, and Melbourne: Cambridge University Press, 1995.

Grivas, Theodore. *Military Governments in California, 1846–1850, with a Chapter on Their Prior Use in Louisiana.* Glendale, Calif.: Arthur H. Clark Company, 1963.

Hacker, Barton C. "The United States Army as a National Police Force: The Federal Policing of Labor Disputes, 1877–1898." *Military Affairs* 33(1969): 255–64.

Hall, Martin H. "The Campbell-Sherman Diplomatic Mission to Mexico." *Bulletin of the Historical and Philosophical Society of Ohio* 13(1955): 254–70.

Halleck, H. Wager. *Elements of Military Art and Science; or, Course of Instruction in Strategy, Fortification, Tactics of Battles, &c.; Embracing the Duties of Staff, Infantry, Cavalry, Artillery, and Engineers. Adapted to the Use of Volunteers and Militia.* Reprint, Westport, Conn.: Greenwood Press, n.d.

Halstead, Murat. "Recollections and Letters of General Sherman." *Independent* 5 (1899): 1610–13.

Harlow, Neal. *California Conquest: War and Peace in the Pacific, 1846–1850.* Berkeley, Los Angeles, and London: University of California Press, 1982.

Harris, Joel Chandler. *Life of Henry W. Grady, Including the Writings and Speeches.* New York: Cassell Publishing Company, 1890.

Harris, Moses. "The Old Army." *War Papers.* MOLLUS, Wisconsin Commandery 2 (1896): 331–44.

Harrison, Lowell H. "The Civil War in Kentucky: Some Persistent Questions." *Kentucky Historical Society Register* 76 (1978): 1–21.

——. and James C. Klotter. *A New History of Kentucky.* Lexington: University Press of Kentucky, 1997.

Hawley, Joseph R. *Major General Joseph R. Hawley, Soldier and Editor (1826–1905).* Edited by Alfred D. Putnam. n.p.: Connecticut Civil War Centennial Commission, 1964.

Hazen, William B. *A Narrative of Military Service.* Boston: Ticknor & Co., 1885.

Heth, Henry. *The Memoirs of Henry Heth.* Edited James L. Robertson, Jr., Westport, Conn.: Greenwood Press, 1974.

Hickenlooper, Andrew. "The Battle of Shiloh, in Two Parts." *Sketches of War History, 1861–1865.* MOLLUS, Ohio Commandery 5(1896): 402–83.

——. *Sherman: General Hickenlooper's Address at the Twenty-third Meeting of the Society of the Army of the Tennessee.* Cincinnati: Press of F.W. Freeman, 1893.

Hinkley, Julian. *A Narrative of Service with the Third Wisconsin Infantry.* Madison: Wisconsin Historical Commission, 1912.

Hirshson, Stanley P. *The White Tecumseh: A Biography of William T. Sherman.* New York: John Wiley & Sons, 1997.

Hitchcock, Henry. "General William T. Sherman." *War Papers and Personal Reminiscences, 1861–1865.* MOLLUS, Missouri Commandery 1(1892): 416–29.

——. *Marching with Sherman: Passages from the Letters and Campaign Diaries of Henry Hitchcock, Major and Asst. Adjutant General of Volunteers, Nov. 1864–May, 1865.* Edited by M.

A. DeWolfe Howe. New Haven, Conn.: Yale University Press, 1927.

Hitt, Michael D. *Charged with Treason: Ordeal of 400 Mill Workers During Military Operations in Roswell, Ga., in 1864–1865.* Monroe, N.Y.: Library Research Associates, 1992.

Holden, Edward S., and W. E. Ostrander, comps. *Centennial History of the United States Military Academy at West Point, New York.* 2 vols. Washington, D.C.: Government Printing Office, 1904.

Holliday, J. S. *The World Rushed In: The California Gold Rush Experience.* New York: Simon & Schuster, 1981.

Hood, John Bell. *Advance and Retreat: Personal Experiences in the United States and Confederate States Armies.* 1880; reprint, New York: Konecky & Konecky, n.d.

Hosmer, James Kendall. *The Last Leaf: Observations during Seventy-Five Years of Men And Events in America and Europe.* New York & London: G.P. Putnam's Sons, 1912.

Hough, Alfred Lacy. *Soldier of the West: The Civil War Letters of Alfred Lacy Hough.* Edited by Robert G. Athern, with an Introducton by John Newbold Hough. Philadelphia: University of Pennsylvania Press, 1957.

Howard, Oliver Otis. *Autobiography of Oliver Otis Howard, United States Army.* 2 vols. New York: Baker & Taylor Company, 1907.

Hurd, E. O. "The Battle of Colliersville." *Sketches of War History.* MOLLUS, Ohio Commandery 5(1903): 243–54.

Huston, James A. "Logistical Support of Federal Armies in the Field." *Civil War History* 7 (1961): 36–47.

Ingersoll, L. D. *A History of the War Department of the United States with Biographical Sketches of the Secretaries.* Washington, D.C.: Francis B. Mohun, Successor to Mohun Brothers, 1879.

Jacob, Katherine Allamong. *Capital Elites: High Society in Washington, D.C., after the Civil War.* Washington, D.C.: Smithsonian Institution Press, 1994.

——. *Testament to Union: Civil War Monuments in Washington, D.C.* Photographs by Edwin Harlan Remsberg. Baltimore and London: Johns Hopkins University Press, 1998.

James, Joseph B. "Life at West Point One Hundred Years Ago." *Mississippi Valley Historical Review* 31, no. 12(June 1944): 21–40.

Jenkins, Walworth. *Q.M.D. or Book of Reference for Quartermasters.* Louisville, Ky.: John P. Morton & Co., 1865.

Jenney, William Lebaron. "With Sherman and Grant from Memphis to Chattanooga–A Reminiscence." *Military Essays and Recollections.* MOLLUS, Illinois Commandery 4(1907): 193–214.

Johnson, Andrew. *The Papers of Andrew Johnson.* Edited by Leroy P. Graf. 16 vols. Nashville: University of Tennessee Press, 1967–2000.

Johnson, R. W. *A Soldier's Reminiscences in Peace and War.* Philadelphia: J.B. Lippincott Company, 1886.

Johnson, Stephen M. *Humanizing the Narcissistic Style.* New York: W.W. Norton & Company, 1987.

Johnston, Joseph E. *Narrative of Military Operations, Directed, During the Late War Between the States.* New York: D. Appleton & Company, 1874.

Johnson, W. Fletcher. *Life of Wm. Tecumseh Sherman, Late Retired General, U.S.A.* N.p.: Edwood Pub. Co., 1891.

Jones, Archer. *Civil War Command and Strategy: The Process of Victory and Defeat.* New York: Free Press, 1992.

Jones, John C., and Winona C. Jones. *The Prairie Pioneers of Western Kansas and Eastern Colorado.* Boulder, Colo.: Johnson Publishing Company, n.d.

Jones, Virgil Carrington. *The Civil War at Sea, March 1862–July, 1863. The River War.* New York: Holt, Rinehart & Winston, 1961.

"Justitia." "To the Honorable Mr. Hawes." *Army-Navy Chronicle* 25 (June 1836): 386–9.

Kelly, R. M. "Holding Kentucky for the Union." *Battles and Leaders of the Civil War* 1 (1887), 373–92.

Kerr, Charles D. "General William T. Sherman." *Glimpses of the Nation's Struggle.* MOLLUS, Minnesota Commandery 3 (1893): 495–502.

Keyes, Erasmus D. *Fifty Years' Observations of Men and Events, Civil and Military.* New York: Charles Scribner's Sons, 1884.

Kiper, Richard L. "John Alexander McClernand and the Arkansas Post Campaign." *Arkansas Historical Review* 56 (1979): 56–79.

Kurz, Wilbur G. "A Federal Spy in Atlanta." *Atlanta Historical Bulletin* 10 (1957): 14–19.

Lancaster, Jane F. "William Tecumseh Sherman's Introduction to War, 1840–1842: Lesson for Action." *Florida Historical Quarterly* 72 (July, 1993): 56–72.

LeDuc, William G. *Recollections of a Civil War Quartermaster. The Autobiography of William G. LeDuc.* St. Paul, Minn.: North Central Publishing Company, 1963.

Leonard, Thomas C. "Red, White and Army Blue: Empathy and Anger in the American West," in Peter Karsten, ed., *The Military in America: From the Colonial Period to the Present.* New rev. ed. New York: Free Press, 1986, 226–38.

Lewis, Lloyd. *Sherman: Fighting Prophet.* New York: Harcourt, Brace and Company, Inc., 1932; reprint, New York:: Konecky & Konecky, n.d.

Liddell Hart, B. H. *The Liddell Hart Memoirs.* Vol. 1, *1895–1938.* New York: G.P. Putnam's Sons, 1965.

——. *Paris, or the Future of War.* New York: E.P. Dutton, 1925.

——. *Sherman: Soldier, Realist, American.* New York: Dodd, Mead & Co., 1930.

Lincoln, Abraham. *The Collected Works of Abraham Lincoln.* Edited by Roy P. Basler. 8 vols. New Brunswick, N. J.: Princeton University Press, 1953.

Littell, John S. *The Commerce and Industries of the Pacific Coast of North America.* San Francisco: A.L. Bancroft & Co., Publishers, 1882.

Longacre, Edward G. *From Union Stars to Top Hat: A Biography of the Extraordinary James Harrison Wilson.* Harrisburg, Pa.: Stackpole Books, 1972.

Lothrop, John A. "Ropes' Attack upon General Sherman." *War Sketches and Incidents.* MOLLUS, Iowa Commandery 2 (1898): 488–507.

Lucas, Marion Brunson. *Sherman and the Burning of Columbia.* College Station and London: Texas A & M University Press, 1976.

Lusk, William Thompson. *War Letters of William Thompson Lusk, Captain, Assistant Adjutant General, United States Volunteers, 1861–1863.* New York: N.p., 1911.

Lyman, Theodore. *Meade's Headquarters 1863–1865. Letters of Colonel Theodore Lyman from the Wilderness to Appomattox.* Edited by George R. Agassiz. Boston: Atlantic Monthly Press, 1922.

Lyons, W. F. *Brigadier-General Thomas Francis Meagher. His Political and Military Career; with Selections from his Speeches and Writings.* New York: D & J Sadler & Company, n.d.

McAlexander, U. G. *History of the Thirteenth Regiment, United States Infantry.* N.p.: Regimental Press, Thirteenth Infantry, 1905.

McCallister, Anna. *Ellen Ewing: Wife of General Sherman.* New York: Benziger Brothers, 1936.

McClellan, George B. *The Civil War Papers of George B. McClellan: Selected Correspondence, 1860–1865.* Edited by Stephen W. Sears. New York: Ticknor & Fields, 1989.

McClure, Alexander K. *Recollections of Half a Century.* Salem, Mass.: Salem Press, 1902.

McConnell, Stuart. *Glorious Contentment: The Grand Army of the Republic, 1865–1900.* Chapel Hill: University of North Carolina Press, 1992.

McCormick, Robert W. "Challenge of Command: Worthington vs Sherman." *Timeline: A Publication of the Ohio Historical Society* 8, no. 3(June-July 1991): 28–39.

McCrory, W. M. "Early Life and Personal Reminiscences of General William T. Sherman." *Glimpses of the Nation's Struggle* MOLLUS, Minnesota Commandery 3(1893): 310–46.

McCulloch, Hugh. *Men and Measures of Half a Century: Sketches and Comments.* New York: Charles Scribner's Sons, 1888.

McDonough, James Lee. *In Hell before Night.* Knoxville: University of Tennessee Press, 1977.

McFeely, William S. *Grant: A Biography.* New York and London: W.W. Norton & Company, 1981.

McKelvey, Blake. *Rochester in the Civil War.* Rochester Historical Society Publications, vol. 22. Rochester, N.Y.: Rochester Historical Society, 1944.

McKnight, W. Mark. *Blue Bonnets o'er the Border: The 79th New York Cameron Highlanders.* Shippensburg, Pa.: White Mane Publishing Company, Inc., 1998.

McMurry, Richard. *John Bell Hood and the War for Southern Independence.* Lexington: University of Kentucky Press, 1982.

McPherson, James M. *Oxford History of the United States.* Vol. 6, *The Battle Cry of Freedom: The Civil War Era.* New York and Oxford: Oxford University Press, 1988.

———. *For Cause and Comrades: Why Men Fought in the Civil War.* New York and Oxford: Oxford University Press, 1997.

Mahan, Dennis Hart. *Advanced-Guard, Out-Post and Detachment Service of Troops, and the Manner of Posting and Handling Them in Presence of the Enemy.* New York: John Wiley, 1853.

Mahon, Samuel. "The Forager in Sherman's Last Campaigns." *War Sketches and Incidents.* MOLLUS, Iowa Commandery 2(1898): 188–200.

Major, Duncan K., and Roger S. Fitch. *Supply of Sherman's Army During the Atlanta Campaign.* Fort Leavenworth, Kans.: Army Service School Press, 1911.

Marcy, Henry O. "Sherman's Campaign in the Carolinas." *Civil War Papers.* MOLLUS, Massachusetts Commandery 3(1900): 331–48.

Marshall-Cornwall, Sir James. *Grant as Military Commander.* New York: B.T. Batsford, Ltd.; reprint, New York: Van Nostrand Reinhold, 1970.

Marszalek, John F. "Celebrity in Dixie: Sherman Tours the South, 1879." *Georgia Historical Quarterly* 66(1982): 366–82.

———. "Philatelic Pugilists." *Cump and Company: A Newsletter for Friends and Fanciers of General William T. Sherman* 6, no.2(Oct. 1999): 28–35.

———. *Sherman: A Soldier's Passion for Order.* New York: Free Press, 1993.

———. *Sherman's Other War: The General and the Civil War Press.* Memphis: Memphis State University Press, 1981.

———. "William T. Sherman: Myth and Reality." *Journal of the Georgia Association of Historians* 15(1994): 1–13.

Masterson, James F. *The Search for the Real Self: Unmasking the Personality Orders of Our Age.* New York: Free Press, 1988.

Matloff, Maurice, et al. *American Military History.* Washington, D.C.: Office of the Chief of Military History, United States Army, 1969.

Meagher, Thomas Francis. *The Last Days of the 69th in Virginia: A Narrative in Three Parts.* New York: Office of the "Irish American," 1861.

Michie, Peter S. *The Life and Letters of Emory Upton, Colonel of the Fourth Regiment of Artillery, and Brevet Major-General, U.S. Army.* New York: D. Appleton and Company, 1885.

Miller, Charles C., ed. And comp. *History of Fairfield County, Ohio and Representative Citizens.* Chicago: Richmond-Arnold Publishing Co., 1912.

Miller, Ernest Smith. *The Francis Blair Family in Politics.* 2 vols. New York: Macmillan Company, 1933.

Moore, John Hammond. "Sherman's 'Fifth Column': A Guide to Unionist Activity in Georgia." *Georgia Historical Quarterly* 68 (1984): 382–409.

Morgan, George W. "The Assault on Chickasaw Bluffs." *Battles and Leaders of the Civil War* 3, 462–70.

Morrison, Andrew P. *Shame: The Underside of Narcissism.* Hillsdale, N.J.: Analytic Press, 1997.

Morrison, James L., Jr. *"The Best School in the World:" West Point, the Pre-Civil War Years, 1833–1866.* Kent, Ohio: Kent State University Press, 1986.

Moulton, C. W. *The Review of General Sherman's Memoirs Examined Chiefly in Light of its Own Evidence.* Cincinnati: Robert Clarke & Co., Printers, 1875.

Mruch, Armin E. "The Role of Railroads in the Atlanta Campaign." *Civil War History* 7(1961): 264–71.

Nenninger, Timothy K. *The Leavenworth Schools and the Old Army: Education, Professionalism, and the Officers' Corps of the United States Army, 1881–1918.* Westport, Conn.: Greenwood Press, 1978.

Nevins, Allan. *Frémont: The West's Greatest Adventurer.* 2 vols. New York & London: Harper & Brothers Publishers, 1928.

——. *Hamilton Fish: The Inner History of the Grant Administration.* New York: Dodd, Mead & Company, 1936.

Nichols, George Ward. *The Story of the Great March, from the Diary of a Staff Officer.* New York: Harper & Brothers, 1865.

Nichols, Roy F., ed. "William Tecumseh Sherman in 1850." *Pennsylvania Magazine of History,* 75(1951): 424–35.

Nunis, Doyce B., ed. *The San Francisco Vigilance Committee of 1856: Three Views: 1. William T. Coleman. 2. William T. Sherman. 3. James O'Meara.* Los Angeles: Los Angeles Westerners (Publication no. 103), 1971.

Park, Roswell. *A Sketch of the History and Topography of West Point and the Military Academy.* Philadelphia: Henry Perkins, 1840.

Parks, J. H. "A Confederate Trade Center under Federal Occupation: Memphis 1862 to 1865." *Journal of Southern History* 7(1941): 289–314.

——. *Joseph E. Brown of Georgia.* Baton Rouge: Louisiana State University Press, 1977.

Parrish, William E. *Frank Blair: Lincoln's Conservative.* Columbia and London: University of Missouri Press, 1998.

Partridge, Samuel Selden. "Letters of S.S. Partridge." Edited by McKelvey Blake. *Rochester in the Civil War.* Rochester Historical Society Publications, vol. 22. Rochester, N.Y.: Rochester Historical Society, 1944.

Paul, Rodman W. *California Gold.* Cambridge, Mass.: Harvard University Press, 1947.

Pepper, George W. *Personal Recollections of Sherman's Campaigns in Georgia and the Carolinas.* Zanesville, Ohio: Hugh Dunne Co., 1866.

——. *Under Three Flags; or, the Story of my Life as Preacher, Captain in the Army, Consul, with Speeches and Interviews.* Cincinnati: Printed for the Author by Curts & Jennings, 1899.

Pitzman, Julius. "Vicksburg Campaign Reminiscences." *The Military Engineer* 15 (Mar.-Apr. 1923): 112.

Porter, Admiral [David Dixon]. *Incidents and Anecdotes of the Civil War.* New York: D.Appleton and Company, 1885.

Porter, Horace. *Campaigning with Grant.* New York: Century, 1897.

Putnam, Douglas. "Reminiscences of the Battle of Shiloh." *Sketches of War History.* MOLLUS, Commandery of Ohio 3. (1888–1890): 197–211.

Rable, George C. "William T. Sherman and the Conservative Critique of Radical Reconstruction." *Ohio History* 93 (1984): 147–63.

Reardon, Carol. *Soldiers and Scholars: The U.S. Army and the Uses of Military History, 1865–1890.* Lawrence: University Press of Kansas, 1990.

Reed, Germaine M. *David French Boyd, Founder of Louisiana State University.* Baton Rouge: Louisiana State University Press, 1977.

Reese, William J. *A Sketch of the Life of Charles R. Sherman.* n.p. n.d.

Reid, Harvey. *The View from Headquarters: Civil War Letters of Harvey Reid.* Frank L. Byrne, Ed. Madison: State Historical Society of Wisconsin, 1965.

Reid, Whitelaw. *A Radical View: The "Agate" Despatches of Whitelaw Reid, 1861–1865.* Edited by James G. Smart. 2 vols. Memphis: Memphis State University Press, 1988.

——. *Ohio in the War. Her Statesmen and Soldiers.* 2 vols.
Columbus: Eclectic publishing Company, 1893.

Reston, James, Jr. *Sherman's March and Vietnam.* New York: Macmillan, 1984.

Rodman, W. Paul. *California Gold.* Cambridge: Harvard University Press, 1947.

Ropes, John Codman. "General Sherman." *Atlantic Monthly* 68(1891): 191–204.

Royster, Charles. *The Destructive War: William Tecumseh Sherman, Stonewall Jackson, and the Americans.* New York: Alfred A. Knopf, 1991.

Rusling, James F. *Men and things I Saw in Civil War Days.* New York: Eaton & Maines, 1899.

Russ, William A. Jr. "Administrative Activities of the Union Army During and After the Civil War." *Mississippi Law Journal* 17(1945): 71–89.

Sanborn, John B. "Remarks on a Motion to Extend a Vote of Thanks to General Marshall." *Glimpses of the Nation's Struggle.* MOLLUS, Minnesota Commandery 4(1898): 615–22.

Sanchez, N. V. "Grafting Romance on a Rose Tree: The True Story of Doña Maria Bonifacio and General Sherman at Monterey." *Sunset,* Apr. 1916: 36–40.

Sanderson, George. *A Brief History of the Early Settlement of Fairfield County.* Lancaster, Ohio: Published by Thomas Wetzler, 1851.

Schofield, John M. *Forty-Six Years in the Army.* New York: the Century Company, 1897.

Schurz, Carl. *The Intimate Papers of Carl Schurz, 1841–1869.* Edited by Joseph L. Schaefer. New York: Da Capo Press, 1970.

——. *The Reminiscences of Carl Schurz.* Edited by Frederick Bancroft and William A. Dunning. 3 vols. The McClure Co., 1908.

Schwarzkopf, H. Norman, with Peter Petrie. *It Doesn't Take a Hero: General Norman H. Schwarzkopf, the Autobiography.* New York: Bantam, 1992.

Scott, Hervey. *A Complete History of Fairfield County, Ohio, 1795–1876.* Columbus, Ohio: Siebert & Lilley, 1877.

Scott, H. L. *Military Dictionary: Comprising Technical Definitions; Information on Raising and Keeping Troops; Actual Service, Including Makeshifts and Law, Government, Regulation, and Administration Relating to Land Forces.* 1861; reprint, New York: Greenwood Press, 1968.

Shanks, W. G. F. "Recollections of General Sherman." *Harper's New Monthly Magazine* 30(1865): 640–46.

Sheridan, Philip. *Personal Memoirs of P. H. Sheridan.* 2 Vols. New York: Charles Webster & Co., 1888.

"Sherman's Winter Campaign through Georgia." *United States Service Magazine* 3(1865): 164–69.

Sherman, Janan. "The Jesuit and the General: Sherman's Private War." *Psychohistory Review* 21(1998), 255–94.

Sherman, John. *Battle of Pittsburg Landing-Volunteers of Ohio. Remarks of Hon. John Sherman of Ohio, in Senate of the United States, May 9, 1862.* Washington, D.C.: Scammell & Co., 1862.

——. "In Commemoration of General William Tecumseh Sherman, April 6, 1892." *Personal Recollections of the War of the Rebellion.* MOLLUS, New York Commandery 2(1897): 56–63.

——. *Recollections of Forty Years in the House, Senate and Cabinet.* 2 vols. Chicago, London & Berlin: Werner Company, 1895.

Sherman, Minnie (Maria) Ewing. "My Father's Letters." *Cosmopolitan* 12 (Nov. 1891): 64–69.

Sherman, Thomas Townsend. *Sherman Genealogy, including Families of Essex, Suffolk, and Norfolk, England. Some Descendants of Immigrants Captain John Sherman, Reverend John Sherman, Edward Sherman and the Descendants of Honorable Roger Sherman and Honorable Charles R. Sherman.* New York: Tobias A. Wright, Printer and Publisher, 1920.

Sherman, William Tecumseh. *Address to the Graduating Class of the Military Academy, West Point, June 15th, 1869.* New York: D. Van Nostrand, Publisher, 1869.

——. "Address to the New England Society, December 28, 1871." *Army & Navy Journal* January 7, 1873.

——. *A Letter of Lieutenant William T. Sherman Reporting on Conditions in California in 1848.* Carmel, California: N. p., 1947.

——. *A Letter from Sherman to Colonel Stevenson, with an Introduction by Wilbur Smith and a Facsimile of the Original Letter from the Collections of the University of California at Los Angeles.* Los Angeles: N.p., 1960.

——. "Annotations by General Sherman." Edited by David Fitzgerald, *Journal of the Military Service Institution of the United States.* Sept. 1893: 978–79.

——. *Correspondence between General W. T. Sherman, U.S. Army and Major General W. S. Hancock, U.S. Army.* Saint Paul, Minn.: n.p., 1871.

——. "The Grand Strategy of the War of the Rebellion." *Century Magazine* 13(1888): 582–98.

——. *Memoirs of General W. T. Sherman: Written by Himself.* 2 vols. New York: D. Appleton and Co., 1875; 2nd ed., 1885.

——. "The Militia." *Journal of the Military Service Institution of the United States* 6(1885), 1–15.

——. "Our Army and Militia." *North American Review* 151 (1890): 129–45.

——. "Some Letters by General W. T. Sherman, U.S.A. Chiefly Concerning Shiloh." Edited by William L. Marshall. *Glimpses of the Nation's Struggle.* MOLLUS, Minnesota Commandery 4(1898): 605–15.

——. *Two Letters from Lieutenant W. T. Sherman, U.S.A.* N.p., 1964.

——. "Unpublished Letters of General Sherman." *North American Review* 152(1891): 371–75.

Skelton, William B. *An American Profession of Arms: The Army Officer Corps, 1784–1861.* Manhattan: University of Kansas Press, 1993.

Slocum, H. W. "Final Operations of Sherman's Army," *Battles and Leaders of the Civil War* 4, 754–58.

Smalley, E. V. "General Sherman." *Century Magazine*(1884): 450–62.

Smith, Mark A. "Sherman's Unexpected Companions: Marching through Georgia with Jomini and Clausewitz." *Georgia Historical Quarterly* 81(1997), 1–24.

Smith, Sherry L. *The View from Officer's Row: Army Perceptions of Western Indians.* Tucson: University of Arizona Press, 1990.

Smith, William Ernest. *The Francis Preston Blair Family in Politics.* 2 vols. New York: Macmillan Company, 1933.

Spaight, J. M. *War Rights on Land.* London: Macmillan and Co., 1911.

Stampp, Kenneth M. *America in 1857: A Nation on the Brink.* New York and Oxford: Oxford University Press, 1990.

Stohlman, Robert F., Jr. *The Powerless Position: The Commanding General of the Army of the United States, 1864–1903.* Manhattan, Kans.: Military Affairs, 1975.

Stone, Henry. "The Atlanta Campaign." *Papers of the Military Historical Society of Massachusetts* 8(1910): 341–92.

Symonds, Craig L. *Joseph E. Johnston: A Civil War Biography.* New York and London: W.W.Norton & Company, 1992.

Taylor, John T. "Reminiscences of Service as an Aide-de-Camp with General William Tecumseh Sherman." *War Talks in Kansas.* MOLLUS, Kansas Commandery(1892,): 128–42.

Thomas, Benjamin P. and Harold Hyman. *Stanton: Life and Times of Lincoln's Secretary of War.* New York: Alfred A. Knopf, 1962.

Todd, William. *The Seventy-Ninth Highlander Volunteers in the War of the Rebellion 1861–1865.* Albany, N.Y.: Press of Brandow, Barton & Co., 1886.

Turner, Henry Smith. *The Original Journals of Henry Smith Turner: With Stephen Watts Kearney to New Mexico and California.* Norman: University of Oklahoma Press, 1966.

U'Ren, Richard C. *Ivory Fortress: A Psychiatrist Looks at West Point.* Indianapolis and New York: Bobbs-Merrill Company, Inc., 1974.

Van Duser, John. "The John Van Duser Diary of Sherman's March from Atlanta to Hilton Head." Edited by Charles Brockman, Jr. *Georgia Historical Quarterly* 53(1969): 220–40.

Villard, Henry. *Memoirs of Henry Villard, Journalist and Financier.* 2 vols. Boston & New York: Houghton Mifflin & Co., 1904.

Walters, John Bennett. "General William T. Sherman and Total War." *Journal of Southern History* 14 (1948): 447–80.

Warner, Ezra J. *Generals in Blue: Lives of the Union Commanders.* Baton Rouge and London: Louisiana State University Press, 1992.

Weigley, Russell F. *The American Way of War: A History of United States Military Strategy and Policy.* New York: Macmillan Publishing Co., 1973.

——. *History of the United States Army.* Bloomington: Indiana University Press, 1984.

Weatherford, John "Sherman Liked the South Once." *Manuscripts* 8 (Winter 1956): 73–79.

Wheaton, Henry. *Elements of International Law: With a Sketch of the History of the Science.* Philadelphia: Carey, Lea and Blanchard, 1836.

Wheeler, Richard. *We Knew Sherman.* New York: Thomas Crowell, 1977.

Widney, Lyman S. "Campaigning with Uncle Billy." *Neale's Monthly* 2(August, 1913): 131–43.

Wilkie, Franc B. *With Pen and Powder.* Boston: Ticknor & Company, 1888.

Williams, Alpheus S. *From the Cannon's Mouth: The Civil War Letters of Gen. Alpheus S. Williams.* Edited by Milo M. Quaife. Detroit: Wayne State University Press, 1959.

Williams, Mary Floyd. *History of the San Francisco Vigilante Committee of 1851: A study of Social Control on the California Frontier in the Days of the Gold Rush.* 1931; reprint, New York: Da Capo Press, 1969.

Williams, T. Harry. *McClellan, Sherman and Grant.* New Brunswick, N.J.: Rutgers University Press, 1962.

Wilson, James Harrison. *The Life of John A. Rawlins. Lawyer, Major General of Volunteers, and Secretary of War.* New York: Neale Publishing Co., 1916.

——. *Under the Old Flag.* 2 vols. New York: D. Appleton, 1912.

Wiseman, C. M. L. *Centennial History of Lancaster, Ohio and Lancaster People.* Lancaster: C.M.L. Wiseman Publishers, 1898.

——. *Pioneer Period and Pioneer People of Fairfield County, Ohio.* Columbus: F. J. Heer Printing Co., 1901.

Wistar, Isaac Jones. *Autobiography of I. J. Wistar, 1827–1905: Half a Century in War and Peace.* Philadelphia: Wistar Institute of Anatomy and Biology, 1937.

Woodhull, Maxwell Van Zandt. "A Glimpse of Sherman Fifty Years Ago." MOLLUS, Commandery of the District of Columbia 4(1914): 452–68.

Worthington, Thomas. *Colonel Worthington Vindicated. Sherman's Discreditable Record at Shiloh on his Own and Better Evidence.* Washington, D.C.: T. McGill & Co. Printers, 1878.

——. *Shiloh. The Only Correct Military History of U.S. Grant and of the Missing Army Records for which he is Alone Responsible to Conceal his Organized Defeat of the Union Army at Shiloh, April 6, 1862.* Washington, D. C: N.d.

Wright, Benjamin C. *Banking in California, 1849–1910.* San Francisco: H.S. Crocker Company, Printers, 1910.

Wyn Koop, Henry M., comp. and ed. *Picturesque Lancaster, Past and Present.* Lancaster: Republican Printing Co., 1897.

Young, Jared W., Ed. "General Sherman on his own Record: Some Unpublished Comments." *Atlantic Monthly* 108(1911): 289–300.

——. "General Sherman's Puritan Heritage." *Eugenical News* 13(1928): 106–9.

## PUBLISHED DOCUMENTS

*Annual Report of the Board of Visitors of the United States Military Academy, West Point, New York, June, 1839.* Washington, D.C.: A.B. Claxton, 1839.

*Military Orders of General William T. Sherman.* N.p, n.d.

*Regulations Established for the Organization and Government of the Military Academy, at West-Point, by Order of the President of the United States; to which is added the Regulations for the Internal Police of the Institution; with an Appendix, Containing the Rules and Articles of War, and Extracts from the General Regulations of the Army Applicable to the Academy.* New York: Wiley and Putnam, 1839.

*Revised Regulations for the Army of the United States, 1861.* By Authority of the War Department. Philadelphia: J.G.L. Brown, 1861.

*A Digest of the Military and Naval Laws of the Confederate States from the Commencement of the Provisional Congress to the End of the First Congress under the Permanent Constitution.* Edited by W. W. Lester and W. J. Brown. Columbia, S.C.: Evans and Cogswell, 1864.

*The War of the Rebellion: A Compilation of the Official Records of the Union and Confederate Armies.* 128 vols. Washington, D.C.: Government Printing Office, 1880–1901

## NEWSPAPERS

*Alta California* (San Francisco)

*Atlanta Journal*

*Atlanta Constitution*

*Baltimore Gazette*

*Boston Daily Advertiser*

*Buffalo Enquirer*

*The Capital* (Washington, D.C.)
*Chicago Times*
*Cincinnati Commercial*
*Cincinnati Gazette*
*Charleston News and Courier*
*Daily Evening Bulletin* (San Francisco)
*Daily Herald* (San Francisco)
*Daily Missouri Democrat*
*Dubuque Telegraph*
*Evening Star* (Philadelphia)
*Indianapolis Daily Journal*
*Ironton Register* (Ohio)
*Louisville Daily Journal*
*Louisville Tribune*
*Memphis Bulletin*
*Memphis Daily Argus*
*Memphis Union Appeal*

*Montreal Daily Witness*
*National Intelligencer* (Washington, D.C.)
*National Tribune* (Washington, D.C.)
*New Orleans Bulletin*
*New York Advertiser*
*New York Herald*
*New York Times*
*New York Tribune*
*News and Courier* (Charleston, S.C.)
*Pointer View* (West Point)
*Portland Oregonian*
*San Francisco Daily Herald*
*San Francisco*
Sunday Capital (Washington)
*Sunny South* (Atlanta)
*World* (New York)

## THESES AND DISSERTATIONS

Acker, Dudley, Jr. "Nantan Lupan: George Crook on America's Frontiers". Ph.D. diss., Northern Arizona University, 1995.

Andrews, Richard Allen. "Years of Frustration: William T. Sherman, the Army, and Reform, 1869–1883. Ph.D diss., Northwestern University, 1968.

Crites, William Ralph. "The Development of Infantry Tactical Doctrine, 1861–1865." M.A. thesis, Duke University, 1968.

Griess, Thomas E. "Dennis Hart Mahan: West Point Professor and Advocate of Military Professionalism, 1830–1871." Ph.D. diss., Duke University, 1968.

Hooper, Ernest Walter, "Memphis, Tennessee: Federal Occupation and Reconstruction, 1862–1870." Ph.D. diss., University of North Carolina, 1957.

Hughes, Michael Anderson. "The Struggle for Chattanooga, 1862–1863." Ph.D. diss., University of Arkansas, 1990.

Hutton, Paul Andrew II. "General Philip Sheridan and the Army of the West, 1867–1888." Ph.D. diss., Indiana University, 1980.

Losson, Christopher Thomas. "Jacob Dolson Cox: A Military Biography," Ph.D. diss., University of Mississippi, 1993.

McNeill, William James. "The Stress of War: The Confederacy and William Tecumseh Sherman during the Last Year of the Civil War." Ph.D. diss., Rice University, 1973.

De Montravel, Peter R. "The Career of Lieutenant General Nelson A. Miles from the Civil War through the Indian Wars." Ph.D diss. St. John's University, 1983.

Moore, Daniel Haas. "Chiefs, Agents, and Soldiers: Conflict on the Navajo Frontier, 1866–1880." Ph.D. diss., University of Northern Arizona, 1988.

Shiman, Philip Lewis. "Engineering Sherman's March: Army Engineers and the Management of Modern War, 1862–1865." Ph.D. diss., Duke University, 1991.

Skirbunt, Peter Daniel. "Prologue to Reform: the Germanization of the United States Army, 1865–1898." Ph.D diss., Ohio State University, 1983.

Toll, Larry A. "The Military Community and the Western Frontier, 1866–1898." Ph.D. diss., Ball State University, 1990.

Waghelstein, John David. "Preparing for the Wrong War: The United States Army and Low-Intensity Conflict, 1775–1890." Ph.D. diss., Temple University, 1990.

Wooster, Robert Allen. "The Military and United States Indian Policy,1865–1903." Ph.D. diss., University of Texas, Austin, 1985.

# INDEX

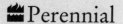
# Perennial

**Books by Lee Kennett:**

## MARCHING THROUGH GEORGIA
*The Story of Soldiers and Civilians During Sherman's Campaign*
ISBN 0-06-092745-3 (paperback)

A gripping history of Sherman's legendary march through Georgia that spelled the beginning of the end for the Confederacy. Beginning with the opening skirmish at Buzzard Roost Gap and continuing all the way to Savannah ten months later, Lee Kennett analyzes General Sherman, and the military action that changed the country politically, historically, and socially. By capturing the experiences of the soldiers and civilians who witnessed the bloody siege, Kennett brings this remarkable march to life.

## SHERMAN
*A Soldier's Life*
ISBN 0-06-017495-1

A major new biography of William Tecumseh Sherman based on considerable fresh archival material; and focusing, more than any current biography, on Sherman as soldier and warrior—not just the Civil War years but his eventful decades as Army officer and leader. Sherman enjoyed a glorious post-Civil War period of 26 years, during which he ruled over large parts of the Western frontier, wrote his classic *Memoirs*, was commanding General of the Army, and spurned repeated invitations to run for the presidency.